# WindEurope Annual Event 2024

AA007500

Journal of Physics: Conference Series Volume 2745

Bilbao, Spain
20 - 22 March 2024

**Editor:**

**Amir R. Nejad**

ISBN: 978-1-7138-9733-0
ISSN: 1742-6588

**Printed from e-media with permission by:**

Curran Associates, Inc.
57 Morehouse Lane
Red Hook, NY 12571

**Some format issues inherent in the e-media version may also appear in this print version.**



Institute of Physics
Dirac House, Temple Back
Bristol BS1 6BE UK

Phone: 44 1 17 929 7481
Fax: 44 1 17 920 0979

techtracking@iop.org

**Additional copies of this publication are available from:**

Curran Associates, Inc.
57 Morehouse Lane
Red Hook, NY 12571 USA
Phone: 845-758-0400
Fax: 845-758-2633
Email: curran@proceedings.com
Web: www.proceedings.com

# TABLE OF CONTENTS

# PREFACE

This Special Issue presents selected contributions from the **WindEurope Annual Event 2024 Conference** that took place 20-22 March in Bilbao. A total of 23 papers are included, addressing one (or more) of the conference themes:

- Resource assessment
- Turbine technology
- Wind farm: installation and operations
- Floating offshore wind
- Digitalisation
- Environmental sustainability and social acceptance
- Grid and system integration
- Renewable hydrogen and Power-to-X.

**Conference Committee for the Scientific Program**

- Amir R. Nejad, Professor, NTNU
- Nicolaos Cutululis, Professor, Technical University of Denmark (DTU)
- Simon Watson, Professor, Delft University of Technology
- Julia Gottschall, Chief Scientist, Fraunhofer IWES
- A. Mehrad Kazemi Amiri, Research Associate, Strathclyde University
- Jan Helsen, Professor, Vrije Universiteit Brussel

We thank all, more than 30, reviewers for their engagement and effort, without which these proceedings would not have been possible. As with previous WindEurope events, the European Academy of Wind Energy (EAWE) has been responsible for organizing the review process for scientific contributions and has contributed to the development of the conference sessions. WindEurope has been responsible for the organization and logistics of the conference, and we thank their highly professional staff and their associates for the excellent collaboration.

**Proceedings editor**

Amir R. Nejad, Professor, Norwegian University of Science and Technology (NTNU)

# Peer Review Statement

All papers published in this volume have been reviewed through processes administered by the Editors. Reviews were conducted by expert referees to the professional and scientific standards expected of a proceedings journal published by IOP Publishing.

- **Type of peer review:** Single Anonymous
- **Conference submission management system:** Morressier
- **Number of submissions received:** 24
- **Number of submissions sent for review:** 24
- **Number of submissions accepted:** 23
- **Acceptance Rate (Submissions Accepted / Submissions Received × 100):** 95.8
- **Average number of reviews per paper:** 3
- **Total number of reviewers involved:** 27
- **Contact person for queries:**
  **Name:** WindEurope Conference Team
  **Email:** conference@windeurope.org
  **Affiliation:** WindEurope

# On Building Predictive Digital Twin Incorporating Wave Predicting Capabilities: Case Study on UMaine Experimental Campaign - FOCAL

**Yuksel R. Alkarem[1], Kimberly Huguenard[1], Richard W. Kimball[2], Babak Hejrati[2], Ian Ammerman[2], Amir R. Nejad[3], Jacob Fontaine[4], Reza Heshami[4], Stephan Grilli[4]**

[1]Civil and Environmental Engineering Department, University of Maine, 35 Flagstaff Road, Orono, Maine 04469, USA

[2]Mechanical Engineering Department, University of Maine, 35 Flagstaff Road, Orono, Maine 04469, USA

[3]Norwegian University of Science and Technology (NTNU), Jonsvannsveien 82, 7050 Trondheim, Norway

[4]Ocean Engineering, University of Rhode Island, Sheets Building, 15 Receiving Road, Narragansett, RI 02882 – USA

E-mail: yuksel.alkarem@maine.edu

**Abstract.** The response of floating wind turbines (FWT) are susceptible to stochastic wave variations. For the optimal operation of FWT, a comprehensive understanding of the phase-resolved wave dynamics and the consequential system response is crucial for real-time monitoring and control. A multi-variate, multi-step, long short term memory (MLSTM), a type of recurrent neural network (RNN) is used to capture complex system dynamics for real-time application. Results indicate that the integration of a wave prediction-reconstruction (WRP) model substantially enhances prediction accuracy by 50% on average relative to the baseline model. The improvement is consistent across various wave extremity and prediction horizons, thereby significantly broadening the scope for timely and precise predictive capabilities.

## 1. Introduction

Current global efforts aim to increase offshore wind capacities. Europe aims to deploy 60 GW of offshore wind capacity by 2030 and 300 GW by 2050 [1]. The Floating Offshore Wind Energy Shot in the United States seeks to reduce the cost of floating offshore wind energy technologies by over 70%, targeting $45 per megawatt-hour by 2035 [2]. To drive down the cost of wind energy systems is to reduce uncertainty in predicting wave-induced loadings. Accurately forecasting such loadings constitutes the initial phase in their mitigation, thereby prolonging the operational lifespan of these system. However, unlike fixed-bottom wind turbines, FWT suffer additional fatigue loads induced by the floaters degrees of freedom (DOF). For instance, the sea-to-land fatigue ratio under standard operations with respect to the tower base fore-aft bending moment can reach up to 7.0 for barge-type FWT [3]. Given the stochastic nature of wave excitations, physics-based modelling is utilized to model the system's dynamics [4, 5, 6, 7, 8]. Yet, for real-time applications, data-driven approach for wind/wave predictive schemes [9, 10], and low-fidelity based digital twin solutions [11, 12] are computationally efficient albeit lack the accuracy the higher-fidelity models provide.

Experimental investigation were conducted on a 1:70 scale model of the FOCAL hull [13], based on the UMaine VolturnUS-S reference floating platform [14], which supported a fully operational model of the IEA 15MW reference turbine [15], as reported in Ref. [10]. During the test, the hull underwent both gradual and sudden sea-state variations. Results showed that controllers with access to future data were more successful at mitigating platform motion. Therefore, it is recommended for real-world applications that control systems should be designed to assimilate both present and near-future wave conditions. Typically, LiDAR camera-based wave reconstruction prediction (WRP) models have a forecast time horizon of $10s$ to $20s$ [16]. Furthermore, accuracy in wave predictions decreases as the forecasted time is further from the present moment. Consequently, the WRP-based predictive digital twin must rapidly compute the influence of predicted waves on the system's dynamics. The updated information can be used based on the level of the digital twin (DT) which varies between mere descriptive to full control autonomy. While contemporary medium-fidelity models such as the open source code OpenFAST [17], can operate faster than real-time, they often lag behind the rapid changes in predicted wave conditions. Moreover, updating the wave field usually mandates simulation re-initialization, a time-consuming process. A potential solution may involve the the employment of a linear state-space model of the system, corrected to physical data through methodologies like Kalman filtering to account for system's non-linearities [18]. This framework, however, lacks the capacity to foresee future system states. This limitation is addressed in this paper and a novel methodology that incorporates WRP models to predict the system's state in the near future is presented.

The remainder of this paper is organized as follows: the first section provides a description of the experimental setup and methodology used for the predictive model and its limitations. Then, results of the methodology are discussed covering the accuracy of the predictive model and sensitivity analysis when considering prediction uncertainty. The paper is concluded with some remarks for future studies.

## 2. Methodology

The aim of this research is to propose a predictive model that leverages the past/present data (either physical or numerical through virtual sensors) and wave data extracted from WRP models to estimate the phase-resolved, short-term future state of the system. In this section, the experimental setup is briefly discussed. Then, the proposed methodology is explained along with the scope of this research and general assumptions considered.

### *2.1. Experimental setup*

The experimental campaign took place at the Harold Alfond wind-wave ($W^2$) basin situated at the Advanced Structures and Composites Center at the University of Maine [19]. The platform tested is the 1:70 scale model of the UMaine VolturnUS-S semi-submersible platform supporting a performance-matched wind turbine with a simulated rotor-nacelle assembly (RNA) mass as illustrated in Figure 1. The experimental setup is described in detail in Ref. [20]. The wave field is generated using 16 wave paddles at one end of the basin. The paddles are calibrated to generate long-crested incident waves. For the purpose of this study, wave only conditions are conducted. The authors are planning to test the model's predictability for coupled wind-wave excitations in the future.

Characterization tests are carried out to determine the system properties. For instance, free decay and hammer tests are conducted to quantify the systems natural frequency, and horizontal offset tests to determine the hydrostatic stiffness of the mooring system setup. These results are reported in Ref. [10, 20]. Then, wave tests on the hull took place. Wave fields are uni-directional

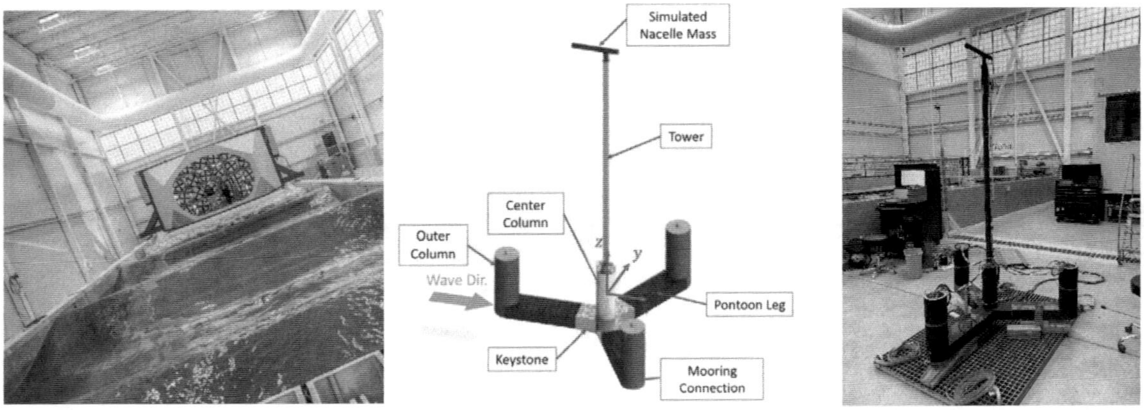


**Figure 1.** wind/wave basin at the University of Maine (left), the 1:70 scale model of the VolturnUS-S offshore wind turbine semi-submersible platform model [19] (center), system characterization test setup (right)

and are characterized using the JONSWAP spectrum, examining four wave spectra with varying extremity ($3.1 < H_s[m] < 10.9$ and $8.96 < T_p[s] < 14.2$).

*2.2. Wave reconstruction prediction models*

Various techniques can be used to sample the ocean surface waves. While wave buoys are accurate, they only provide a one-point measurement in space. To overcome this, ocean waves can be sampled from an optical sensor (such as LiDAR), ensuring a dense wave measurement. One drawback of this method is that the sampling accuracy depends on weather adversity. In stormy weather, where data is most valuable, visual clearance can hinder the capability of optic sensors to correctly intersect the ocean surface and get correct measurements. Additionally, The measured data can be highly non-uniformly distributed in space and time [16], due to the shadowing effect caused by optic rays approaching the ocean surface tangentially, as illustrated in Figure 2. Waves closer to the sensor obstruct upstream waves. However, the ocean wave surface can still be mathematically constructed based on the sampled data using simplified models such as those based on linear wave theory (LWT), which is computationally efficient, or using more complex predictive schemes such as the nonlinear improved Choppy wave model (ICWM) [16, 21, 22]. Figure 2 illustrates the capability of the Choppy model to perform spatio-temporal forecast of ocean waves using numerical tank to generate LiDAR data.

Within the scope of this study, the deployment of a sophisticated continuous wave prediction algorithm is paramount. This algorithm generates wave forecast at the location of the floater, denoted as $x_c$, utilizing observations collected from a distinct location, $x_b < x < x_e$, as depicted in Figure 3. Each timestep prediction yields a forecast that is systematically integrated into a dynamic buffer file that gets updates as time progresses. The boundaries of the nowcast and forecast are determined by the minimum and maximum group velocities of the nowcast wave components, $C_{g,min}$ and, $C_{g,max}$, respectively. It is imperative to approach reconstructed data outside these boundaries with a heightened level of scrutiny due to their diminished physical relevance and the consequent impact on the reliability of the predictions.

*2.3. Near-present mirroring*

The use of models to compute accurate response of the platform in a computationally cost efficient manner is crucial. The non-linear aero-hydro-servo-elastic simulator (OpenFAST) [23],

Journal of Physics: Conference Series **2745** (2024) 012001 doi:10.1088/1742-6596/2745/1/012001

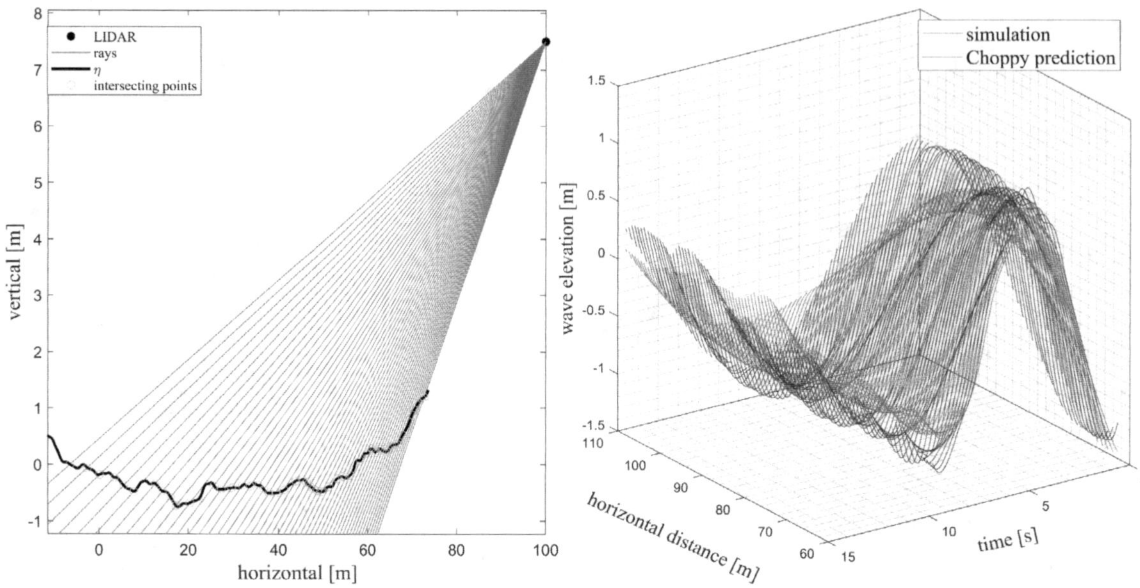


**Figure 2.** LiDAR rays intersecting numerically created ocean wave surface with shadowing effect around 20m and 25m (x and y axes are not to scale) (left). Spatio-temporal wave reconstruction prediction results using Choppy method compared to simulated wave data (right).

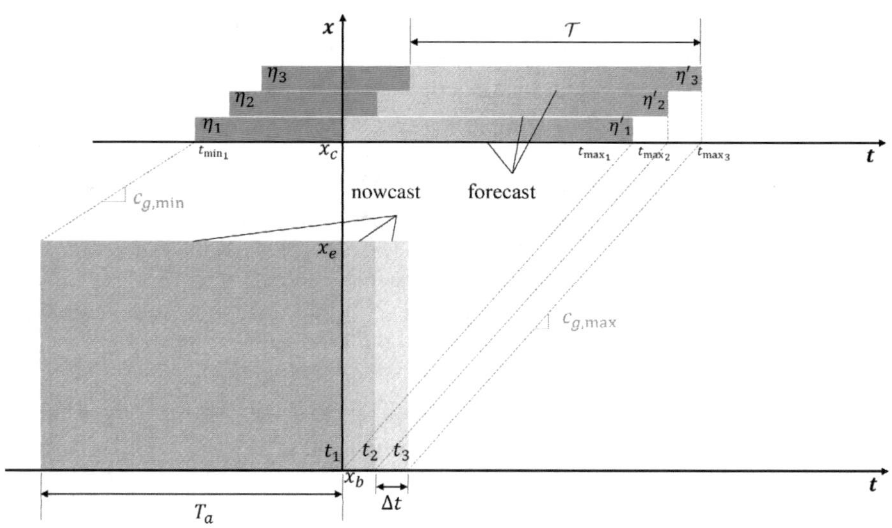


**Figure 3.** Continuous wave prediction algorithm and the differentiation between nowcast and forecast at the location of interest $x_c$, based on wave measurements between $x_b$ and $x_e$ over an acquisition time, $T_a$ (inspired by [22]).

developed by the National Renewable Energy lab (NREL), provides a numerically-generated linear state-space approximation of the full system dynamics [24]. The non-linear model is tuned to match the experimental results in [10] and includes the 6 rigid-body degrees of freedom of the floating platform and the first two tower bending modes in the fore-aft and side-side directions. Low frequency second order wave effects are not included in the simulation. The tower-top mass of the rotor-nacelle assembly (RNA) is included, but the turbine blades are set to zero length and inflowing wind is disabled.

The past and near-present dynamics can be modelled using a state-space model, updated through a Kalman Filter to correct for non-linearities present in the real system. The structure of the Kalman filter validated in [18] is shown in Figure 4, illustrating how a small subset of real-time system measurements is used to provide accurate virtual measurements of system parameters.

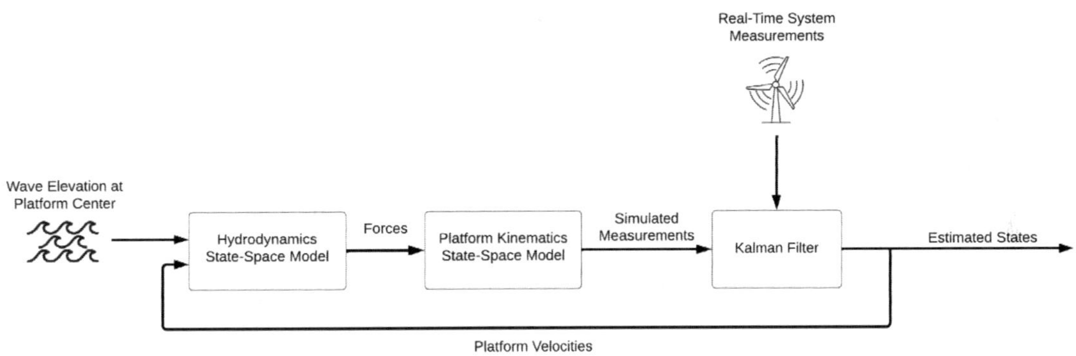


**Figure 4.** Linear model and Kalman filter structure for real-time virtual sensing.

The Kalman filter implemented in [18] utilized five real-time measurements from the FWT system: pitch, roll, and mooring line tensions. Uncertainties from the mooring line tensions are based on the scaled up noise from the load cells used in the tank tests, while the uncertainty in pitch and roll measurements are set based on a commercially available inertial measurement unit (IMU). The magnitude of uncertainty (Q) in the state-space model, used in the Kalman filter as a weight on the predictions, is set as twice the difference in variance between the linear model and experimental results, according to Eq. 1:

$$Q_i = 2||\sigma^2_{experiment} - \sigma^2_{state-space}|| \quad (1)$$

where $Q_i$ represents uncertainty in the prediction of the $i^{th}$ degree of freedom and $\sigma^2$ is the variance.

The Kalman filter is validated in [18] against experimental data collected from wave tank tests conducted on a 1:70 scale model of the UMaine VolturnUS-S semi-submersible FWT platform [10]. Using measurements of the platform pitch/roll and the three mooring line tensions, the Kalman filter is shown to accurately and robustly improve estimates of states, namely surge, heave, and tower base bending moment, across a wide range of sea conditions. Figure 5 presents a sample of the validation results for the surge DOF, illustrating how the Kalman filter is able to correct for the low-frequency platform oscillation caused by second-order wave effects not included in the linear model.

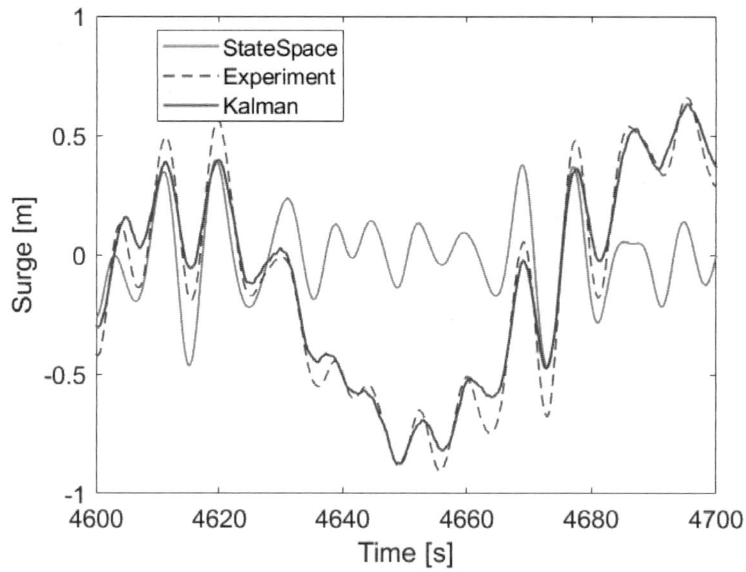


**Figure 5.** Response values of surge measured during wave tests on the 1:70 scale model of the UMaine VolturnUS-S floating wind platform compared with linear state-space (red) and Kalman filter adjusted (purple) values.

*2.4. System dynamics forecast integrated with wave reconstruction prediction model*

Prediction of the floater's motion and forces exerted by the upcoming wave is conducted through training a recurrent neural network (RNN) type and feeding it past and near-present data (either directly from the physical system in case the sensor of the measure of interest exists or from the kalman-corrected linear model in case a virtual sensor is to be used). The RNN used here is a multi-variate long short term memory (MLSTM) that predicts a certain time-series signal $\mathcal{T}$ ahead in time, where $\mathcal{T} > \Delta t$, rendering MLSTM a multi-step model. For simplicity and computational efficiency, the model's labels (output) are the predicted states at the last timestep of the forecast, $\hat{\xi}_j(t + \mathcal{T})$. At each timestep, a new predictive value is available, rendering predictive model continuous after reaching $t >= \mathcal{T}$. The MLSTM has three main gates; a forget gate, an input gate, and an output gate. If the MLSTM has been trained and takes the forecasted wave from the WRP model as input, then the model is named MLSTM-WRP.

The data is managed in the flow shown in Figure 6. The data is first normalized via min-max normalization methodology. Then, the input signals are checked and cleaned for any idleness or irregularities. One can perform pre-processing on the cleaned data to check for data clustering and correlation between sensors using techniques such as principal component analysis (PCA) or confusion matrix, respectively. The normalized data are then restructured to allow for supervised training or testing. The restructuring occurs at every timestep. For instance, Table 1 illustrates the restructuring of data for supervised learning/testing for a given time, $t$ of both MLSTM and MLSTM-WRP. The MLSTM only accepts features (input) of past signals, $(\xi_j)$, where $j = 1, 2, ..., N$ and $N$ is the number of states of interest. Meanwhile, MLSTM-WRP additionally accepts the spatially forecasted wave elevations in past/near-present, $\eta$, and the spatio-temporally forecasted wave elevations, $\eta'$, in continuous pattern as illustrated in Figure 3. If timestep is $\Delta t$, for the features to have equal lengths, and since the acquisition time, $T_a$ (the time the model uses as basis for prediction), does not necessarily equal the prediction horizon, $\mathcal{T}$, $\eta'$ must be interpolated with a different timestep, $\Delta t^* = \mathcal{T}/T_a \times \Delta t$. The RNN model uses

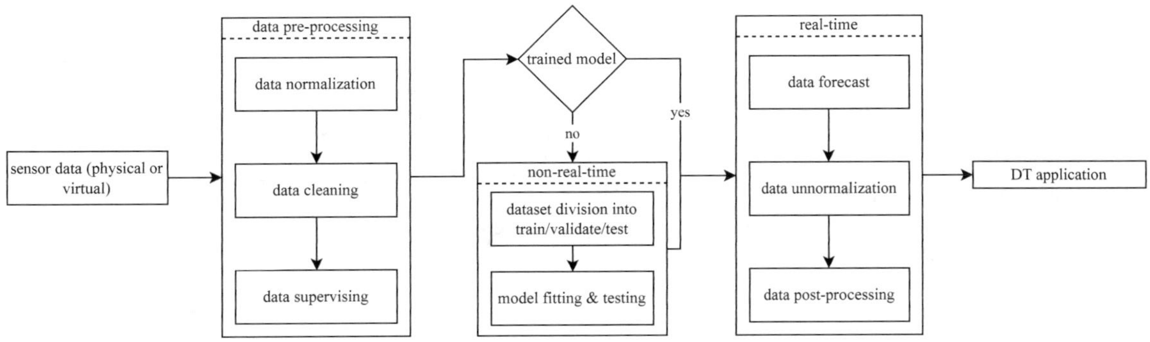


**Figure 6.** Flowchart illustrating data processes from the physical asset to the virtual asset in non-real-time for fitting the model and real-time for DT applications.

| models | | features | | | | | labels |
|---|---|---|---|---|---|---|---|
| MLSTM-WRP | MLSTM | DOF | $\xi_1(t-T_a)$ | $\xi_1(t-T_a+\Delta t)$ | ... | $\xi_1(t-\Delta t)$ | $\hat{\xi}_1(t+\mathcal{T})$ |
| | | | $\xi_2(t-T_a)$ | $\xi_2(t-T_a+\Delta t)$ | ... | $\xi_2(t-\Delta t)$ | $\hat{\xi}_2(t+\mathcal{T})$ |
| | | | $\vdots$ | $\vdots$ | $\vdots$ | $\vdots$ | $\vdots$ |
| | | | $\xi_N(t-T_a)$ | $\xi_N(t-T_a+\Delta t)$ | ... | $\xi_N(t-\Delta t)$ | $\hat{\xi}_N(t+\mathcal{T})$ |
| | | WRP | $\eta(t-T_a)$ | $\eta(t-T_a+\Delta t)$ | ... | $\eta(t-\Delta t)$ | |
| | | | $\eta'(t)$ | $\eta'(t+\Delta t^*)$ | ... | $\eta'(t+\mathcal{T})$ | |

**Table 1.** MLSTM and MLSTM-WRP restructuring for supervised learning/testing at time, $t$, with a timestep, $\Delta t$, acquisition time, $T_a$, and prediction horizon, $\mathcal{T}$. The $\xi$ is the measured state and $\hat{\xi}$ is the predicted one.

$n = T_a/\Delta t$ for indexing past/near-present data and $m = \mathcal{T}/\Delta t$ for indexing future data.

### *2.5. Scope of study and key assumptions*
This research puts forward a framework on conducting continuous, real-time future prediction of floating wind turbine's dynamics based on previously acquired signals from physical or virtual sensors and forecasted wave data from measurements. It is assumed here that sensors provide a continuous data stream and the WRP model provides accurate wave data predictions with error in the reported range in the literature [16]. Results in this paper are focused on using physical sensor data from the experimental campaign and wave elevation data from wave gauges multiplied by an error function to resemble data acquired by an optical sensor. However, a sensitivity analysis is carried out to demonstrate the influence of the accuracy of the WRP on the output of the prediction.

## 3. Results and Discussions
### *3.1. MLSTM v. MLSTM-WRP at various prediction horizons; a generalization study*
This section is dedicated to an in-depth evaluation of the predictive accuracy of two models, the MLSTM and MLSTM-WRP, with respect to various states of the system and a wide range of wave-induced excitations. Both models are subjected to identical configurations and datasets, with a training ratio of 0.5, equivalent to 14,500 seconds of data. Preliminary hyperparameter optimization scheme yielded optimal settings: the ratio of past to future indexing parameter is established at $n/m = T_a/\mathcal{T} = 0.59$, encompassing a single hidden layer with 96 neurons,

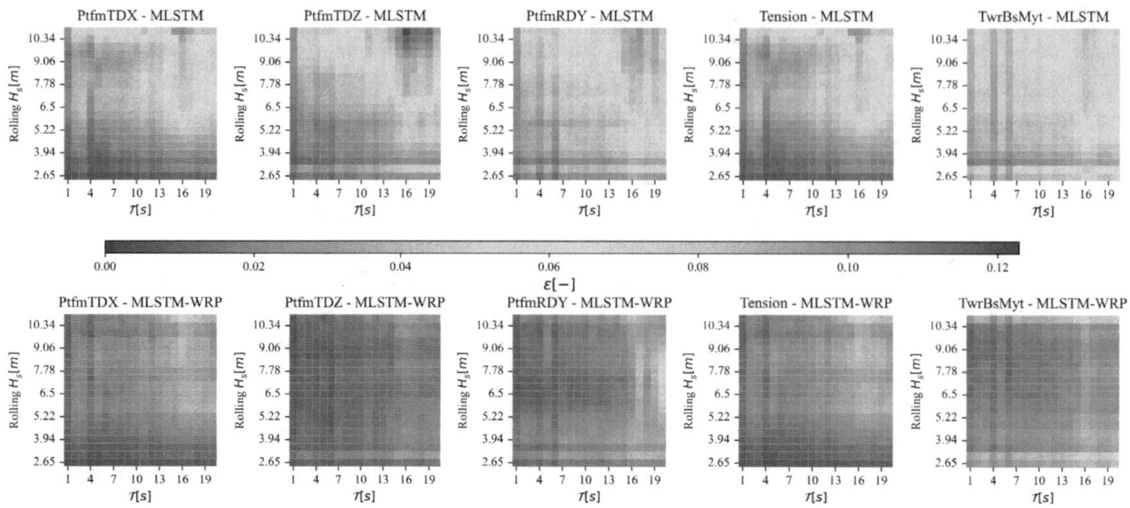


**Figure 7.** Normalized error between measured values and those predicted by MLSTM (first row) and MLSTM-WRP (second row) as a function of prediction horizon and rolling significant wave height for various states

an epoch number of 60, a batch size of 53, utilization of the Adam optimization algorithm, and a mean absolute error (MAE) as the loss function. The training regime for both models encompassed a range of prediction horizons, $1 < \mathcal{T}[s] < 20$. Subsequent predictions are executed on an unseen dataset. The error, $\varepsilon$, normalized by the range of the measured state is calculated at each timestep, comparing the actual, $\xi_j$, and forecasted values, $\hat{\xi_j}$, as delineated in Eq. 2:

$$\varepsilon(t) = \frac{\mid \hat{\xi_j}(t) - \xi_j(t)\mid}{\xi_{jmax} - \xi_{jmin}} \tag{2}$$

The states investigated are the horizontal surge displacement of the floater aligned with the wave direction (PtfmTDX), vertical heave displacement of the floater (PtfmTDZ), pitch rotation in the fore-aft direction (PtfmRDY), the highest loaded mooring line tension (Tension), and the fore-aft tower base bending moment (TwrBsMyt). Figure 7 illustrates the normalized error as a function of the rolling significant wave height, $H_s$, and prediction horizon, $\mathcal{T}$, for the aforementioned states, as predicted by both the MLSTM and MLSTM-WRP models. The rolling significant wave height is determined by averaging the highest one-third of wave heights, obtained using a zero-crossing method over a rolling period of $T_r = 1000s$. The analysis, as depicted in Figure 7, reveals that although the MLSTM model yields relatively satisfactory predictions under moderate wave conditions and within low to medium prediction horizons, its precision diminishes under extreme wave conditions and at longer prediction horizons. Conversely, the MLSTM-WRP model demonstrates a significant 50% reduction in average error. Moreover, the error margins are remarkably consistent, showcasing negligible variance, thereby affirming the superior accuracy and generalizability of the MLSTM-WRP model.

The time-series of the measured signal and the predicted values by the two models and their respective power spectral density (PSD) in the main DOF are given in Figures 8 and 9 for WRP prediction horizon of $\mathcal{T} = 5s$ and $20s$, respectively. For short-term predictions ($\mathcal{T} = 5s$), both models indicate good agreement with the observations except at the low-frequency region in

 2745 (2024) 012001 doi:10.1088/1742-6596/2745/1/012001 

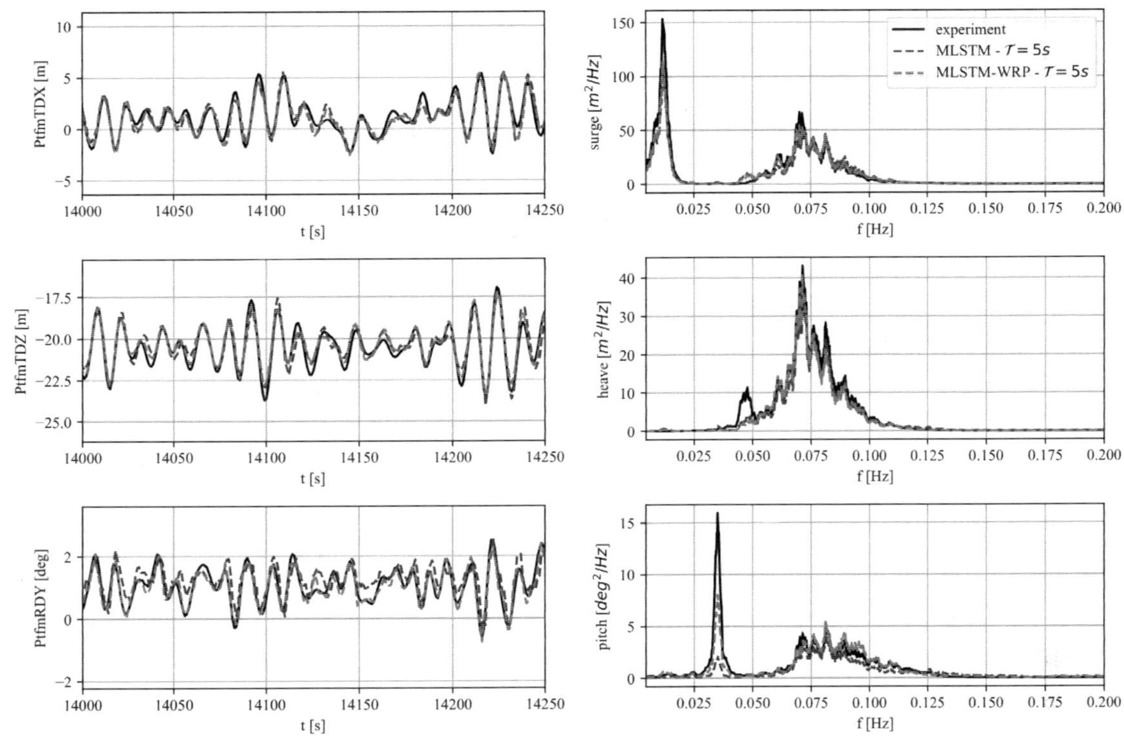


**Figure 8.** Time-series history and the corresponding power spectra density of the floater's motion in surge (PtfmTDX), heave (PtfmTDZ), and pitch (PtfmRDY) and the predicted values using MLSTM and MLSTM-WRP for a prediction horizon of $\mathcal{T} = 5s$.

heave and pitch. The authors believe this is due to the fact that short-term prediction horizon does not cover the full natural period in heave and pitch. The semi-submersible platform under investigation has heave and pitch natural periods in the 20-30 seconds region. For MLSTM-WRP, this discrepancy is eliminated with the longer prediction horizon of ($\mathcal{T} = 20s$) as depicted in Figure 9. MLSTM at longer prediction horizon suffers immensely in most frequency regions. Further analysis on this platform and on different systems is needed to generalize this observation.

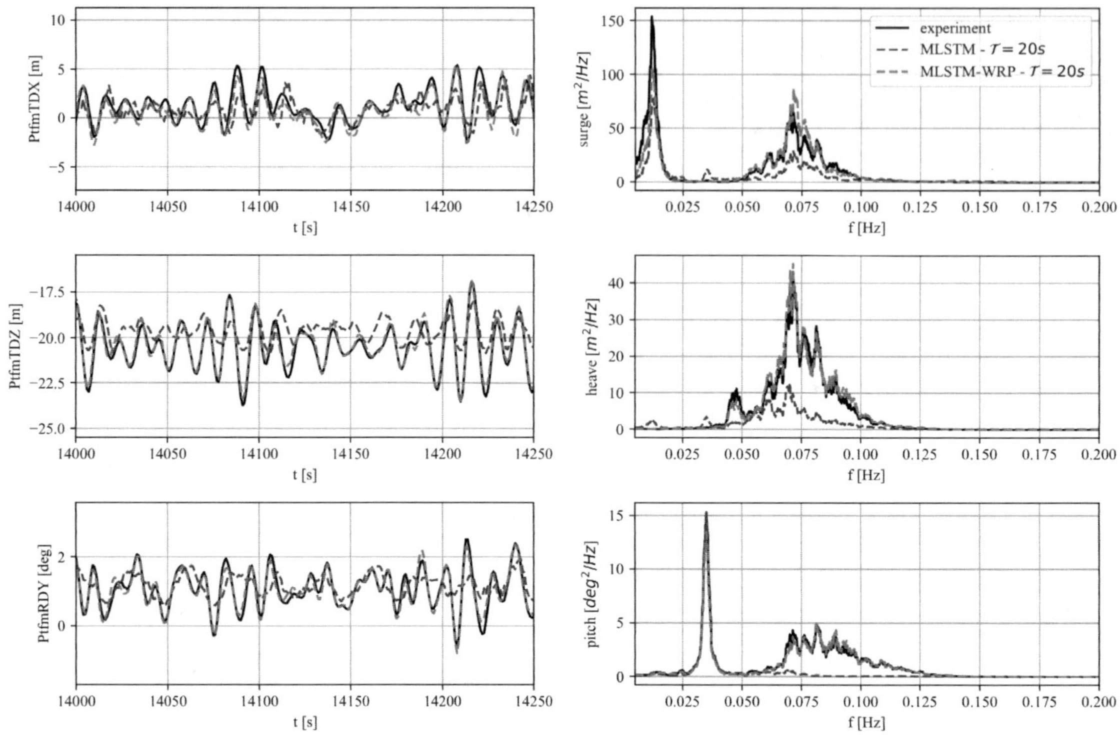


**Figure 9.** Time-series history and the corresponding power spectra density of the floater's motion in surge (PtfmTDX), heave (PtfmTDZ), and pitch (PtfmRDY) and the predicted values using MLSTM and MLSTM-WRP for a prediction horizon of $\mathcal{T} = 20s$.

### *3.2. WRP sensitivity analysis*

The error function used in this study is inspired by the work of [16]. As the ratio of $t/T_p$, where $T_p$ is the instantaneous peak wave period, increases or decreases away from 0.0, the error of the predictor increases. Two coefficients are introduced to modify the time-dependent error dependency; $\alpha$, which is multiplied by the height of the error function to amplify/condense the error throughout the width of the function, and $\beta$ which is multiplied by the width of the error function to shrink/expand the prediction accuracy over the horizon. Ranging these two coefficients in combination ($\alpha \in [1.0, 2.0]$, $\beta \in [1.0, 10.0]$) results in the error functions and the subsequent coefficient of determination, $R^2$, for various time-series signals described in Figure 10. The prediction horizon is set to $\mathcal{T} = 20s$ to create more contrast in the results of the analysis. All three mooring line analysis are added here. It is observed that the model is sensitive to the accuracy of the wave predictor. The models ability to predict the system's states reduce as amplification/shrinkage combination increases. The state with the highest sensitivity is the tower base bending moment.

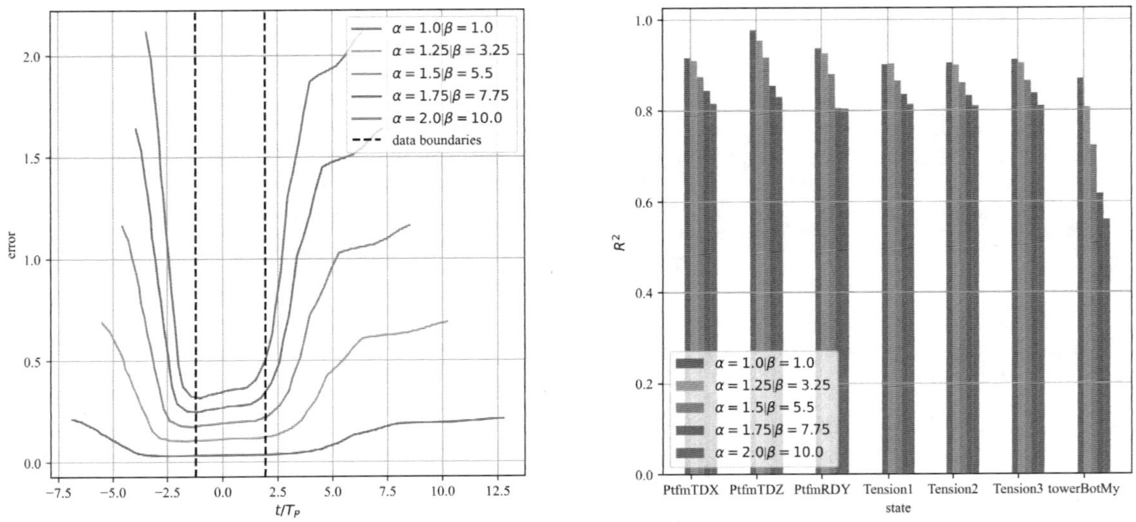


**Figure 10.** Error function amplification and shrinkage as a function of normalized time (left), model accuracy in predicting system's states for varying accuracy of wave predictions

## 4. Conclusions and Remarks for Future Work

For the enhancement and development of offshore wind systems, accurate and quick phase-resolved predictions of the sea-state and its effect on the systems dynamics is the first step to ensure the digital twin is up-to-date and to build better control strategies to enhance the longevity of the system. For instance, a structural controller, when using future wave forecast would significantly improve float control in large waves. Multi-variate long short term memory recurrent neural network, when fitted correctly, is a quick and accurate method to predicting system's states. However, it is notoriously weak when it comes to predicting multiple steps ahead in time as error accumulates.

Integrating WRP capabilities to the MLSTM extend the assurance of obtaining accurate predictions with a prediction horizon governed by the software and hardware capabilities of the wave reconstruction prediction model. Preliminary sensitivity study shows that MLSTM-WRP precision deteriorates with the lowered quality of the wave reconstruction with a tower-base bending moment being more vulnerable to wave prediction accuracy than others.

Future work will extend to this research by applying this method on wind/wave cases to investigate the accuracy of the MLSTM-WRP model. Sensitivity studies will also cover a broad range of faults created by the LiDAR measurement systems. Ultimately, control integration with a WRP application is to be deployed.

## 5. Acknowledgment

The authors would like to acknowledge the Department of Energy for funding this research under project name: Ocean_193360_1, project grant #: DE-SC0022103.

## References

[1] Antonio LJ. 2020. Communication from the Commission to the European Parliament, The Council, The European Economic and Social.

[2] Department of Energy, U.S. 2023. The Office of Energy Efficiency and Renewable Energy (EERE) `https://www.energy.gov/eere/wind/wind-energy-technologies-office`.

[3] Jonkman, J.M. and Matha, D., 2011. Dynamics of offshore floating wind turbines—analysis of three concepts. Wind Energy, 14(4), pp.557-569.

[4] Robertson, A., Jonkman, J., Musial, W., Popko, W. and Vorpahl, F., 2014. IEA Wind Task 30 Offshore Code Comparison Collaboration Continued.

[5] Robertson, A.N., Wendt, F., Jonkman, J.M., Popko, W., Dagher, H., Gueydon, S., Qvist, J., Vittori, F., Azcona, J., Uzunoglu, E. and Soares, C.G., 2017. OC5 project phase II: validation of global loads of the DeepCwind floating semisubmersible wind turbine. Energy Procedia, 137, pp.38-57.

[6] Robertson, A.N., Gueydon, S., Bachynski, E., Wang, L., Jonkman, J., Alarcon, D., Amet, E., Beardsell, A., Bonnet, P., Boudet, B. and Brun, C., 2020, September. OC6 Phase I: Investigating the underprediction of low-frequency hydrodynamic loads and responses of a floating wind turbine. In Journal of Physics: Conference Series (Vol. 1618, No. 3, p. 032033). IOP Publishing.

[7] Wang, L., Robertson, A., Jonkman, J., Kim, J., Shen, Z.R., Koop, A., Borràs Nadal, A., Shi, W., Zeng, X., Ransley, E. and Brown, S., 2022. OC6 Phase Ia: CFD simulations of the free-decay motion of the DeepCwind semisubmersible. Energies, 15(1), p.389.

[8] Alkarem, Y.R. and Ozbahceci, B.O., 2021. A complemental analysis of wave irregularity effect on the hydrodynamic responses of offshore wind turbines with the semi-submersible platform. Applied Ocean Research, 113, p.102757.

[9] Steele, S.C., Albertson, S., Fontaine, J., Dahl, J.M., Grilli, S.T., Hashemi, R.M., Alkarem, Y.R., Kimball, R.W. and Hejrati, B., 2023. Active Control of the Wave-Induced Motions of a Float: Real-Time Simulations with a Digital Twin and Experimental Validation. In Proc. 33rd Intl. Offsh. and Polar Engng. Conf. ISOPE.

[10] Alkarem, Y.R., Huguenard, K., Kimball, R.W., Hejrati, B., Goupee, A., Albertson, S., Fontaine, J., Grilli, S.T., Steele, S. and Dahl, J., 2023, June. Digital Twin Model Development for Mitigating Floating Offshore Wind Turbine Motions Due to Wave Actions. In ISOPE International Ocean and Polar Engineering Conference (pp. ISOPE-I). ISOPE.

[11] Branlard, E., Jonkman, J., Brown, C. and Zang, J., 2023. A digital-twin solution for floating offshore wind turbines validated using a full-scale prototype. Wind Energy Science Discussions, 2023, pp.1-34.

[12] Branlard, Emmanuel, Jason Jonkman, Scott Dana, et al. (2020). "A digital twin based on Open- FAST linearizations for real-time load and fatigue estimation of land-based turbines". In: Journal of Physics: Conference Series. Vol. 1618. 2. IOP Publishing, p. 022030.

[13] Kimball, R., Robertson, A., Fowler, M., Mendoza, N., Wright, A., Goupee, A., Lenfest, E. and Parker, A., 2022, May. Results from the FOCAL experiment campaign 1: turbine control co-design. In Journal of Physics: Conference Series (Vol. 2265, No. 2, p. 022082). IOP Publishing.

[14] Allen, C., Viscelli, A., Dagher, H., Goupee, A., Gaertner, E., Abbas, N., Hall, M. and Barter, G., 2020. Definition of the UMaine VolturnUS-S reference platform developed for the IEA wind 15-megawatt offshore reference wind turbine (No. NREL/TP-5000-76773). National Renewable Energy Lab.(NREL), Golden, CO (United States); Univ. of Maine, Orono, ME (United States).

[15] Gaertner, E., 2020. Definition of the IEA wind 15-megawatt offshore reference wind turbine Tech. Rep. NREL/TP-5000-75698 National Renewable Energy Laboratory Golden, CO.

[16] Desmars, N., Bonnefoy, F., Grilli, S.T., Ducrozet, G., Perignon, Y., Guérin, C.A. and Ferrant, P., 2020. Experimental and numerical assessment of deterministic nonlinear ocean waves prediction algorithms using non-uniformly sampled wave gauges. Ocean Engineering, 212, p.107659.

[17] Jonkman, J., 2015. Fast simulation tool containing methods for predicting the dynamic response of wind turbines (No. FAST). National Renewable Energy Lab.(NREL), Golden, CO (United States).

[18] Ammerman I., Hejrati, B., Alkarem, Y., Huguenard, K., Kimball, R., 2023. Experimental Validation of a Kalman Observer using Linearized OpenFAST and a Fully-instrumented 1:70 Model. Wind Energy

[19] Wang, L., Bergua, R., Robertson, A., Jonkman, J., Ngo, T., Das, T., Sarker, D., Flavia, F.F., Harries, R., Fowler, M. and Lenfest, E., 2023, October. Experimental Validation of Models of a Hull-Based Tuned Mass Damper System for a Semisubmersible Floating Offshore Wind Turbine Platform. In Journal of Physics: Conference Series (Vol. 2626, No. 1, p. 012067). IOP Publishing.

[20] Fowler, M.J., 2023. Floating Offshore-wind and Controls Advanced Laboratory Program: 1: 70-scale Testing of a 15 Mw Floating Wind Turbine.

[21] Kim, I.C., Ducrozet, G., Bonnefoy, F., Leroy, V. and Perignon, Y., 2023. Real-time phase-resolved ocean wave prediction in directional wave fields: Enhanced algorithm and experimental validation. Ocean Engineering, 276, p.114212.

[22] Kim, I.C., Ducrozet, G., Leroy, V., Bonnefoy, F., Perignon, Y. and Bourguignon, S., 2024. A real-time wave prediction in directional wave fields: Strategies for accurate continuous prediction in time. Ocean Engineering, 291, p.116445.

[23] National Renewable Energy Lab. OpenFAST. `https://openfast.readthedocs.io/en/main/`

[24] Jonkman, J.M., Wright, A.D., Hayman, G.J. and Robertson, A.N., 2018, November. Full-system linearization for floating offshore wind turbines in OpenFAST. In International Conference on Offshore Mechanics and Arctic Engineering (Vol. 51975, p. V001T01A028). American Society of Mechanical Engineers.

Journal of Physics: Conference Series **2745** (2024) 012002 doi:10.1088/1742-6596/2745/1/012002

# Improving data sharing in practice – power curve benchmarking case study

**Sarah Barber[1], Yu Ding[2]**

[1]Institute for Energy Technology, OST - Eastern Switzerland University of Applied Sciences, Oberseestrasse 10, CH-8640 Rapperswil-Jona.

[2]H. Milton Stewart School of Industrial and Systems Engineering, Georgia Institute of Technology, 755 Ferst Drive NW, Atlanta, GA 30332-0205, USA

E-mail: `sarah.barber@ost.ch`

**Abstract.** In this work, a data sharing case study is presented, aiming to investigate and demonstrate how data sharing can be improved in practice in the wind energy sector. The case study is part of the WeDoWind framework, which creates tangible incentives to motivate different types of people to actually share data in practice. For this, a WeDoWind "challenge" on the topic of wind turbine power curve benchmarking is created and implemented within the framework. The results allow five different data-driven power curve prediction methods to be compared. The best method reduces the model error by as much as 70% in terms of mean absolute error and 45% in terms of root mean squared error compared to the standard industry method of binning. The results of a survey filled out by the participants show that data sharing could be improved compared to previous WeDoWind case studies, by providing more clear comparison and evaluation criteria, as well as by better integrating students into the WeDoWind framework. Overall, we find that "challenge"-based collaborations can help the industry become more innovative, by providing a motivation and basis for sharing data, as well as for comparing and benchmarking different methods. Finally, our experience in doing this as part of the present work allows us to make some suggestions for improving data sharing in practice.

## 1. Introduction

A lack of data sharing is a widely-recognised challenge in the wind energy sector today, as demonstrated in a recent review paper [1]. In this paper, the main challenges related to digitalisation in wind energy were recently established to be (1) creating FAIR[1] data frameworks, (2) connecting people and data to foster innovation, and (3) enabling collaboration and competition between organisations. Some attempts have already been made to address these challenges, including the Sharewind metadata registry [2], data marketplaces including the Greenbyte marketplace for wind data[2] and the IntelStor Market Intelligence Ecosystem[3], comparison and benchmarking activities including IEA Wind Task 31[4], IEA Wind Task 30 (OC6) WP3 Benchmark[5] and CREYAP: Comparison of Resource and Energy Yield Assessment

[1] Findable, Acessible, Interoperable, Reusable

[2] `https://www.greenbyte.com/marketplace/wind-only` (accessed on 3 August 2022)

[3] `https://www.intelstor.com/` (accessed on 3 August 2022)

[4] `https://iea-wind.org/task31/` (accessed on 3 August 2022)

[5] `https://iea-wind.org/task30/` (accessed on 29.02.2024)

Journal of Physics: Conference Series **2745** (2024) 012002 doi:10.1088/1742-6596/2745/1/012002

Procedures [3], open-source tools such as the Brightdata app[6], OpenOA[7] and the Data Science for Wind Energy R Library[8]. More information can be found in [4].

There are many different types of collaboration in which data sharing is necessary, including the sharing of design data between wind turbine manufacturers and component providers, between wind farm planners and buyers, between operators and owners, as well as between researchers and industry. In order to improve data sharing between researchers and industry, the WeDoWind framework has recently been developed and demonstrated in two use cases related to wind turbine fault detection [4, 5]. The main innovation of this framework is the way it creates tangible incentives to motivate different types of people to actually share data in practice. It helps people and organisations solve the most pressing problems collaboratively based on "challenges" and data provided by the industry. This improves data sharing in the sense that new data sets are made available to researchers, and/or data sets are made more accessible and usable to researchers. It was shown that it is possible to motivate asset owners to share new data sets with groups of innovators, and that the resulting solutions to the "challenges" provide added value to the asset owners. A recent assessment of participants' surveys filled out after the two case studies also allowed several obstacles related to data sharing in the context of WeDoWind to be identified, including the need to provide more clear comparison and evaluation criteria, to better integrate students into the process, to well-describe the data through standardised metadata and to prepare sufficient data for a given task [6].

The goal of the present work is to carry out a further data sharing case study, aiming to further investigate and demonstrate how data sharing between researchers and industry can be enhanced in practice in the wind energy sector. In particular, we address how to (1) provide more clear comparison and evaluation criteria and (2) better integrate students into the data sharing process. The topic of this case study is wind turbine power curve benchmarking, which is interesting and important for the wind energy sector. While there have been some comparison studies of different power curve methods, see for example, Table 1 in [7] or Table 5.7 in [8], there has not been a public blind study in which different teams trained their respective models without knowing the power output in the test dataset.

The work presented in this paper was carried out in the following steps: Design and implementation of the case study (Section 2); Summary and comparison of the results (Section 3); Implementation and evaluation of a participants' survey (Section 4); Discussion of how data sharing can be improved in practice (Section 5). These steps are described in the remainder of this paper, followed by the conclusions in Section 6.

## 2. Design and implementation of the case study

WeDoWind is a framework for bringing asset owners together with researchers and model developers in a "win-win" situation, whereby asset owners get easy access to state-of-the-art data analytics and model developers get access to relevant asset data to train and validate their models. For this case study, a new WeDoWind "space" was created together with the co-author of this paper, Yu Ding, author of the book, *Data Science for Wind Energy* [8]. A new WeDoWind "challenge" was then created in this "space". The goal of this "challenge" was to benchmark and discuss data-driven power curve models. We addressed the above-mentioned data-sharing obstacles of how to (1) provide more clear comparison and evaluation criteria and (2) better integrate students into the process as described below.

[6] `https://www.brightwindanalysis.com/brightdata/` (accessed on 3 August 2022)
[7] `https://github.com/NREL/OpenOA` (accessed on 3 August 2022)
[8] `https://github.com/TAMU-AML/DSWE-Package/` (accessed on 3 August 2022)

*2.1. How to provide more clear comparison and evaluation criteria*

In order to provide more clear comparison and evaluation criteria compared to previous WeDoWind "challenges", we firstly downloaded the open data from the Kelmarsh and Penmanshiel wind farms [9, 10]. Kelmarsh has six sets of wind turbine data and Penmanshiel has 14 sets. The datasets contain 10-minute averaged wind turbine SCADA data. The data was pre-processed by firstly eliminating negative power or zero power when there is a valid wind speed. Next, the datasets were separated into training and test datasets. After this step, the training dataset for each turbine spanned one calendar year. The remaining valid data corresponded to 75% to 90% of the original 52,560 records per turbine, depending on the turbine.

The **training dataset** has the following 14 columns for each data record—one is the time when the data was recorded, one is the active power measured, i.e., the response, and the remaining 12 are the environmental inputs, as follows:

- Wind speed
- Wind speed sensor 1
- Wind speed sensor 1 standard deviation
- Wind speed sensor 2
- Wind speed sensor 2 standard deviation
- Air density adjusted wind speed
- Wind direction
- Nacelle position
- Wind direction standard deviation
- Nacelle position standard deviation
- Ambient temperature
- Turbulence intensity

For the **test dataset**, the parameter to be predicted ("active power") and the timestamp were removed, which ensured that the test data was not easily traceable in the original dataset. Two additional columns were added indicating (a) whether it is a daytime measurement (between 6am and 6pm local time) or it is a nighttime measurement (from 6pm to 6am of the next day) and (b) which month the measurement was taken. This was done on request of the participants because this information would be known in practice. The job of the participants was to make a prediction on the missing (withheld) active power in the test datasets. In order to do this, an upload link was provided, and participants had to upload their results with the "active power" column completed with their predicted values. This ensured that the results could be easily processed and the evaluation criteria calculated in a consistent way. The test datasets are from the same wind farms and same wind turbines but different years. The test datasets are also of one year in duration. However, the data is not of one natural year but assembled from multiple years; for instance, the January data may be from 2016 but the February data is from 2019 and so on. This was done on order to mix up the data so that it was not traceable back to the original dataset. However, the data within each month remained temporally sequenced.

After this, two performance criteria were required in the solutions: the root mean squared error (RMSE) and the mean absolute error (MAE). For a given turbine, $y$ denotes the withheld active power and $\hat{y}$ denotes a participant's prediction. For turbine $t$, its RMSE is defined as:

$$\mathrm{RMSE}^{(t)} = \sqrt{\frac{1}{n_T}\sum_{i=1}^{n_T}\left(y_i^{(t)} - \hat{y}^{(t)}(\boldsymbol{x}_i^{(t)})\right)^2}, \tag{1}$$

where $n_T$ is the number of data points associated with this specific turbine in the test set and $\hat{y}^{(t)}(\boldsymbol{x}_i^{(t)})$ signifies that the prediction is made for turbine $t$, when taking $\boldsymbol{x}_i$ of turbine $t$ as its input. Using the same notations, the MAE for turbine $t$ is defined as

$$\mathrm{MAE}^{(t)} = \frac{1}{n_T}\sum_{i=1}^{n_T}\left|y_i^{(t)} - \hat{y}^{(t)}(\boldsymbol{x}_i^{(t)})\right|. \tag{2}$$

These values were calculated by the "challenge" provider for each individual turbine using the actual, withheld power measurements and the predicted power submitted by the participants. The results were ranked in a "leaderboard", which was available to the participants, by averaging the RMSE and MAE for all the wind turbines in each wind farm. We ranked the performance of a submission based on the grand average of RMSEs and MAEs of all turbines in the test datasets, i.e.,

$$\overline{\mathrm{RMSE}} \equiv \frac{1}{m}\sum_{t=1}^{m}\mathrm{RMSE}^{(t)} \quad \text{and} \quad \overline{\mathrm{MAE}} \equiv \frac{1}{m}\sum_{t=1}^{m}\mathrm{MAE}^{(t)}, \tag{3}$$

where $m$ is the total number of turbines and in this "challenge," $m = 20$.

### *2.2. How to better integrate students into the process*

We attempted to better integrate students into the data sharing process by providing teaching material and video recordings of a graduate-level machine learning course alongside the monthly meetings and interim presentation webinars for the participants. This allowed participants to learn about different machine learning techniques as they progressed with the challenge. Even though the space was available to anyone in the world with an email address and an internet connection, the students of a machine learning course at Texas A&M University were also required to take part in this "challenge" as part of their studies.

## 3. Summary and comparison of the results

The "challenge" was launched on January 9, 2023 and concluded on June 14, 2023. Monthly webinars allowed participants to discuss the progress and discuss any questions or ideas with the "challenge" provider. There was an interim presentation on March 22. Participants were required to register and thus receive a participant ID. They were allowed to submit their predictions at anytime. The final presentation and conclusion of the "challenge" took place on June 14.

In total, 11 teams registered to take part, and five teams submitted full results, among which three were the student teams from the Texas A&M University machine learning class mentioned above. When a team submitted multiple results, their submission is marked with a "ref00" after their participation ID. A baseline using the IEC binning method [11] was provided by the "challenge" provider. Additionally, one participating team submitted results for one turbines using the IEC standard method. Since this was done only for a single turbine, the result was not included in the leaderboard. However, the results matched well with the baseline.

The leaderboard includes two lists, based on the grand average of RMSE and MAE, respectively. The two rankings do not necessarily agree with each other. As mentioned earlier, in order for a submission to be included in the leaderboard, the submission must include active power prediction for all 20 turbines on the two wind farms. A submission including predictions only for a subset of turbines is deemed "incomplete" and its result is not included in the leaderboard.

Table 1 summarises the submitted machine learning methods used by each team. If one team submitted multiple results, usually only the best result is included, with the only exception

**Table 1.** Methods used by different teams

| Team | Method | Note |
|---|---|---|
| ID_5 | Multi-layer neural network | No. 6 on the MAE list and No. 8 on the RMSE List |
| ID_7 Method 1 (Ref006) | XGBoost | No. 1 on the MAE list and No. 3 on the RMSE list |
| ID_7 Method 2 (Ref007) | TempGP. Use the `R` package [13]. TempGP was recently published [12] | No. 5 on the MAE list and No. 2 on the RMSE list |
| ID_8 | Multi-layer neural network | Multiple submissions with improved tuning of the baseline method. The best result is No. 7 on the MAE list and No. 5 on the RMSE list |
| ID_9 | XGBoost | Four submissions with different tuning strategies. Two of the four submissions are competitive. They are No. 2 and No 3 on the MAE list, and No. 6 and No. 7 on the RMSE list. |
| ID_11 | Multi-layer neural network | No. 4 on the MAE list and No. 1 on the RMSE list. |

for ID_7. Two of the ID_7's methods are included, as they are different methods, rather than different fine tuning of the same basic method, and also the two included methods both show sufficient competitiveness in performance. The different methods included multi-layer neural networks, XGBoost and TempGP. Multi-layer neural networks are neural networks with more than one layer of artificial neurons or nodes. XGBoost (eXtreme Gradient Boosting) is an open-source software library which provides a regularising gradient boosting framework. TempGP [12] is a Gaussian process model designed for data with temporal auto-correlation.

Tables 2 and 3 present the performance comparison of each solution based on RMSE and MAE, respectively, which are normalised with the rated power (1 = rated power). The last column of each table indicates the percentage increase of the method in that row relative to the best result in the same table. A larger number in this column indicates a worse performance for the said method.

It can be seen that the ranking based on MAE does not fully agree with the ranking based on RMSE. This is not surprising. These two performance criteria are not mathematically identical. On the other hand, they are related, so that a method competitive on the MAE list tends to do well on the RMSE list, too. One can clearly observe that the machine learning methods can achieve a significant reduction in modelling errors while learning a turbine-specific power curve. The best method reduces the model error by as much as 70% in terms of MAE and 45% in terms of RMSE. It is interesting to note that both best results were obtained by Texas A&M University student teams, who did not have prior experience with wind turbine power curve modelling.

We would like to stress the following. The top five methods on the MAE list have a MAE value within 2.5% of the best MAE, whereas the top three methods on the RMSE list have an RMSE value within 3.1% of the best RMSE. These relatively small differences are probably not enough to differentiate these methods in terms of their general performance in future datasets, considering that the current results were obtained from two wind farms. It would be more prudent to treat the methods in the top group all as competitive and use additional datasets to evaluate them further.

## 4. Implementation and evaluation of a participants' survey

Following the completion of the "challenge", a survey was sent out to the participants. The goal of this survey was to understand if data sharing had been improved compared to previous WeDoWind "challenges" following our attempts to (1) provide more clear comparison and evaluation criteria and (2) better integrate students into the data sharing process. The survey was split into several sections related to motivation, value added and suggestions for improvements. Some sections involved quantifying the participants' experiences on a scale of one to five, and some were qualitative. We did not include specific questions related to the attempted improvements in this study, because we wanted to directly compare survey answers with the two previous challenges, which had already been completed at the start of the current project.

In general, the feedback was positive. Figure 1 shows a summary of the quantitative answers compared to a previous study for two other "challenges" (called "EDP" and "WinJi") as presented in [6]. In this summary, answers of 1 and 2 have been given the label "Disagree", 3 = "Neutral", 4 and 5 = "Agree". In the survey section "General benefits", the percentage of participants answering "Agree" increased in comparison to the previous projects for all of the questions except "Participation in this challenge resulted in some useful new contacts". This indicates that providing teaching material and video recordings of a graduate-level machine learning course and requiring the students of the course to take part in this "challenge" as part of their studies enabled students to be better integrated into the process (obstacle number (2) described in Section 2.2). However, not as many contacts could be made due to the high proportional of students participating.

Furthermore, in the survey section "The data and the challenge", 100% of the participants answered "Agree" to all of the questions. This demonstrates that we were successfully able to provide more clear comparison and evaluation criteria in this case study (obstacle number (1) described in Section 2.1). Not only was the "challenge" topic perceived to be realistic and interesting, but the provided data was of sufficient quality and quantity for solving the challenge, and it was sufficiently documented and labelled. In the "Process" section, the digital platform was found to be useful for documenting the "challenge"; however, its ease of use and technical support was once again a weak point that needs working on. The results of the "Evaluation" section confirms that the comparison and evaluation criteria were more clear– a lower proportion of participants said that they would have preferred a formal evaluation method to have been defined by the "challenge" providers at the beginning (obstacle number (1) described in Section 2.1).

Finally, the "Future potential" section shows a promising future for this type of collaboration– 100% of the participatns thought that WeDoWind could help the wind energy sector benchmark and evaluate ML methods for various applications in the future, that WeDoWind could provide a central knowledge hub for wind energy in the future, they intend to take part in a WeDoWind "challenge" in the future and they would like future WeDoWind "challenges" to include more collaboration components.

In summary, the survey results provide a strong indicator that data sharing was improved compared to previous WeDoWind "challenges" by (1) providing more clear comparison and evaluation criteria and (2) better integrating students into the data sharing process. It should be noted that these results are based on a fairly low number of participant responses (nine for "EDP", seven for "WinJi" and six for the present study), and therefore the results should be treated with case. Further case studies with more participants are underway in order to further investigate how to improve data sharing in the sector.

## 5. Discussion of how data sharing can be improved in practice

Data sharing is a complex problem with an array of different obstacles, including confidentiality, cybersecurity, data privacy, fear of losing competitive advantage, organisational and sector-wide culture. There are many different types of collaboration in which data sharing is necessary, including the sharing of design data between wind turbine manufacturers and component providers, between wind farm planners and buyers, between operators and owners, as well as between researchers and industry. Fostering data sharing is therefore not easy, and requires a range of different strategies. In this work, we focus on fostering data sharing between industrial partners and academia, with the hope that the learning outcomes and strategies developed will be transferable to other groups of people and sectors.

This work indicates that data sharing between academia and industry can be improved by (1) providing more clear comparison and evaluation criteria and (2) better integrating students into the data sharing process. Furthermore, participants indicated that "challenge"-based collaborations can help the industry become more innovative, by providing a motivation and basis for sharing data, as well as for comparing and benchmarking different methods. Our experience in doing this as part of the present work allowed us to make some suggestions for improving data sharing in practice:

- Data owners are more likely to share data if a concrete benefit is visible. A benchmarking project with well-defined comparison metrics can help to achieve this.
- Students can be encouraged to use the shared data by integrating a "challenge" into their course of studies.
- Including students as participants can increase the quality and innovation level of the resulting solutions.
- A digital platform is useful to enable documentation of the "challenges" and communication between the participants and "challenge" providers. However, it is challenging to achieve the required level of usability of this platform.

Further work and discussion of the learning outcomes related to data sharing is required across and beyond the wind energy sector, in order to address the complex problem of improving data sharing. This requires further data sharing projects to be carried out and solutions to the different data sharing obstacles to be developed collaboratively.

## 6. Conclusions

In this work, a data-sharing case study was carried out, aiming to investigate and demonstrate how data sharing can be improved in practice in the wind energy sector. The case study was part of the WeDoWind framework, which creates tangible incentives to motivate different types of people to actually share data between academia and industry in practice. For this, a WeDoWind "challenge" on the topic of wind turbine power curve benchmarking was created and implemented within the framework. The results allowed five different data-driven power curve prediction methods to be compared. The best method reduced the model error by as much as 70% in terms of MAE and 45% in terms of RMSE compared to the standard industry method. The results of a survey filled out by the participants showed that data sharing could be improved compared to previous WeDoWind case studies, by providing more clear comparison and evaluation criteria, as well as by better integrating students into the WeDoWind framework. Overall, we found "challenge"-based collaborations can help the industry become more innovative, by providing a motivation and basis for sharing data, as well as for comparing and benchmarking different methods. Our experience in doing this as part of the present work allowed us to make some suggestions for improving data sharing in practice.

**Table 2.** Performance comparison of participating teams based on normalised MAE

| Participant_ID | Submission_ID | MAE | RMSE | MAE increase relative to the lowest MAE |
|---|---|---|---|---|
| 7 | ID_7_ref006 | 0.0146 | 0.0262 | 0% |
| 9 | ID_9_ref003 | 0.0148 | 0.0286 | 1.4% |
| 9 | ID_9_ref001 | 0.0149 | 0.0287 | 1.5% |
| 11 | ID_11_ref001 | 0.0150 | 0.0254 | 2.2% |
| 7 | ID_7_ref007 | 0.0150 | 0.0259 | 2.5% |
| 5 | ID_5_ref001 | 0.0153 | 0.0288 | 4.5% |
| 8 | ID_8_ref003 | 0.0158 | 0.0284 | 7.9% |
| 7 | ID_7_ref003 | 0.0160 | 0.0281 | 9.3% |
| 7 | ID_7_ref001 | 0.0167 | 0.0295 | 14.1% |
| 7 | ID_7_ref005 | 0.0171 | 0.0289 | 16.8% |
| 9 | ID_9_ref004 | 0.0178 | 0.0316 | 21.3% |
| 9 | ID_9_ref002 | 0.0178 | 0.0316 | 21.6% |
| 8 | ID_8_ref001 | 0.0195 | 0.0338 | 33.2% |
| 8 | ID_8_ref002 | 0.0196 | 0.0320 | 33.9% |
| 7 | ID_7_ref002 | 0.0199 | 0.0325 | 35.9% |
| IEC Binning | Challenge provider | 0.0250 | 0.0370 | 70.8% |
| 7 | ID_7_ref004 | 0.0264 | 0.0416 | 80.3% |

**Table 3.** Performance comparison of participating teams based on normalised RMSE

| Participant_ID | Submission_ID | MAE | RMSE | RMSE increase relative to the lowest RMSE |
|---|---|---|---|---|
| 11 | ID_11_ref001 | 0.0150 | 0.0254 | 0% |
| 7 | ID_7_ref007 | 0.0150 | 0.0259 | 1.9% |
| 7 | ID_7_ref006 | 0.0146 | 0.0262 | 3.1% |
| 7 | ID_7_ref003 | 0.0160 | 0.0281 | 10.6% |
| 8 | ID_8_ref003 | 0.0158 | 0.0284 | 11.8% |
| 9 | ID_9_ref003 | 0.0148 | 0.0286 | 12.7% |
| 9 | ID_9_ref001 | 0.0149 | 0.0287 | 13.0% |
| 5 | ID_5_ref001 | 0.0153 | 0.0288 | 13.4% |
| 7 | ID_7_ref005 | 0.0171 | 0.0289 | 13.8% |
| 7 | ID_7_ref001 | 0.0167 | 0.0295 | 16.1% |
| 9 | ID_9_ref004 | 0.0178 | 0.0316 | 24.2% |
| 9 | ID_9_ref002 | 0.0178 | 0.0316 | 24.5% |
| 8 | ID_8_ref002 | 0.0196 | 0.0320 | 26.0% |
| 7 | ID_7_ref002 | 0.0199 | 0.0325 | 27.9% |
| 8 | ID_8_ref001 | 0.0195 | 0.0338 | 33.0% |
| IEC Binning | Challenge provider | 0.0250 | 0.0370 | 45.6% |
| 7 | ID_7_ref004 | 0.0264 | 0.0416 | 63.7% |

| | EDP | | | WinJi | | | Power Curve | | |
|---|---|---|---|---|---|---|---|---|---|
| Percentage of answers | "Agree" | "Neutral" | "Disagree" | "Agree" | "Neutral" | "Disagree" | "Agree" | "Neutral" | "Disagree" |
| **General benefits** | | | | | | | | | |
| Participation in this challenge gave me some new insights or ideas I can use for my research | 57% | 29% | 14% | 40% | 40% | 20% | 67% | 33% | 0% |
| Participation in this challenge resulted in some useful new contacts | 43% | 29% | 29% | 60% | 20% | 20% | 33% | 67% | 0% |
| Participation in this challenge increased my visibility in the wind energy sector | 43% | 29% | 29% | 20% | 60% | 20% | 67% | 33% | 0% |
| Participation in this challenge helped me understand the challenges of using real data from the industry | 72% | 0% | 29% | 80% | 20% | 0% | 100% | 0% | 0% |
| Participation in this challenge helped me compare my work with others and evaluate its success | 72% | 14% | 14% | 100% | 0% | 0% | 100% | 0% | 0% |
| **The data and the challenge** | | | | | | | | | |
| The provided data was of sufficient QUALITY for solving the challenge | 43% | 14% | 43% | 0% | 20% | 80% | 100% | 0% | 0% |
| The provided data was of sufficient QUANTITY for solving the challenge | 57% | 29% | 14% | 0% | 0% | 100% | 100% | 0% | 0% |
| The provided data was sufficiently documented and labelled | 43% | 43% | 14% | 0% | 20% | 80% | 100% | 0% | 0% |
| The challenge topic was realistic and interesting | 86% | 14% | 0% | 100% | 0% | 0% | 100% | 0% | 0% |
| **Support** | | | | | | | | | |
| The challenge providers were active and supportive during the process | 43% | 0% | 57% | 20% | 80% | 0% | 100% | 0% | 0% |
| The WeDoWind operators were active and supportive during the process | 86% | 0% | 14% | 100% | 0% | 0% | 100% | 0% | 0% |
| **The process** | | | | | | | | | |
| The digital platform was useful for summarising and centralising the knowledge relevant for the challenge | 57% | 29% | 14% | 60% | 20% | 20% | 67% | 0% | 33% |
| The webinars were a useful way of sharing ideas | 57% | 29% | 14% | 100% | 0% | 0% | 67% | 33% | 0% |
| The digital platform was easy to use | 0% | 43% | 57% | 80% | 0% | 20% | 0% | 0% | 100% |
| The tech support of the digital platform was sufficient | 43% | 14% | 43% | 40% | 20% | 40% | 0% | 67% | 33% |
| The regular email updates were motivating | 71% | 29% | 0% | 100% | 0% | 0% | 67% | 33% | 0% |
| **Evaluation** | | | | | | | | | |
| I would have preferred a formal evaluation method to have been defined by the challenge providers at the beginning | 86% | 14% | 0% | 80% | 20% | 0% | 67% | 0% | 33% |
| I would have preferred to have defined a formal evaluation method together with the other participants at the beginning | 71% | 29% | 0% | 80% | 20% | 0% | 33% | 33% | 33% |
| **Future potential** | | | | | | | | | |
| WeDoWind could help the wind energy sector benchmark and evaluate ML methods for various applications in the future | 71% | 0% | 29% | 100% | 0% | 0% | 100% | 0% | 0% |
| WeDoWind could provide a central knowledge hub for wind energy in the future | 57% | 29% | 14% | 100% | 0% | 0% | 100% | 0% | 0% |
| I intend to take part in a WeDoWind challenge in the future | 83% | 0% | 14% | 80% | 20% | 0% | 100% | 0% | 0% |
| I would like future WeDoWind challenges to include more collaboration components | 71% | 29% | 0% | 80% | 20% | 0% | 100% | 0% | 0% |

**Figure 1.** Summary of survey results ("Power Curve") compared to previous projects ("EDP" and "WinJi")

## 7. References

[1] Clifton A, Barber S, Bray A, Enevoldsen P, Fields J, Sempreviva A M, Williams L, Quick J, Purdue M, Totaro P and Ding Y 2023 *Wind Energy Science* **8** 947–974 URL `https://wes.copernicus.org/articles/8/947/2023/`

[2] Maria S A, Allan V, Christian B, Robert V D, Gregor G, Kjartansson D H, Pilgaard M L, Mattias A, Nikola V, Stephan B, Javier S R, Pawel G, Inge R T, Christian B H, Willem W J and W H K 2017 Taxonomy and metadata for wind energy Research &Development Integrated Research Programme on Wind Energy, IRPWIND Grant agreement no 609795, , collaborative project, Start date: 01st December 2013 Work Package 2 - Deliverable D2.3 Lead Beneficiary: DTU Delivery date: 12 December 2017 Dissemination level: PU The research leading to these results has received funding from the European Union Seventh Framework Programme under the agreement 609795. URL `https://doi.org/10.5281/zenodo.1199489`

[3] Mortensen N, Nielsen M and Ejsing Jørgensen H 2015 Comparison of resource and energy yield assessment procedures 2011-2015: What have we learned and what needs to be done? *Proceedings of the EWEA Annual Event and Exhibition 2015* (European Wind Energy Association (EWEA)) paper for poster presentation; EWEA Annual Conference and Exhibition 2015 ; Conference date: 17-11-2015 Through 20-11-2015

[4] Barber S, Lima L A M, Sakagami Y, Quick J, Latiffianti E, Liu Y, Ferrari R, Letzgus S, Zhang X and Hammer F 2022 *Energies* **15** ISSN 1996-1073 URL `https://www.mdpi.com/1996-1073/15/15/5638`

[5] Barber S, Izagirre U, Serradilla O, Olaizola J, Zugasti E, Aizpurua J I, Milani A E, Sehnke F, Sakagami Y and Henderson C 2023 *Energies* **16** ISSN 1996-1073 URL `https://www.mdpi.com/1996-1073/16/8/3567`

[6] Barber S, Hammer F and Henderson C 2023 *Journal of Physics: Conference Series* **2507** 012003 URL `https://dx.doi.org/10.1088/1742-6596/2507/1/012003`

[7] Ding Y, Barber S and Hammer F 2022 *Frontiers in Energy Research, section Wind Energy* **10** 1050342

[8] Ding Y 2019 *Data Science for Wind Energy* (Boca Raton, FL, USA: Chapman & Hall)

[9] Plumley C 2023 Kelmarsh wind farm data URL `https://doi.org/10.5281/zenodo.8252025`

[10] Plumley C 2023 Penmanshiel wind farm data URL `https://doi.org/10.5281/zenodo.8253010`

[11] International Electrotechnical Commission (IEC) 2005 *IEC TS 61400-12-1 Ed. 1, Wind Turbines – Part 12-1: Power Performance Measurements of Electricity Producing Wind Turbines* (Geneva, Switzerland: IEC)

[12] Prakash A, Tuo R and Ding Y 2023 *Technometrics* **65** 70–82

[13] Kumar N, Prakash A and Ding Y 2020 *DSWE: Data Science for Wind Energy,* `R` *Package* (The Comprehensive `R` Archive Network (`CRAN`) https://cran.r-project.org/web/packages/DSWE/index.html)

# Challenges on prognostics and health management for wind turbine components

**Jokin Cuesta**[1,2], **Urko Leturiondo**[1], **Yolanda Vidal**[2,3], **Francesc Pozo**[2,3]

[1] Artificial Intelligence and Data Area. Ikerlan Technology Research Centre, Basque Research and Technology Alliance (BRTA). Pº J. Mª. Arizmendiarrieta, 2. 20500 Arrasate/Mondragón, Spain

[2] Control, Data and Artificial Intelligence (CoDAlab), Department of Mathematics, Escola d'Enginyeria de Barcelona Est (EEBE), Campus Diagonal-Besòs (CDB), Universitat Politècnica de Catalunya (UPC), Eduard Maristany 16, 08019 Barcelona, Spain

[3] Institute of Mathematics (IMTech), Universitat Politècnica de Catalunya (UPC), Pau Gargallo 14, 08028 Barcelona, Spain

E-mail: `uleturiondo@ikerlan.es`

**Abstract.** This paper explores the applicability of prognostics and health management (PHM) for wind turbines (WTs), presenting the PHM approach along with challenges and opportunities in the context of WT components. First, the PHM framework is introduced, consisting of three blocks: observation, analysis, and action. Critical components and failure modes for WTs are identified, and data acquisition strategies using supervisory control and data adquisition (SCADA) and condition monitoring (CM) data are discussed. Prognostics, specifically remaining useful life (RUL) estimation, employs physics model-based, data-driven, and hybrid models. Finally, challenges and opportunities related to data, analysis and CM, and developing RUL prediction models have been found. Data challenges include data standardization, limited public datasets, and data quality issues. Analysis and CM challenges address new sensorless and non-intrusive techniques, as well as the fusion of data sources. Prognostics model challenges involve uncertainty management, interpretability issues, and the need for online updates. Addressing challenges requires incorporating physical knowledge, utilizing transfer learning, and improving online RUL prediction methods.

## 1. Introduction

To achieve the European Commission's objective of climate neutrality by 2050, the reduction of the levelized cost of energy is a key point. According to the International Renewable Energy Agency [1], the operation and maintenance (O&M) costs, comprising fixed, and variable components are typically made up between 10% and 30% of the levelized cost of energy for the majority of wind industry projects in 2022. Designing an effective maintenance plan is crucial for cost reduction and enhanced productivity. Two primary strategies, corrective and proactive, are employed to achieve maintenance objectives. Corrective actions address failures reactively, restoring normal operations promptly but with uncertain expenses and safety concerns. Proactive actions, including preventive and predictive activities, are conducted before failures. Preventive measures reduce the probability of unplanned failures but may not fully utilize the equipment's lifespan, leading to the possibility of over-maintenance. Predictive actions leverage historical data and condition monitoring (CM) for early fault detection, offering efficient maintenance planning with benefits in downtime, costs, reliability, and production quality. This shows that maintenance tasks for wind turbines (WTs) should be optimized, moving from corrective and preventive maintenance to a predictive maintenance approach, where these tasks are scheduled depending on the actual and future condition of the components. In this context, the engineering field of prognostics and health management (PHM), which covers the group of techniques for monitoring component wear evolution, has a crucial role. It has raised interest in several fields, including electric vehicle batteries, aerospace/aviation, energy, civil, chemical, process and industrial engineering, and transportation and manufacturing [2–6]. Thanks to PHM, it is possible to predict the remaining useful life (RUL) of components and assets from historical and actual operating data [7]. Some approaches have been done on renewable energy systems [8, 9], and, notably, PHM has been instrumental in enhancing the reliability of WT components, such as bearings [10, 11], gearboxes [12, 13], and blades [14, 15]. This work explores the applicability of PHM to WT components, emphasizing both the challenges and opportunities inherent in its implementation.

The structure of this paper is designed to provide a comprehensive exploration of the PHM framework applied to WTs. In Section 2, we lay the foundation by introducing the key components of the PHM framework, encompassing observation, analysis, and action. Moving forward, Section 3 delves into the specific application of PHM to WTs, critical components of WTs, exploring failure modes, data sources, and prognosis methodologies are explored. Section 4 addresses the challenges and opportunities, categorized into three blocks: data, analysis and CM, and prognosis models. Finally, in Section 5, conclusions are drawn.

## 2. Prognostics and health management framework

PHM stands as a maintenance strategy within the area of machinery and equipment management, strongly related to predictive maintenance. It is defined as a computer-based analysis method to work with information and data related to how structure, system or components (SSCs) function and are maintained [6]. According to the International Organization for Standardization (ISO), PHM addresses three major tasks: detection of problems (anomalies are detected), diagnosis of the faults and their causes, and prognosis of future fault progression (prediction of RUL is done) [16]. The final aim is to reduce O&M costs, optimizing the maintenance periods based on predicted degradation, and extending the life cycle of a SSC [17].

To achieve these goals, a PHM framework is shown in Figure 1, slightly different from the provided in the literature [6, 9, 17, 18]. The introduced PHM framework consists of ten modules divided into three blocks: observation block, analysis block and action block, which presents two modifications from the previous works presented. First, it reorders the decision layer and human-machine interface layer, emphasizing the important role of the visual information provided by the dashboard in decision-making. Second, a last module is introduced, dedicated to the analysis of the economic impact of the decisions made, to optimize the economic aspects associated with the required investment to apply PHM and the cost reduction obtained by its application.

It is noteworthy that, although the PHM figure displays modules in a chronological clockwise order, in the action block, visualization, decision-making, and economic impact analysis are grouped together. This arrangement is designed to facilitate the identification of the best maintenance action.

### *2.1. Observation block*

The first block, named the observation block, lays the foundation for effective PHM implementation. Understanding the system involves a deep examination of the system's design, function, and operation, considering the varying importance of subsystems and components according to the potential failure modes. Critical aspects include prioritizing components most crucial for PHM, evaluating their consequences to prioritize maintenance efforts.

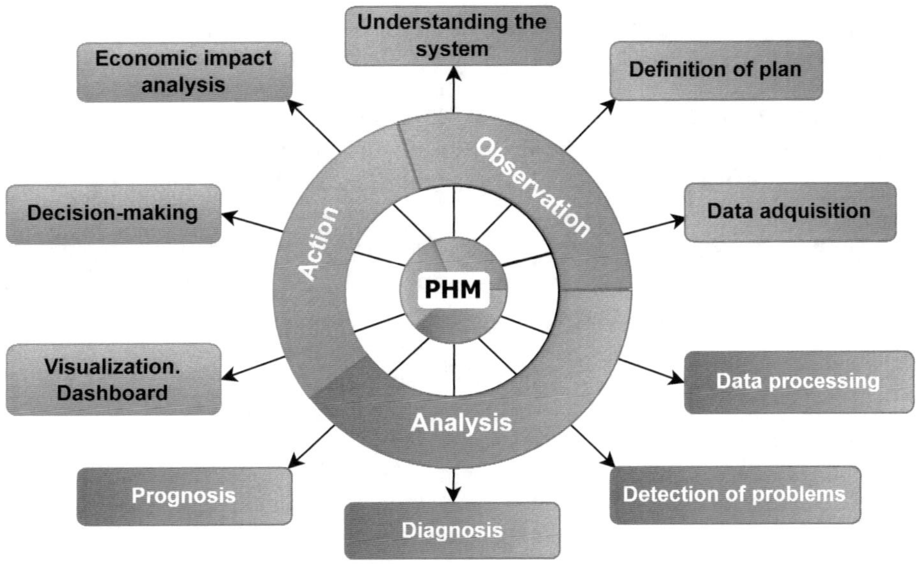


**Figure 1.** PHM framework.

The next step involves planning, building upon the research from the previous module. This phase includes the definition of a maintenance plan based on the criticality of the components, selecting appropiate sensors and outlining data capture protocols. Then, depending on these maintenance policies, the necessary data to support each approach will be acquired, deploying the strategically placed sensors in key SSCs and utilizing measurement hardware and software to capture relevant parameters. This phase goes beyond mere data collection; it includes evaluating the collected data to ensure it meets the quality standards defined in the initial plan, ensuring the reliability of data.

### *2.2. Analysis block*

The second block, called the analysis block, is built upon the significant data collected in the previous block. It focuses on data processing, detection of anomalies, fault diagnosis, and prognosis. This is where the heart of PHM resides, as advanced methods are employed to analyze fault evolution patterns from collected data.

The initial module of the block focuses on efficient data processing methods of the time-series: filtering, anomaly cleaning, or data imputation techniques. It also incorporates the use of merged condition indicators through techniques like principal component analysis (PCA) [19], investigating the extraction of features from various sources, such as vibration signals or acoustic emissions, in different domains like time, frequency, and time-frequency. These features, including basic statistical indicators, spectral characteristics, and time-frequency representations, are crucial for future problem detection, diagnosis, and prognosis. Diagnosis, as per ISO 13372:2012 [20], involves analyzing symptoms to determine fault characteristics and severity, providing insights into root causes. On the other hand, prognosis, also according to ISO 13372:2012 [20], involves forecasting future conditions and RUL, utilizing advanced techniques to construct potential failure timelines for effective maintenance planning based on component degradation.

### *2.3. Action block*

The third block, denominated action block, focuses on the shift from data analysis to action. Maintenance actions will be performed based on the information received from the previous steps. This information will be displayed in visual dashboards, and, together with an economic impact analysis, the decisions will be made.

First, it entails the development of visual dashboards, which serve as a user interface that provides insights into the condition of the SSC and helps maintenance teams with subsequent decision-making. These could include time-series plots, trend lines, heatmaps, and other easy-interpretable data representations. Moreover, key performance indicators that evaluate critical information about the SSC's performance should be shown in the dashboards, including, among others, fault severity scores or RUL estimations. Additionally, it delves into economic impact analysis, the optimization of which will directly influence decision-making. It involves evaluating the costs and benefits associated with different maintenance actions, not only the direct costs but also the potential cost savings of the extended equipment life or the reduced downtime. In essence, this analysis will help in the prioritization of the most urgent actions, as mentioned in the first block.

## 3. Prognostics and health management approach to wind turbines

This section explores the application of PHM strategies in the context of WTs. The discussion is structured across three integral sub-sections: firstly, an examination of critical components and failure modes, emphasizing the identification of high-failure-rate components; secondly, an in-depth analysis of data sources, and thirdly, focusing on prognostics and models of RUL estimation, categorizing them into into physics model-based, data-driven, and hybrid models. Each of the three subsections will contribute to the development of a robust framework for implementing PHM strategies in the field of wind energy.

### *3.1. Critical WT components and failure modes*

To apply the PHM framework in the wind sector, the target components of WTs must be considered. In onshore WTs, components with higher failure rates include electrical and control systems, blades, pitch systems, and the yaw system, as reported in studies [21, 22]. Similarly, offshore WTs exhibit the highest failure rates in the pitch system, followed by the generator [21]. A higher average failure rates can be observed in offshore WTs compared to onshore, which can be attributed to the severe operating conditions inherent to offshore environments, such as elevated mean wind speeds and exposure to corrosive saltwater. In relation to downtime, the gearbox, generator, blades emerge as the most critical subsystems for both onshore and offshore WTs [21]. Additionally, failures associated with the gearbox, rotor blades, yaw system, and generator represent a higher expenditure, in that specified order [23]. Santelo et al. [24] divided the failure periods into two main categories: premature failure period and full time working period. They concluded that the pitch system, converter, generator, and yaw system are the most critical components in the infant mortality period (up to 6 years from the beginning).

Identifying potential failure modes is essential for developing effective PHM strategies. Every component within a WT has its unique set of failure modes, and the associated downtime contributions are carefully documented and analyzed. The main failure modes of these components are presented in Table 1 [22, 24–32].

WT components are susceptible to various failure modes that can impact their functionality and performance. In the pitch system, issues such as oil leaks and battery failures can disrupt the precision of blade adjustments [25–27]. Frequency converter failures, including inverter and crowbar issues, may lead to electrical disruptions affecting power conversion [26, 28]. Gearbox problems, such as bearing and gear failures, can result in mechanical inefficiencies [29, 30]. Generator-related failures, encompassing wearing, electrical issues, and overheating, can affect

**Table 1.** WT component failure modes

| WT component | Failure mode | Description | References |
|---|---|---|---|
| Pitch system | Oil issues | Leaks, unscheduled oil changes, and unscheduled oil top ups | [25–27] |
| | Battery failure | Unexpected battery failure. | |
| | Motor failure | Wear or electrical issues | |
| | Motor converter failure | Electrical faults | |
| | Valve issues | Blockages or malfunctions | |
| | Accumulator problems | Leaks or pressure issues | |
| | Internal leakage | Internal leakage of proportional valve, internal leakage of solenoid valve, hydraulic cylinder leakage | |
| | Sensor Failure | Innaccurate feedback for pitch control | |
| Frequency converter | Inverter failure | Generator-side or Grid-side inverter failure | [26, 28] |
| | Crowbar failure | Uncontrolled voltage spikes | |
| | Cooling Failure | Pump failures, blockages, or leaks in the cooling system | |
| | Control Board failure | Electric issues/ component wear | |
| | Open/short circuit | Resistors, capacitors, and electronic switches | |
| Gearbox | Bearing failure | Axial cracking, spalling (contact fatigue), pitting (surface fatigue), and brinelling (fretting) | [29, 30] |
| | Gear failure | Abrasive wear, pitting, cracking, scuffing | |
| | Oil leakage | Resulted from worn seals or damaged components | |
| Generator | Wearing | Wear and tear on generator components | [30, 31] |
| | Electrical problems | Insulation failure and electrical imbalance | |
| | Winding damage | Due to electrical faults, overheating | |
| | Rotor asymmetries | Caused by imbalances | |
| | Bar break | Due to material fatigue or overloading | |
| | Overheating | Caused by electrical faults, inadequate cooling, or manufacturing defects | |
| Main shaft | Misalignment | Causing increased friction, vibration, and potential damage | [22] |
| | Crack | Caused by material fatigue, manufacturing defects, or excessive stress due to varying wind conditions | |
| | Corrosion | Caused by chemical reactions with the environment | |
| | Coupling failure | Caused by caused by wear, misalignment, or overloading | |
| Yaw system | Abrasions of gear teeth | Caused by lack of greasing and bearing breakage | [24, 32] |
| | Brake failure | Caused by brake pad wear, brake pad contamination and brake disc wear | |
| | Hydraulic failure | Hydraulic oil leaks and unstable braking forces | |
| | Electrical failure | Failures in the drive and drive motor, angular transducer, and abrasions in cables | |
| Blades | Crack fatigue | Fatigue cracks in the blades. | [32] |
| | Corrosion | Corrosion on the blades. | |
| | Aerodynamic imbalance/asymmetry | Due to manufacturing defects, non-uniform accumulation of ice, dirt, moisture or damage by lightning | |

power generation [30, 31]. Yaw system failures, such as gear abrasions and brake issues, may compromise the WT's ability to align with the wind [24, 32]. Blade failures, including crack fatigue and corrosion, pose structural risks and aerodynamic imbalances [32].

### *3.2. Data sources*

Regarding the data to be used to monitor and predict faults in the aforementioned components, supervisory control and data adquisition (SCADA) data and CM data are the main ones [33–35]. Based on the study of Wang et al. [36], parameters obtained from a wind farm SCADA system can be classified into wind parameters (e.g., wind speed, deviation), performance parameters (e.g., power output, rotor speed, or blade pitch angle), vibration parameters (e.g., tower acceleration or drivetrain vibration), and temperature parameters (e.g., bearing temperature). Even if SCADA data are appropriate for predictive maintenance, their low sampling frequency (typically 10-minute averages) limits the usage of many conventional CM techniques (e.g., spectral analysis) [37].

In contrast, it has been demonstrated that CM data are efficient for early fault diagnosis, but their real-world exploitation has some limitations because of the difficulty to isolate faulty vibration signatures acquired in such complex systems like WTs, considering their highly predominant non-stationary conditions [38]. The measurement obtained from condition monitoring systems (CMSs) may include strains, vibrations, acoustic emissions, oil debris counts, oil condition measurements, electric currents, and voltages [39]. One of the main kinds of signals used in predictive maintenance is vibration signatures, specially in rotating machinery like drive trains and pitch bearings [40]. Moreover, new health indicators based on different approaches compared to the classical ones (based on time- and frequency-domain) have been recently developed, like entropy-based indicators [41].

### *3.3. Prognostics. RUL estimation*

The RUL of a SSC is a key element in prognostics, as its estimation is a valuable input to establish effective predictive maintenance plans and, subsequently, for the optimization of the O&M of WTs thereof. There are different approaches used for RUL prediction [22, 42–44]. The ones presented in this paper include physics based-models, data-driven models, and hybrid models.

Physics model-based strategies refer to techniques grounded solely in the mathematical representation of a system and its failure modes. These methods encompass physics of failure (PoF) and system modeling [2]. PoF focuses on understanding failure patterns influenced by life cycle loading conditions, considering the geometry and material properties of the target component to analyze potential failure modes and evaluate RUL. The initial phase involves a detailed failure modes, mechanisms, and effects analysis. To apply this approach to real systems, the identification of relevant loading conditions is essential. Subsequently, failure criteria or indicators are mapped as functions of loading conditions, material properties, and component geometry. In-situ data collection provides loads, and pertinent features are extracted for use in PoF models to estimate damage and RUL. System modeling employs mathematical functions such as differential equations to accurately represent the system. These models include Paris Law, oil viscosity models [45], and Miner´s rule.

Data-driven techniques use historical data, e.g., sensor readings, operational history, or SCADA data to construct a model. They are commonly used when there is a lack of physical understanding or where the model exhibits a complex non-linear behavior. There is no consensus on the classification of data-driven methods for RUL calculation. For example, Ramezani et al. [46] consider that these techniques include deep neural networks, generative models, kernel-based learning algorithms, probabilistic models, and fuzzy logic techniques. In contrast, Gordan et al. [47] consider two main groups of algorithms: prediction and classification-based data mining algorithms (e.g., artificial neural networks, deep learning (DL), fuzzy logic, PCA, clustering analysis, support vector machines, regression analysis, decision trees, Bayesian analysis), and optimization-based data mining algorithms (e.g., genetic algorithms, particle swarm optimization, ant colony optimization). It should be considered that, at the same time, some of these big groups of algorithms are composed of a large amount of specific methods. That is the case of DL techniques, where time-series data can be analyzed using algorithms like multilayer perceptron neural networks, backpropagation neural networks, convolutional neural networks, recurrent neural networks, long short-term memory, restricted Boltzmann machines, deep belief networks, generative adversarial networks, and auto-encoders [48], among others.

Hybrid models represent a combination of the aforementioned approaches. According to Zhang et al. [8], these combinations can be classified in three main types: machine learning (ML)-stats model, stats-physics model and other hybrid models. They get the positive aspects of each type of model while minimizing their respective limitations. Results obtained from hybrid models have shown higher interpretability and intuitiveness compared to data-driven models [8], which are often considered as *black boxes* due to their lack of explainability. In this context, explainable artificial intelligence algorithms have raised to explain the predictions, such as local interpretable model-agnostic explanations or Shapley additive explanation methods, which help hybrid models to gain explainability [49]. Moreover, hybrid models effectively handle uncertainties, yet their high algorithmic complexity and reliance on physical modeling for degradation phenomena can be resource-intensive [9].

## 4. Challenges and opportunities

This section outlines the challenges and opportunities associated with the application of PHM in the context of WTs. Although PHM brings many benefits, it also faces challenges that need to be considered. Issues found in the literature are divided into challenges related to data; encompassing aspects like quality, lack of standardized protocols, and the need for publicly available datasets; analysis and CM, which involves exploring novel sensorless techniques and addressing challenges in data collection; and models for prognostics, addressing issues such as uncertainty, interpretability or consideration of variable operational conditions.

### *4.1. Data challenges*

Implementing PHM in WTs faces data challenges that could significantly impact the effectiveness of the monitoring and maintenance strategies, specially in its acquisition and processing. As mentioned in section 3.2, the most used data sources are SCADA and CMS data. However, these are not always available, or their quality is not as desired.

One of the primary challenges is the absence of standardized protocols for WTs SCADA data. The lack of uniformity in data formats, parameters, and naming conventions across different WT systems makes difficult the integration and analysis of information, limiting the development of cohesive PHM solutions [4, 37]. The lack of uniformity in data formats makes it challenging to create algorithms and models that can work seamlessly across diverse WT systems, as well as robust analytics tools for performance monitoring and optimization.

Data quality is another key factor to consider [4]. One of the major drawbacks to using SCADA data is its low-frequency resolution, in numerous instances, a 10-minute sampling period [37]. This negatively affects the PHM capabilities, possibly hiding short-lived events [50]. The current data enhancement methods for incomplete data cannot directly generate balanced and usable high-quality training datasets, and the highly imbalanced condition between healthy and unhealthy data is still one of the problems to be solved urgently in the research of equipment life prediction [51]. PHM often deals with a large volume of data gathered under healthy conditions and a limited number of data points that indicate faulty states [46].

Another significant constraint in the development of models for WTs PHM is the scarcity of publicly available datasets [9, 52]. It restricts researchers from accessing information vital for developing, training, and validating RUL estimation models. The implementation and approval of a benchmark database would enable the evaluation of various PHM algorithms based on their effectiveness, interpretability, scalability, and dependability [46]. Generating synthetic data may be found as a solution to the problem, either using mathematical models, simulation tools or test rigs. However, these are not always capable of generalizing effectively real-world conditions [35]. Moreover, the datasets at hand might not sufficiently capture the wide spectrum of WT configurations, environmental conditions, and operational scenarios. Dealing with miscellaneous data with multifarious characteristics is one of the main challenges [53].

Additionally, enterprises often exhibit reluctance to share data with external entities for constructing predictive and prognostic models. To address this issue, the emerging field of federated learning has gained prominence. Federated learning allows the training of ML models on decentralized clients without necessitating the sharing of raw data [54]. In this setup, instead of gathering data from clients, the model is distributed among enterprises and trained with their local data. Then, the updated model properties, such as weights and biases, are transmitted to the server for aggregation. Through multiple iterations of this process, a global master model can be generated without directly accessing the underlying data [54]. Some federated learning approaches to WTs have been found in the literature [55–57].

### *4.2. Analysis and CM challenges*

Analysis CM challenges focus on Traditional sensor-based CMSs costs, development of smart sensors and industrial internet of things (IIoT), and data sources fusion techniques.

Traditional sensor-based CMSs bring increased costs associated with sensor deployment (susceptibility to environmental factors), communication, high-frequency data acquisition, storage systems and potential points of failure in the sensors themselves. Additionally, retrofitting existing turbines with numerous sensors can be logistically challenging and may disrupt normal operations during installation. To address this issue, new emerging trends, such as virtual sensing or drone-captured monitoring, have surfaced. Virtual sensing involves utilizing available data from real sensors with a model of the system to derive estimated data from unmeasured points [58]. Moreover, for diagnostic purposes, it has been shown that drone-captured images offer an alternative method for detecting faults in offshore WT blades [59].

The development of smart sensors and IIoT technologies has given rise to an increased high-quality data scarcity problem, characterized by noise and uncertainty [60]. The sensor readings involved in this context may encompass incomplete data, misreadings, unlabeled data, signal interference, disconnections, imbalanced classes, and unseen classes, highlighting the complexity of the data landscape [46]. Moreover, due to time-varying environmental loads, offshore WTs typically operate in dynamic conditions, resulting in vibration signals displaying robust non-stationarity [61]. In non-stationary conditions, fault characteristic frequencies may fluctuate with speed variations, causing blurred frequency lines in spectra.

Equally, it becomes crucial to consider varying operating conditions, incorporating environmental factors like wind speed and ambient temperature into monitoring methods. This ensures the applicability of the techniques to real-world scenarios. Additionally, investigating component interactions is essential for improving prognosis capabilities [44]. Utilizing advanced signal processing procedures can aid in differentiating signals related to specific component faults, particularly bearings, thereby refining the precision of prognostics tasks and contributing to the overall enhancement of WT maintenance and operational strategies.

On the other hand, for an overall assessment of a WT at the system level, a CM-based performance assessment can be addressed. This contributes to the expanding field of performance-based maintenance by introducing a novel approach rooted in CM, specifically focusing on non-binary failures [19]. No comprehensive knowledge is required for the interpretation of the results, as a consequence, it could be used as a simple tool for monitoring the performance of WTs [61].

Even though any of all these sources have been already analyzed, the fusion of SCADA and CM data is conceived, which implies a challenge, as mentioned before, because of the large difference between their sampling frequencies (10 minutes in the former, even milliseconds in the latter), and the data acquisition rates (continuous in the case of SCADA, and periodically scheduled in CM data, usually three times per day with a duration of several minutes). It should be highlighted that the innovative aspects of fusing both types of data are clearly shown due to the tiny number of references in the literature [62, 63].

### *4.3. Prognostics models challenges*

The challenges associated with prognostics models include uncertainty management, the lack of interpretability, physical knowledge embedded modeling, applying transfer learning, and lack of online updates.

One of the most important challenges to prognostics is uncertainty management because most of the techniques are based on approximations [60]. Accurate RUL prediction depends on effective uncertainty estimation and management. There have been several papers in this field, but still there is a need for additional investigation [42]. This uncertainty may arise from sensor noise (as mentioned in 4.2), data processing, filtering, estimation methods, model uncertainty, working condition fluctuations, prediction uncertainty, and others [42, 64]. Uncertainty is inherent, but techniques such as Bayesian neural networks have arisen to reduce it [64–68].

In the context of PHM, as commented in subsection 3.3, using DL has been one of the popular approaches. Nevertheless, conventional end-to-end DL methodologies are often characterized by a lack of interpretability [35,61]. The inherent *black box* nature of these approaches restricts their applicability in industrial scenarios [69], so exploring the interpretability of DL is particularly beneficial for engineers to make more convincing decisions [51]. Generally, interpretable ML model training is critical for improving the transparency and trustworthiness of ML models [49]. One way to improve this interpretability is to include physical knowledge into the model [4, 70], converting them into hybrid models [44]. The parallel fusion structure, integrating physics models and ML modules, provides the advantage of leveraging the strengths of both approaches simultaneously, playing the role of compensators in improving accuracy, robustness, and interpretability while enabling a comprehensive understanding of systems' degradation behavior [69]. Kobayashi et al. [49] presented inherently interpretable ML models: rectified linear unit-deep neural network, explainable boosting machine, fast interpretable greedy-tree sums, and decision tree surrogates. This model interpretation explains how the input and output variables of a ML model are related to the decision-making process. However, not many papers studying interpretability methods applied to RUL of WTs, so there is a need on further investigation on this.

To improve the performance of the models, the integration of the physical knowledge of the system is becoming an option. In pursuit of this objective, an emerging field known as physics informed neural networks (PiNNs) is raising. PiNNs can assimilate features aligned with scientific principles, thereby progressing towards the development of interpretable and more accurate deep neural models [70]. For instance, well-known physical fatigue damage models can be combined with data-driven layers to model degradation processes in bearings, finally merging into a cumulative damage model for RUL prediction [71]. This approach stands out in prognostics due to its robustness, offering heightened accuracy when the underlying physics of the models remain consistent throughout the life cycle of the component. Additionally, it demands less data compared to purely data-driven methods, as noted by Rezamand et al. in their work [44].

There are some shortages and limitations in existing algorithms, including a lack of online updates, vessel failures, extreme weather conditions, numerous constraints, and limited flexibility for evolving environments [4, 72]. The majority of models are currently offline, and transitioning them to online or onboard introduces additional challenges in constructing the DL model. So, comes the need for developing new algorithms and hardware architectures to cope with continuous online learning of non-stationary sequence data as well as keeping the pre-trained model [60]. Bayesian networks have been used in an attempt to solve online RUL prediction tasks. The modeling strategy provides promising solutions for online RUL prediction in industrial scenarios by updating the model parameters with sequential-arrival data. At present, online RUL prediction is at the initial research stage, and it is urgent to develop more online RUL prediction methods to help engineers implement maintenance in advance [51]. Finally, achieving optimal efficiency and reliability needs not only the standardization of data but also the generalization of performance evaluation metrics [9, 46].

To adapt the models to the variable operational conditions, utilizing transfer learning is highlighted as another opportunity in the predictive maintenance of WTs [4, 46, 51, 60, 73]. Transfer learning gains insights into the performance of a system and subsequently applies that acquired knowledge to other applications [46], implementing it on different datasets [51]. In the field of PHM, it has been proven to be an effective tool for solving typical rolling bearing problems such as fault detection and fault diagnosis [51], as most of these methods are used for classification and not regression [73]. As in practical PHM applications the target domain labels are often not easy to obtain, unsupervised transfer learning methods have been proposed [4]. However, not all extracted features are transferable, so RUL prediction based on transfer learning cannot yet guarantee the accuracy of predictions for different types of equipment with large differences in working conditions [51].

In the context of offshore WTs, an effective prognostics strategy should address challenges related to maintenance, given the remote locations and increased downtime resulting from the considerable distance between wind farms and ports or shores [72]. The complexity inherent in various foundation types, exposure to adverse weather conditions, and the necessity for motion-compensated gangways and specialized vessels must be integral considerations in devising a robust prognostics strategy. Furthermore, the strategy should encompass the heightened maintenance costs associated with offshore operations, accounting for specific equipment requirements and the impact of severe working conditions on failure rates [72].

## 5. Conclusions

This paper has introduced the applicability of PHM for WTs, while highlighting the current challenges and opportunities in this domain, regarding data, analysis and CM, and models used for RUL estimation. The data challenges encompass issues related to the acquisition and processing of information, particularly with the absence of standardized protocols for SCADA data, and the scarcity of publicly available datasets. The data quality, low-frequency resolution, and challenges in handling data on different degradation stages further complicate the development of effective PHM solutions. CM faces challenges in the deployment of hardware and software tools for data gathering, leading to increased costs and potential points of failure; the varying operating conditions of WTs, and fusion of data sources. Prognostics models encounter challenges like uncertainty management, lack of interpretability in DL approaches, and the need for online updates. Addressing these challenges involves incorporating physical knowledge into models, utilizing transfer learning, and improving online RUL prediction methods. The models should consider variable operations conditions, specially in offshore WTs. To sum up, all challenges and opportunities are gathered in Table 2.

## Acknowledgments

This study was partially funded by the Spanish Agencia Estatal de Investigación (AEI) — Ministerio de Ciencia e Innovación, the Fondo Europeo de Desarrollo Regional (FEDER) through the research projects PID2021-122132OB-C21, PID2021-122132OB-C22, and TED2021-129512B-I00; and by the Generalitat de Catalunya through the research project 2021-SGR-01044.

**Table 2.** Challenges and opportunities of PHM applied to WTs.

| Categories | Challenge | Description | Opportunities | References |
|---|---|---|---|---|
| Data | Absence of standardized protocols for SCADA | Lack of uniformity in data formats, parameters, and naming conventions across different WT systems, making integration and analysis difficult. | Standardization of protocols for SCADA data. | [4,37] |
| | Data quality | Challenges associated with low-frequency resolution of SCADA data, data enhancement for incomplete data, and dealing with a large volume of healthy data and limited faulty data. | Improvement of data quality, exploration of advanced data enhancement methods. | [4,50,51] |
| | Scarcity of public datasets | Limited access to vital information for developing RUL estimation models due to the scarcity of public datasets. | Creation and approval of benchmark databases, exploration of synthetic data. | [9,46,52] |
| | Reluctance in data sharing | Enterprises' reluctance to share data for predictive and prognostic models. Introduction of federated learning as an emerging field to train models on decentralized clients without sharing raw data. | Adoption of federated learning approaches, addressing concerns of data sharing. | [54–57] |
| CM | Traditional sensor-based CMS costs | Increased costs, susceptibility to environmental factors, and logistical challenges associated with traditional data gathering software and hardware. | Adoption of virtual sensing, exploration of alternative diagnostic methods. | [58,59] |
| | High-quality data scarcity | Challenges in dealing with high-quality data scarcity, characterized by noise and uncertainty of measurements. Consideration of time-varying environmental loads and dynamic operating conditions. | Development of smart sensors, incorporation of environmental factors into monitoring methods. | [60,61] |
| | Fusion of SCADA and CM data | Integration challenges due to differences in sampling frequencies and data acquisition rates between SCADA and CM data. | Exploration of innovative fusion approaches for SCADA and CM data. | [62,63] |
| Models | Uncertainty management | Challenges associated with uncertainty in RUL prediction, arising from sensor noise, data processing, model uncertainty, and working condition fluctuations. | Investigation and application of techniques for effective uncertainty estimation and management. | [9,42,64–68] |
| | Lack of interpretability | Lack of interpretability in conventional end-to-end DL methodologies. | Exploration of interpretability techniques. | [35,51,61,69,70] |
| | Integration of physical knowledge | Improve accuracy and performance of the model by integrating physical knowledge of the system. | Research and application of PiNNs and hybrid models. | [44,70,71] |
| | Variable operational conditions | Challenges in feature transferability and unsupervised transfer learning methods. | Implementation of transfer learning for system performance insights. | [46,48,51,60,73] |
| | Lack of online updates | Challenges in achieving online RUL prediction and updating models with sequential-arrival data. | Development of online RUL prediction methods, exploration of Bayesian networks. | [60,72] |

## References

[1] IREA 2023 Renewable power generation costs 2022
[2] Bhat D, Muench S and Roellig M 2023 *e-Prime-Advances in Electrical Engineering, Electronics and Energy*

**4**

[3] Rezvanizaniani S M, Liu Z, Chen Y and Lee J 2014 *Journal of Power Sources* **256** 110–124

[4] Hu Y, Miao X, Si Y, Pan E and Zio E 2022 *Reliability Engineering and System Safety* **217**

[5] Liu D, Shi J, Liao Z and Guo H 2022 *Journal of Advanced Manufacturing Science and Technology* **2**(4) this article presents a review on PHM methods for gears, presenting four PHM methods (Physical-Based Method, Knowledge model-based method, data driven model-based methods and hybrid methods.

[6] Polverino L, Abbate R, Manco P, Perfetto D, Caputo F, Macchiaroli R and Caterino M 2023 *International Journal of Engineering Business Management* **15**

[7] Ferreira C and Gonçalves G 2022 *Journal of Manufacturing Systems* **63** 550–562

[8] Zhang W, Vatn J and Rasheed A 2022 A review of failure prognostics for predictive maintenance of offshore wind turbines *Journal of Physics: Conference Series* vol 2362 (Institute of Physics)

[9] Saidi L and Benbouzid M 2021 *Electronics (Switzerland)* **10**(22)

[10] Azevedo H D M D, Araújo A M and Bouchonneau N 2016 *Renewable and Sustainable Energy Reviews* **56** 368–379

[11] Wang P, Long Z and Wang G 2020 *Energy Reports* **6** 173–182

[12] Goswami P and Rai R N 2023 *Engineering Failure Analysis* **146**

[13] Shaheen B W and Németh I 2023 *Processes* **11**(1)

[14] Boutrous K, Puig V and Nejjari F 2022 A set-based prognostics approach for wind turbine blade health monitoring *IFAC-PapersOnLine* vol 55 (Elsevier B.V.) pp 402–407

[15] Lee I Y, Roh H D and Park Y B 2022 *Materials and Design* **223**

[16] ISO ISO 133811:2015 Condition monitoring and diagnostics of machines-Prognostics-Part 1 Standard International Organization for Standardization Geneva, CH

[17] Zio E 2022 *Reliability Engineering and System Safety* **218**

[18] Zonta T, da Costa C A, da Rosa Righi R, de Lima M J, da Trindade E S and Li G P 2020 *Computers and Industrial Engineering* **150**

[19] Urmeneta J, Izquierdo J and Leturiondo U 2023 *Renewable Energy* **205** 281–292

[20] ISO 2012 Condition monitoring and diagnostics of machines-vocabulary Tech. Rep. 13372:2012 International Organization for Standardization

[21] Dao C, Kazemtabrizi B and Crabtree C 2019 *Wind Energy* **22**(12) 1848–1871

[22] Gao Z and Liu X 2021 *Processes* **9**(2) 1–19

[23] Tazi N, Châtelet E and Bouzidi Y 2017 *Energies* **10**(3)

[24] Santelo T N, de Oliveira C M R, Maciel C D and Roberto J R B J 2022 *Journal of Control, Automation and Electrical Systems* **33**(2) 505–521

[25] Carroll J, McDonald A and McMillan D 2016 *Wind Energy* **19**(6) 1107–1119

[26] Gayo J B 2011 Final publishable summary of results of project reliawind. URL `www.ReliaWind.eu`

[27] Walgern J, Fischer K, Hentschel P and Kolios A 2023 *Energy Reports* **9** 3273–3281

[28] Liang J, Zhang K, Al-Durra A, Muyeen S M and Zhou D 2022 *Energy Reports* **8** 5341–5369

[29] Owolabi O I, Madushele N, Adedeji P A and Olatunji O O 2023 *Journal of Reliable Intelligent Environments* **9**(4) 399–419

[30] Olabi A G, Wilberforce T, Elsaid K, Sayed E T, Salameh T, Abdelkareem M A and Baroutaji A 2021 *Energies* **14**(17)

[31] Lydia M and Kumar G E P 2023 *Condition monitoring in wind turbines: a review* (Elsevier) pp 229–247

[32] Catelani M, Ciani L, Galar D and Patrizi G 2020 *IEEE Transactions on Instrumentation and Measurement* **69**(9) 6241–6249

[33] Márquez F P G and Gonzalo A P 2022 *Archives of Computational Methods in Engineering* **29**(5) 2935–2958

[34] Maldonado-Correa J, Martín-Martínez S, Artigao E and Gómez-Lázaro E 2020 *Energies* **13**(12)

[35] Stetco A, Dinmohammadi F, Zhao X, Robu V, Flynn D, Barnes M, Keane J and Nenadic G 2019 *Renewable Energy* **133** 620–635

[36] Wang K S, Sharma V S and Zhang Z Y 2014 *Advances in Manufacturing* **2**(1) 61–69

[37] Leahy K, Gallagher C, O'Donovan P and O'Sullivan D T 2019 *Energies* **12**(2)

[38] Liu X, Du J and Ye Z S 2022 *IEEE Transactions on Industrial Informatics* **18** 986–995

[39] Sheng S 2017 *Prognostics and Health Management of Wind Turbines—Current Status and Future Opportunities* (Cham: Springer International Publishing) pp 33–47

[40] Sandoval D, Leturiondo U, Pozo F, Vidal Y and Salgado O 2019 *Dyna* **94** 636–642

[41] Sandoval D, Leturiondo U, Vidal Y and Pozo F 2021 *Sensors* **21** 849

[42] Lei J, Zhang W, Jiang Z and Gao Z 2022 *Journal of Failure Analysis and Prevention* **22**(6) 2119–2137

[43] Do M H and Söffker D 2021 *Renewable and Sustainable Energy Reviews* **145**

[44] Rezamand M, Kordestani M, Carriveau R, Ting D S, Orchard M E and Saif M 2020 *IEEE Transactions on Instrumentation and Measurement* **69**(12) 9306–9328

[45] Zhu J, Zink J, Combution H, He D and Bechhoefer E 2013 *Journal of Chemical Science and Technology* **2**(3) 100–115
[46] Ramezani S B, Cummins L, Killen B, Carley R, Amirlatifi A, Rahimi S, Seale M and Bian L 2023 *IEEE Access* **11** 41741–41769
[47] Gordan M, Sabbagh-Yazdi S R, Ismail Z, Ghaedi K, Carroll P, McCrum D and Samali B 2022 *Measurement: Journal of the International Measurement Confederation* **193**
[48] Thoppil N M, Vasu V and Rao C S 2021 *Journal of Vibration Engineering and Technologies* **9**(6) 1123–1145
[49] Kobayashi K and Alam S B 2024 *Engineering Applications of Artificial Intelligence* **129**
[50] Vidal Y, Pozo F and Tutivén C 2018 *Energies* **11**(11)
[51] Chen J, Huang R, Chen Z, Mao W and Li W 2023 *Mechanical Systems and Signal Processing* **193**
[52] Maldonado-Correa J, Martín-Martínez S, Artigao E and Gómez-Lázaro E 2020 *Energies* **13**(12)
[53] Soleimani M, Campean F and Neagu D 2021 *Quality and Reliability Engineering International* **37**(8) 3746–3778
[54] Bagheri B, Rezapoor M and Lee J 2020 *Manufacturing Letters* **24** 136–139
[55] Cheng X, Shi F, Liu Y, Liu X and Huang L 2022 *Energy* **254**
[56] Jiang G, Fan W P, Li W, Wang L, He Q, Xie P and Li X 2022 *Measurement: Journal of the International Measurement Confederation* **199**
[57] Wang L, Fan W, Jiang G and Xie P 2023 *Energy* **284**
[58] Mora B, Basurko J, Sabahi I, Leturiondo U and Albizuri J 2023 *Sensors* **23**(10)
[59] Yang C, Liu X, Zhou H, Ke Y and See J 2023 *Renewable Energy* **203** 267–279
[60] Elattar H M, El-Brawany M A, Elminir H K, Ibrahim D A and Ramadan E 2023 *Computers & Industrial Engineering* 109605
[61] Gao Z and Odgaard P 2023 *Renewable Energy* **218**
[62] Turnbull A, Carroll J and McDonald A 2021 *Wind Energy* **24** 197–211
[63] Natili F, Daga A P, Castellani F and Garibaldi L 2021 *Applied Sciences* **11**
[64] Wang L, Cao H, Ye Z and Xu H 2023 *Reliability Engineering and System Safety* **238**
[65] Zhu R, Chen Y, Peng W and Ye Z S 2022 *Reliability Engineering and System Safety* **228**
[66] Bai R, Noman K, Feng K, Peng Z and Li Y 2023 *Reliability Engineering and System Safety* **238**
[67] Zhang Y, Zhang C, Wang S, Dui H and Chen R 2023 *Reliability Engineering & System Safety* 109666
[68] Merainani B, Laddada S, Bechhoefer E, Chikh M A A and Benazzouz D 2022 *Renewable Energy* **182** 1141–1151
[69] Deng W, Nguyen K T, Medjaher K, Gogu C and Morio J 2023 *Applied Mathematical Modelling* **124** 325–352
[70] Chen X, Ma M, Zhao Z, Zhai Z and Mao Z 2022 *Journal of Dynamics, Monitoring and Diagnostics* 200–207
[71] Yucesan Y and Viana F 2020 *International Journal of Prognostics and Health Management* **11** 17
[72] Ren Z, Verma A S, Li Y, Teuwen J J and Jiang Z 2021 *Renewable and Sustainable Energy Reviews* **144**
[73] Ye Z and Yu J 2023 *IEEE Systems Journal* **17**(1) 62–71

Journal of Physics: Conference Series **2745** (2024) 012004 doi:10.1088/1742-6596/2745/1/012004

# Scenarios for offshore wind co-existence opportunities and trade-offs

**S S Pettersen[1], Ø Bergh[2], Ø Aarnes[1], C S Benjamin[1], M V Bjordal[1], S S Hjøllo[2], L Landberg[1], B Pretlove[1], K Brinchmann[3], J H Sandberg[4], M L Strømme[2], P A Horneland[2]**

[1] Group Research and Development, DNV, Høvik, Norway
[2] Institute of Marine Research, Bergen, Norway
[3] Mainstream Renewable Power, Fornebu, Norway
[4] Norwegian Fishermen's Association, Trondheim, Norway

Sigurd.pettersen@dnv.com

**Abstract**. This paper introduces the MARCO (MARine CO-existence scenario building) concept for using scenario exploration in stakeholder engagement processes in offshore wind. MARCO builds on spatial analyses using geographic information systems (GIS), and projections over time using system dynamics simulation models. We position the concept within the existing literature on tools for decision support and stakeholder participation, and provide a preliminary status on the spatial baselines, as well as example scenarios for area usage in offshore wind and implications, including risks and co-existence opportunities, on other sectors and nature.

## 1. Introduction

Fossil fuels are by far the largest contributor to global climate change, and there is an urgent need for upscaling of renewable energy production. Despite concerns about the impacts on area usage and habitat loss, the political expectations to the offshore wind industry in Europe have been rising ever higher over the last few years. The 2022 *Esbjerg Declaration* targets were succeeded by the 2023 *Ostend Declaration*, outlining ambitions to grow offshore wind in the North Sea to 300 GW by 2050 [1]. Signed by 100 companies across the value chain, the *Offshore Renewable Industry Declaration* that followed, outlined several action points for governments to realize these ambitions, including inclusion of non-price (e.g. sustainability) criteria in auctions [2]. At COP28 in December 2023, 130 national governments endorsed a pledge to triple renewable generation capacity by 2030 [3].

DNVs *Energy Transition Outlook* (ETO) forecasts more than 360 GW of offshore wind in Europe by 2050, of which 20% will be floating [4]. DNVs *Ocean's Future to 2050: Spatial Competition Forecast* shows a five-fold increase in the area reserved for offshore installations in Europe by 2050, of which 80% will be offshore wind farms (OWFs) [5,6].

At the same time as the ocean industries face risk of increased spatial competition, the world is experiencing climate change, biodiversity loss and other threats to ocean health. The dual objectives of ocean protection and productivity in ocean industries are tightly connected, and policymakers respond on several levels:

1. The *Convention of Biological Diversity* (CBD) sets forth targets to protect 30% of the ocean and land by 2030 [7]
2. Multi-use is increasingly encouraged under the EU *Maritime Spatial Planning Directive* [8]
3. Co-existence related requirements are included in offshore wind tendering [9]
4. Offshore wind developers seek to become nature-positive, halting and reversing nature loss or even restoring nature [10], motivated by financiers and disclosure requirements [5,9]

For offshore wind to gain social acceptance and be established as a sustainable and equitable industry, early stakeholder dialogue is essential to map technical and scientific constraints for projects and enable good planning for development of operation, minimizing negative impacts and making use of co-existence opportunities. There is a clear need for 1) a shared knowledge base on risks and opportunities related to co-existence, and 2) clarity on how risks and opportunities related to co-existence play out under different trajectories for environmental change and industry development.

The objective of this paper is two-fold: First, explore the current knowledge on marine spatial planning (MSP), co-existence, and decision support and stakeholder engagement tools by conducting a literature review. Second, to provide early examples of issues that are being studied in the collaborative innovation project MARCO (*MARine CO-existence scenario building*). The key value of the research being conducted in MARCO is to provide a common knowledge base and a set of tools that allow a joint understanding of risk and opportunity with respect to co-existence.

## 2. Marine spatial planning, consenting processes, and co-existence

Marine spatial planning is defined as "the public process of analyzing and allocating the spatial and temporal distribution of human activities in marine areas to achieve ecological, economic and social objectives that have been specified through a political process" [11]. All EU member states are required to have national level marine spatial plans in place, to be updated at a regular basis. Specific attention is given to the alignment of MSP with sector-specific objectives, such as scaling up renewable energy [8]. Norway has its own ecosystem-based ocean management planning framework, but as a non-member state is not obliged to implement the EU MSP Directive.

More than 100 countries currently have a type of MSP process in place [12]. Stakeholder engagement, consideration of environmental impacts, and consideration of future changes, including climate change and the dynamics of human activity in the ocean, are among several remaining challenges to the MSP approach [13]. While originally rooted in ecosystem-based management [12], some argue that the EU MSP process, at least in its early phases, had too much of an industry focus [13,14]. Gissi et al. reviewed how change and temporality is accounted for in the process, focusing on the dynamics of the socio-ecological system (internal dynamics, vs external drivers like climate change), and conclude that more research is needed on adaptation and flexibility in MSP [15].

Consenting processes for offshore wind developments depend on MSP, as the MSP process results in allocation of space to ocean users. Still, countries vary greatly in their approach, with countries like Denmark and Sweden allowing an open-door approach where the industry can suggest and apply to develop OWFs outside those areas formally opened by the government [9]. Increasingly, government tenders for OWFs call for co-existence, by specifying non-price criteria related to this [9], as recommended under the EU MSP Directive [8].

Co-existence is when several actors make use of or depend on the same place, at the same time or subsequently [16], including co-existence with nature when activities are expected to have little or no negative impact, or even a positive impact on the environment [10]. Schupp et al [16] position co-existence/co-location as one of several modes of multi-use classified by level of system integration. Higher degrees of integration include 1) synergistic multi-use through sharing of peripheral services, logistics, and infrastructures, and 2) multi-purpose platforms.

Co-existence between offshore wind and fisheries is a primary concern in European seas. Fisheries may be negatively affected due to overlaps between fishing grounds and proposed wind farm locations. Power cables and bottom anchorages could restrict the use of commonly used fishing gears such as bottom trawls. Fishing vessels may need to sail longer distances to catch the same amount of fish, as a consequence of loss of access to specific fishing grounds and longer transit routes. Longer transit routes to and from the fishing grounds would result in increasing fishing effort, implying higher fuel consumption per unit of catch, increasing the costs and the environmental footprint of seafood.

Offshore wind could also have site-specific impacts on the marine environment at different phases of development (installation, operation or decommissioning) with cascading effects to species targeted by fisheries [17]. It is still unclear how large the area of influence is, and whether local effects lead to effects at the ecosystem or population level. Frequently discussed negative impacts include noise, changes in hydrodynamics, alteration of habitat and electromagnetic fields (EMF). Impacts from noise range from physical damage, physiological stress to behavioral changes in some species [18]. Wind power generation reduces the wind speed in the vicinity which translates to possible changes in turbulence, mixing, stratification and hydrodynamics in general [19]. Effects may be highly variable, but these changes could also lead to changes in primary production and onwards to the rest of the food chain [20].

Wind farm's fundaments will also introduce hard substrates that can attract many species. They can have an artificial reef effect [21] but also increases the risk of serving as vectors for invasive species to flourish [22]. Fish species could also interact with EMFs through subsea cables, and these could disrupt the migration or foraging cues that are critical for some fish species [23]. Production of offshore renewable energy represents a new sector, and a new "challenge" for fisheries which have used these areas for generations. So how do we ensure these industries are valued on equal terms?

Related to co-existence but at higher level of system integration [16], multi-use concepts are being explored to derive more value from every square kilometer of ocean area. Tests of offshore aquaculture co-located with OWFs have been conducted in both the North Sea and the Baltic Sea [24], in areas that normally are restricted to other ocean uses due to safety risks and corresponding liability concerns [25]. A recent study by Maar et al. [24] quantify potential harvests (for food), nutrient removals, and carbon removals for blue mussels and sugar kelp grown within wind farms. They find that the food production potential from these systems were much higher in the North Sea, but that low-trophic aquaculture could play an important role for nutrient removal in the Baltic.

As seen from the discussion above, good co-existence results as a combination of measures during project planning, installation, operation, and decommissioning. These measures include 1) processes that contribute to reducing stakeholder conflicts and identification of new synergies during planning, 2) specific considerations to minimize user conflicts and environmental impacts related to site selection, field layout and system design, and 3) specific measures aimed at deriving more value from each hectare through co-location or multi-use of ocean space.

## 3. Participatory scenario building to understand co-existence opportunities and trade-offs

MARCO *(MARine CO-existence scenario building)* is an innovation project aiming to build model-based scenarios for co-existence to improve stakeholder inclusion and early-stage decision making. The approach (see Figure 1) combines spatial analyses by GIS with system dynamics to address the temporal dynamics of industry-industry, and industry-environment interactions. It addresses the need for tools that handle the temporal dynamics of the spatial planning and consenting process, while allowing for a participatory approach.

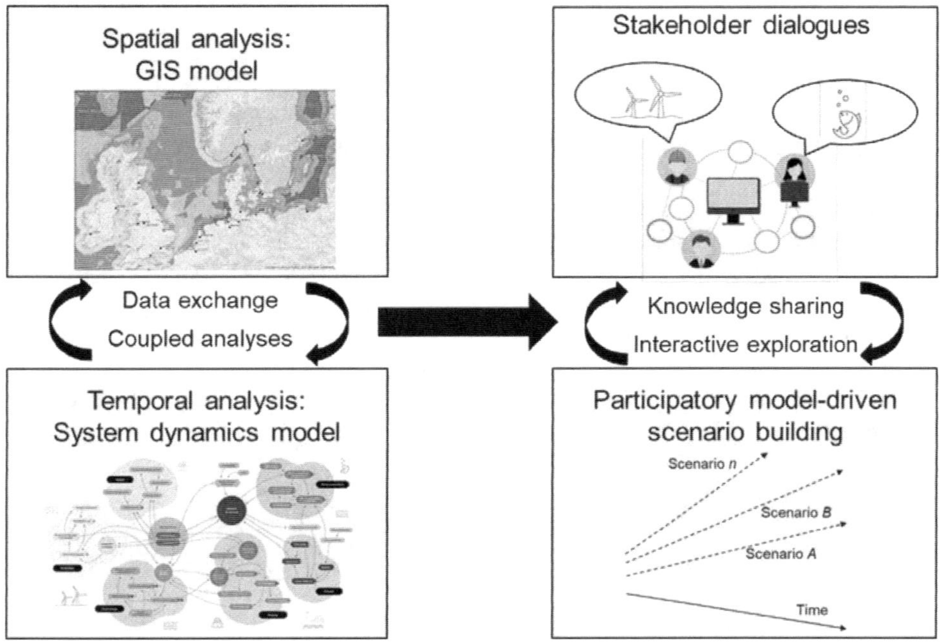


**Figure 1.** The main methodology of the MARCO project [5].

Key steps include 1) identification of key uncertainties related to space use, co-existence, and impacts with stakeholders, 2) parametrization of uncertainties as scenario variables, 3) implementation of scenario variables in system models, 4) simulation of alternative future scenarios defined with stakeholders, 5) quantification of co-existence synergies and impacts, and 6) visualization of results in maps and dashboards for exploration by stakeholders.

Exploration of alternative pathways for what *can* happen allows for a stakeholder-inclusive approach, where scenarios revealing key trade-offs, e.g., between energy production, seafood production, and environmental protection, are to be jointly defined through dialogue between actors that need to co-exist in the ocean.

We now review some key contributions to the literature on decision support methods relevant to understanding the trade-offs that need to be made during offshore wind project development to ensure successful co-existence in the ocean space.

### *3.1. Spatial decision support tools*

Pinarbasi et al. reviewed a total of 34 tools for decision support in MSP in 2016-2017 [26], following an earlier review by Stelzenmüller et al. [27] focused on MSP tools developed between 1998 and 2009. Pinarbasi et al. found that two-thirds of all tools were spatially explicit. The most common purposes of MSP tools were site identification, environmental impact assessment, communication, and use conflict analysis. Few tools included future projections, socio-economic analysis, and stakeholder engagements [26]. Since that publication, several new GIS-based tools have been introduced, e.g., providing spatial interaction analysis [28-30] and cumulative impact assessments [29-32].

Bergh et al. [28] presented a participatory GIS-approach where they introduced the GIS-based SEAGRID tool applied to coastal aquaculture in Western Norway and elicited spatial interaction scores for all pairs of marine uses and natural areas through stakeholder surveys. Spatial interaction scores provided an overview of whether there are synergies (positive score) or conflicts (negative score) between specific uses of the ocean.

Hammar et al. [29] introduced scenarios for cumulative impact assessments (CIA) in the Swedish MSP context to allow for ecosystem-based MSP. They compared a situation without an MSP process to two MSP scenarios for 2030. First, they quantified environmental impact under a *negotiated* plan, which are the Swedish MSPs after extensive stakeholder involvement. Second, they quantified environmental impact under an *eco-alternative* plan, which gave a higher priority to ensuring good environmental status. Also following the CIA methodology, Bonnevie et al. [30] introduced the SEANERGY tool, taking a geospatial indexing approach to describing the relative impact of multiple human pressures on the marine environment. They included an expert-based scoring system for assessing potential synergies and conflicts among uses to facilitate co-existence.

To inform spatial analyses such as cumulative impact assessments, the risk of impact from industry sectors on ecosystem components can be assessed semi-quantitatively, as for example done for Norwegian waters [31] using the ODEMM (Options for Delivering Ecosystem-based Marine Management) approach. Solutions for reducing impacts on marine ecosystems should also be evaluated in an integrated perspective to ensure efficiency when considering cumulative impacts across sectors [32]. Operationalizing the ODEMM risk assessments has the potential to integrate all relevant knowledge and provide risk-based decision support for the industry and management bodies.

### 3.2. *Optimization-based methods*

Optimization applied to siting of offshore wind considers objectives related to energy production [33-40], variability and intermittency challenges [33, 34], and cost minimization/profit maximization [35-38]. When optimizing offshore wind, availability of space, considerations of other users, physical characteristics impacting technical feasibility [5,6], and avoidance of ecologically sensitive areas [29-32] are often seen as constraints.

For spatial allocation of offshore wind, energy system level challenges like the intermittency in renewable power production, will impact optimal siting. Hjelmeland and Nøland [33] showed that the impact of direct grid connections from Norwegian offshore wind to European energy markets introduce correlation with wind patterns in other European countries. They concluded in favor of locating Norwegian offshore wind in the Norwegian Sea, rather than the North Sea to avoid resulting price cannibalization. Solbrekke et al. [34] studied the role of a hypothetical grid between five hypothetical wind farms in the Norwegian Sea to reduce intermittency challenges due to siting.

After siting, design problems like layout can be approached using optimization methods. Micrositing of offshore wind is extensively covered in operations research, e.g. by Cazzaro et al. [35] who presented a combined layout optimization and cable routing problem, maximizing lifetime energy revenues. Their combined optimization approach sacrifices revenue made from energy production, which would be achieved by larger spacing between and thus reduced wake effects, for shorter cable routing that reduces investment costs. This then improves *both* net present value and area utilization.

Spatial planning looks beyond offshore wind, and specialized tools developed for ecosystem-based management, e.g. MARXAN, have been applied to spatially optimizing offshore wind siting when considering multiple objectives, for instance by Göke et al. [36]. Basirati et al. [37] developed a generic multi-objective integer program which maximizes the weighted sum of the value of an area to a set of stakeholders in shipping, environmental protection, and offshore wind.

### 3.3. *Multi-criteria decision-making methods*

Looking beyond optimally allocating offshore wind, spatial planning problems are multi-objective, and can be formulated using an objective function that weighs energy output against seafood production, interests of shipping, and environmental impacts. Collaborative and participatory trade-off exploration is a useful way of getting a common understanding of knowledge basis and goal conflicts during planning and stakeholder dialogues thereby enabling co-existence.

Virtanen et al. [38] presented an analysis based on a spatial prioritization approach for Finnish marine areas, using 150 spatial data layers to perform a suitability assessment. They compared the main objective of the wind farm developer, namely to maximize profits, against solutions that would also take into consideration societal and biodiversity objectives. By handling societal and environmental issues as goals rather than constraints, their method was able to find economically profitable solutions that compromised social and environmental factor to a very small degree. Solbrekke et al. [39] used the analytical hierarchy process to model the decision-making problem arising when there are several actors like offshore wind investors, fishermen, and environmentalists with partially conflicting preferences regarding the siting of wind farms offshore. They constructed a hierarchical suitability score around wind resource attractiveness, techno-economical aspects, met-ocean constraints, social acceptance, and environmental considerations. Kyriazi et al. [40] introduced a game-theoretic framework for allocation of marine space among users. They applied this process to a three-actor case study where wind farm developers, fishermen, and conservationists identify a unique, fair allocation of space at the Doggerbank, and demonstrated that coalitions may form through the process. Pinarbasi et al. [41] developed a Bayesian Belief Network to integrate the ecosystem perspective into feasibility assessments of offshore wind. They included both environmental and technical feasibility criteria, as well as considerations of visibility from shore and conflicts with other marine uses.

From the perspective of consenting processes, the inclusion of non-price criteria in tendering will drive the industry to consider multiple objectives. Next, we look at literature that emphasizes the multi-actor nature of the planning situation, and views it as a learning process, rather than a well-defined problem to be analyzed.

### *3.4. Simulation and participatory scenario planning for stakeholder engagement*

Simulation methods can be powerful tools to engage stakeholders, visualize and explore outcomes of diverse policies. Vermeulen-Miltz et al. [42] used system dynamics to ensure user involvement in an MSP process in South Africa. While their approach did not include offshore wind, they considered key interlinkages between maritime, seafood, tourism, discharge from land, and marine life.

Abspoel et al. [43] reported on the MSP Challenge, a gaming approach to ensuring mutual understanding and communicating MSP, implemented as a role-playing game, a board game and as an interactive digital simulation platform. The tool has been developed through several EU projects and demonstrated in several high-level MSP fora. Serious gaming approaches and use of simulation could play an important role in social learning processes, as exemplified by the Dutch community of practice aimed at contributing to co-existence and multi-use between offshore wind and other ocean stakeholders [44].

Scenario analysis is another commonly used approach which facilitates stakeholder participation. Couture et al. [45] reviewed scenario literature on MSP for aquaculture, another industry whose development is dependent on alignment with marine spatial plans. They found that *projections*, what will likely happen, and *exploration,* what can happen, are commonly used in that sector. *Normative* scenarios, what should happen, have mostly been conducted on the national level. As an enabler of the energy transition, offshore wind is often viewed through the *normative* lens, with concrete ambitions outlining a desired growth trajectory. A recent example is Maar et al [24], who studied the potential for multi-use of offshore wind and low-trophic aquaculture under three distinct offshore wind scenarios, labelled *optimistic, realistic,* and *pessimistic.*

Several of the approaches listed earlier also have an element of participatory modelling, particularly GIS-contributions [28-30, 36, 37] and multi-objective methods [38-40]. However, these are often not temporally explicit scenarios, but focus on scenarios based on impacts of decisions made at one point in time.

## 4. Scenarios for area usage – implications for co-existence

Having reviewed core contributions on MSP and co-existence, as well as specific decision support and stakeholder engagement tools, we now turn to present some early examples of scenarios developed around area usage for offshore wind.

This section introduces 1) the main geographical area of interest for the MARCO project, a smaller part of which will be explored more thoroughly through a dedicated use case, and 2) documents an example of scenarios for area usage in the North Sea.

### *4.1. North Sea area of interest*

The core area of interest for offshore wind development in Norwegian waters is the North Sea. As shown in Figures 2 – 5, this is an area where there is already significant activity, and where there are ecologically valuable areas. This provides a current baseline for understanding the area needs of different sectors, and serves as a starting point for identifying whether there are specific co-existence opportunities and/or challenges, for instance between offshore wind and protected areas.

In the maps below (Figures 2 – 5), we show the proposed development areas for offshore wind as suggested by Norwegian Water Resources and Energy Directorate (NVE) [46], and offshore aquaculture areas proposed by the Norwegian Directorate of Fisheries [47]. Figure 2 shows these areas along with particularly valuable and vulnerable areas (SVOs) [48], spawning areas that are seasonally closed for surveys [49], and vulnerable biotopes like corals, sponges, and seapens [50].

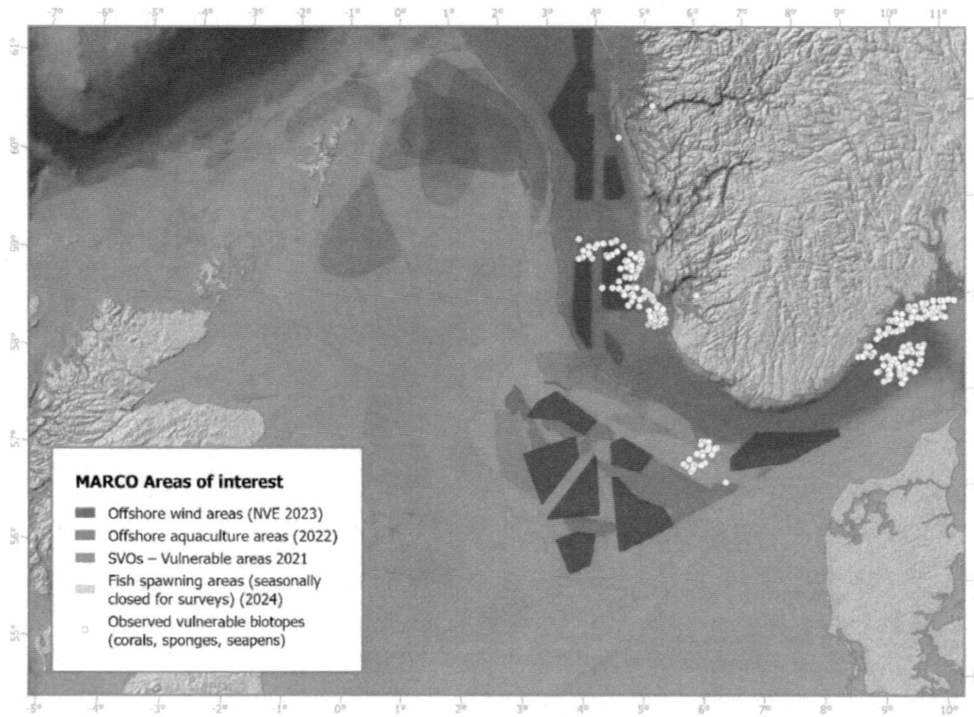


**Figure 2.** Areas of interest for offshore wind [46] and offshore aquaculture [47], with data on particularly valuable and vulnerable areas (SVOs) [48], spawning areas that are seasonally closed for surveys [49], and observed vulnerable biotopes [50]. Note that areas with no vulnerable biotopes cannot be inferred with this map.

Figure 3 shows the fishing done by Norwegian fishing vessels in the North Sea, including all gear types [51]. Most of the proposed offshore wind areas do overlap with fisheries, but there are some examples where the degree of overlap is relatively small. Figure 4 shows the data for shipping, including offshore vessels [52]. Shipping lanes are clearly visible along the coasts of Denmark and Norway, as well as along the coastlines, and have been well-accounted for when proposing the areas. In comparison, there are clear examples from Figure 3, that fisheries activity partially overlaps with one of the suggested areas. One can also clearly identify oil and gas installations from the high intensity traffic of offshore support vessels in the vicinity of these.

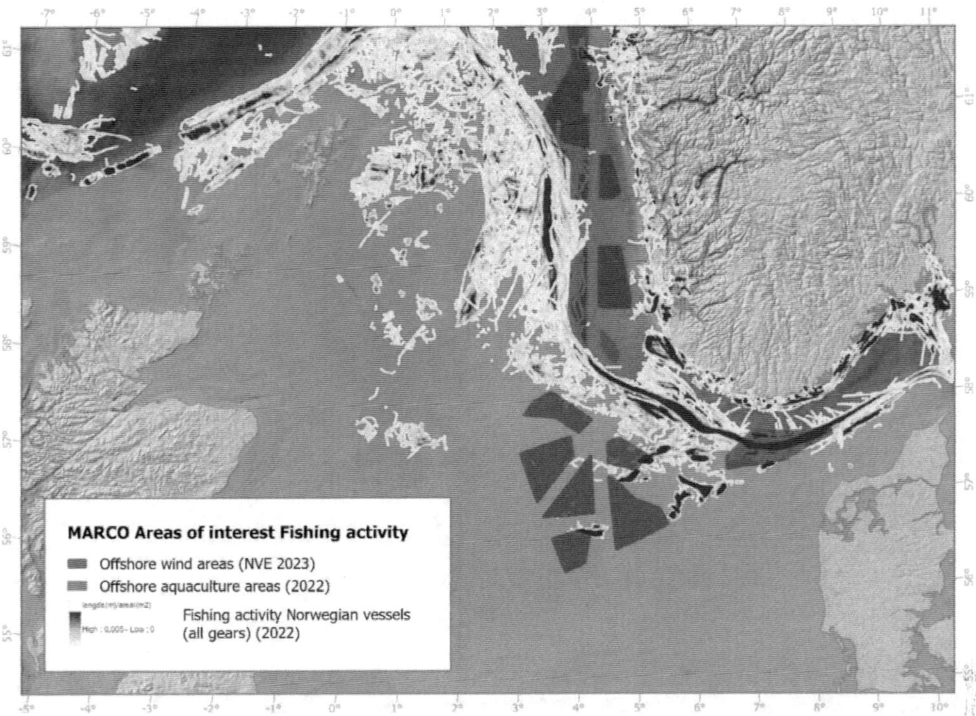


**Figure 3.** Areas of interest for offshore wind [46] and offshore aquaculture [47], with data for Norwegian fishing activity including all gear types in 2022 [51]. Red color indicates high fisheries intensity, while yellow indicates lower intensity.

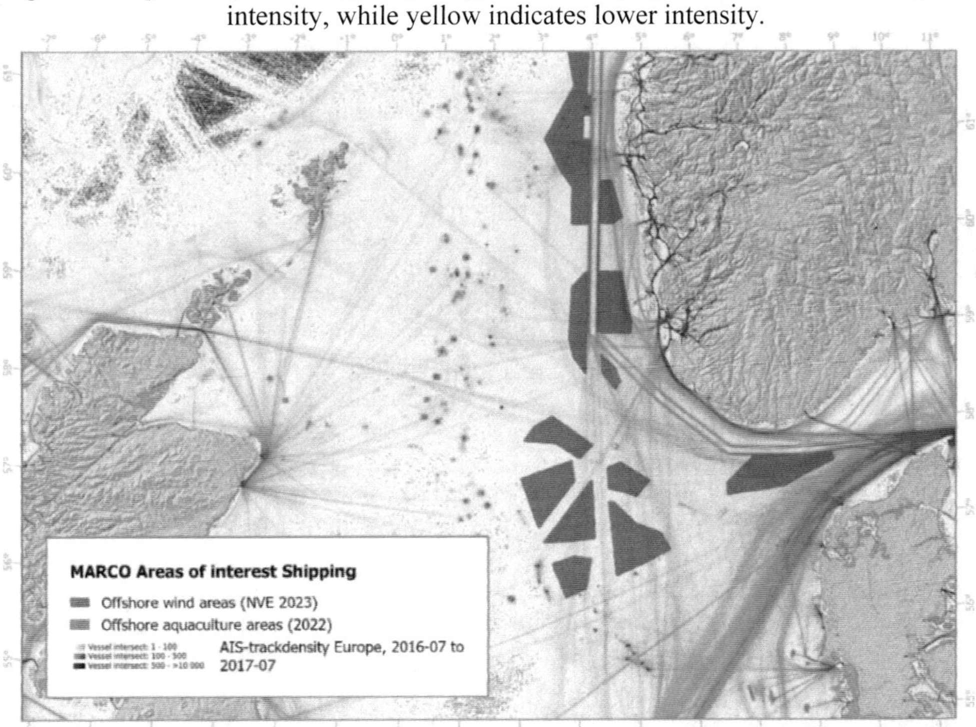


**Figure 4.** Areas of interest for offshore wind [46] and offshore aquaculture [47], with data for maritime traffic from July 2016 to July 2017 (including offshore support vessels) [52]. Dark red color indicates high traffic density, while lighter red indicates lower density.

### *4.2. Identification of key uncertainties and causal structures*

By modelling the causal relationships between underlying drivers in the energy and ocean systems, we establish baselines on which to build scenarios. Figure 5 addresses a wide range of topics, including i) technology and economy, ii) policy and regulations, and iii) nature and climate, where future uncertainty can be a main source of distrust among the stakeholders in the ocean. By investigating the causal drivers, sensitivities, and potential outcomes, the MARCO project will contribute to transparency around the uncertainty and establish a common understanding of the underlying scientific knowledge and knowledge gaps.

Figure 5 shows simplified causal structures that affect the offshore wind sector (e.g., cost developments impacting speed of buildout) and aspects of co-existence with other sectors (e.g., area claims that infringe on fishing areas, change in fish population, sharing of infrastructure). Notably, we also consider factors that are external to offshore wind as such, like policy calls to extend nature protection, e.g. [7], and climate change. Scenario variables in Figure 5 refer to specific levers that can be altered to generate scenarios for exploration. The number of possible combinations for these hint at the flexibility in the approach at creating alternative pathways and outcomes, that can be specified according to stakeholder needs.

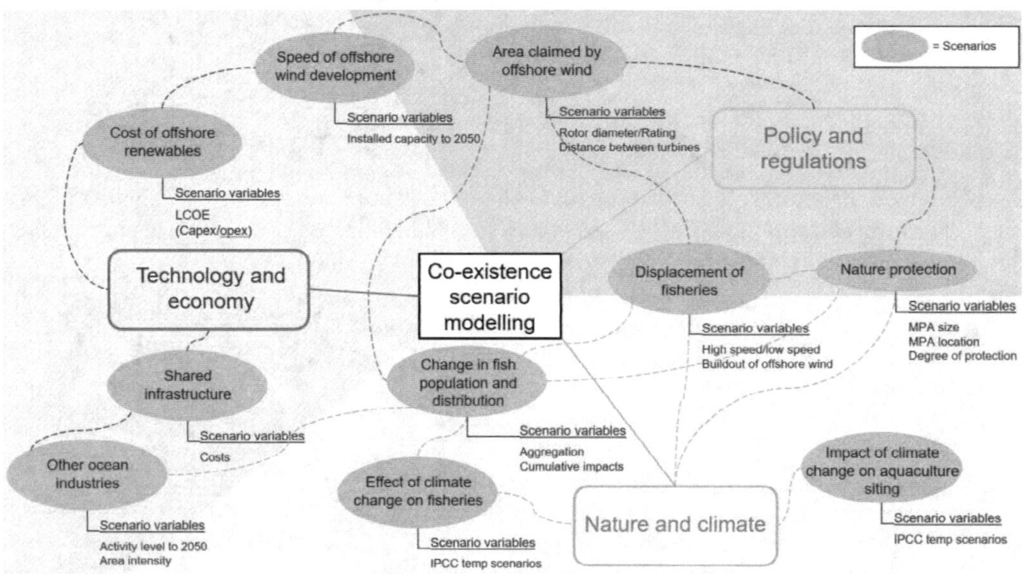


**Figure 5.** Schematic of the causal structures that affect the offshore wind sector for the MARCO-project. Key uncertainties to be addressed by constructing scenarios for co-existence.

### *4.3. North Sea area usage*

To provide a core example on which to build further scenarios, we provide estimates for area usage ("Area claimed by offshore wind" in Figure 5) in the North Sea by varying two parameters: installed capacity and area intensity. For *installed capacity*, we use the North Sea results of DNV's *Spatial Competition Forecast* [5], as a starting point, as based on drivers from DNV's *Energy Transition Outlook* [4]. In that assessment, the installed capacity of the North Sea in 2050 will be 240 GW [5] (compared to more than 360 GW for Europe as a whole [4]). As an alternative trajectory, we use the *Ostend Declaration* target of 300 GW [1]. Note that the example scenarios focus on the long-term trend. In the short-term, numerous issues impact the pace of buildout, such as immature regulatory frameworks, taxation schemes, and contracts for difference. Offshore wind has additionally been hard hit by inflationary pressure and rising material prices, and supply chain bottlenecks in the last few years. Naturally, these developments can also impact the industry willingness to prioritize good co-existence.

Estimation of *power density, capacity density,* or *area intensity*, as we refer to it here, can be conducted in several ways. Either a "bottom-up" process focusing on the layout and turbine design parameters, or through a "top-down" data analysis. The latter assessment could be done by establishing reference modes for area intensity developments over time, or by considering specific tendering requirements. For the opened areas in Norway, the required area intensity is only 3.5 MW/km$^2$ [53]. Lower area intensity results when field layout is optimized for maximum energy yield by reduction of wake losses, while this could also make the wind farm more apt for co-use with active fisheries within the wind farm due to more space between turbines. With reduced spacing between turbines, area intensity increases, thereby excluding active fishing within the wind farm.

The "bottom-up" assessment in DNV's *Spatial Competition Forecast* [5], based on simulated developments in average turbine rating and diameter and distance between turbines, is used as a starting point to create three pathways for area intensity, with a "Low" pathway indicating average area intensities between 3 and 4 MW/km$^2$ for the period to 2050 on average. The "DNV" pathway reflects a "Middle-of-the-road" development for area intensity improving from 5 to 6.8 MW/km$^2$ in 2050 on average. The "High" case reaches almost 10 MW/ km$^2$ on average. In all scenarios, floating wind has approximately 20% higher area intensity, and constitutes around 14% of installed capacity.

Figure 6 shows the resulting six alternative pathways for offshore wind area in the North Sea, considering the "DNV" and "Ostend" pathways for installed capacity, combined with the three area intensity trajectories provided above.

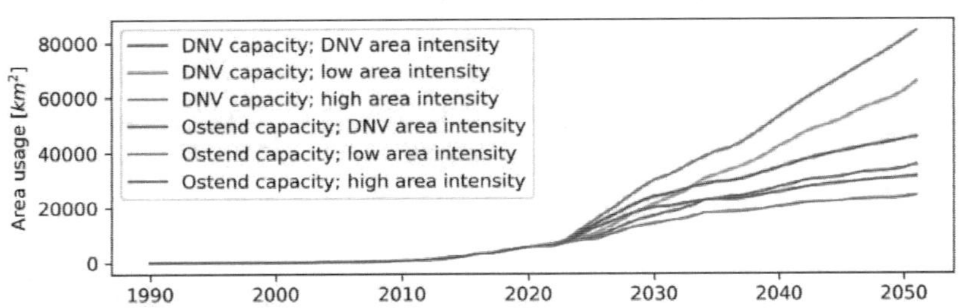


**Figure 6.** Six potential scenarios for area reserved for offshore wind, considering two installed capacity pathways and three alternate pathways for area intensity. "High area intensity" reaches 10 MW/km$^2$, while "low area intensity" is between 3 and 4 MW/km$^2$. "DNV area intensity" improves to 6.8 MW/km$^2$. "DNV capacity" implies 240 GW and "Ostend capacity" implies 300 GW installed in the North Sea by 2050.

The total area of the North Sea is 694,000 km$^2$. Under the six scenarios defined above, the corresponding demand on North Sea area will vary as described in Table 1. In the case of low area intensity, combined with buildout corresponding to the Ostend declaration, there will be high risk of spatial conflicts with other industries and with other wind farms. Additionally, such a development will likely have sea basin-wide cumulative effects, far beyond the local scale.

**Table 1.** 2050 area usage prognosis for the six scenarios above.

| **Scenario** | **Area usage [km$^2$]** | **Share of North Sea area [%]** |
|---|---|---|
| DNV capacity; DNV area intensity | 34,105 | 4.9 |
| DNV capacity; low area intensity | 62,008 | 8.9 |
| DNV capacity; high area intensity | 23,520 | 3.4 |
| Ostend capacity; DNV area intensity | 44,550 | 6.4 |
| Ostend capacity; low area intensity | 81,001 | 11.7 |
| Ostend capacity; high area intensity | 30,724 | 4.4 |

### *4.4. Implications for co-existence with nature and other sectors*

Area efficiency is a key concern for other users of ocean space, as it greatly impacts the area available to other stakeholders, both industry and nature. We now briefly introduce key implications for co-existence with other sectors and identify issues that will be further explored in the MARCO project.

*4.4.1. Co-existence with nature.* The ambition to protect 30% of the ocean by 2030 also comes into play with respect to offshore wind co-existence [7]. The International Union for Conservation of Nature (IUCN) classifications of marine protected areas (MPAs) allow renewable energy generation in categories IV (conservation park), V (habitat protection) and VI (general use). However, an OWF can only be considered a protected area if nature conservation is a priority in case of conflict in goals [54]. This condition might not be achievable given current technologies.

At present, there are OWFs co-located with MPAs in the UK and Germany subject to different regulations for different countries [55]. Both OWFs and MPAs restrict fishing to certain degrees and many species take refuge in the new structures [56]. The idea of OWFs as MPAs is controversial because the species assemblage in artificial structures will never be the same as those in naturally occurring habitats [57]. However, the increase in habitat complexity and recruitment of food sources for higher trophic levels may augment the populations of targeted species [58, 21].

Nature-inclusive design (NiD) measures may also be added towards the goals of fisheries augmentation and mitigation of environmental impacts. Examples include new materials for scour protection and introduction of fish refuge structures [59, 60].

OWFs will never be an alternative to MPAs for strict conservation but co-location may be possible for specific goals such as recovery of benthic habitats in areas that were previously disturbed by bottom trawling. Previous studies have shown that benthic fauna can recover when bottom trawling is stopped [61]. OWFs may also be able to protect and enhance the abundance of specific species with the "artificial reef effect" [21] and this should be viewed at a bigger scale with potential positive consequences such as recruitment subsidy [62] and spillover [63].

Knowledge gaps that wait to be filled are especially related to introduction of the foundations (wake effects on primary production, aggregation/artificial reef effects, habitat use for marine mammals), electromagnetic fields (influence distance, migration routes) and noise [64]. The Norwegian Institute of Marine Research [64] recommends standardized, thorough investigations over 3 years before development, as well as monitoring of both physical and biological changes during operation and after decommissioning the facility. Highlighting the need for rapid expansion of scientific knowledge on the topic, the *International Council for the Exploration of the Sea (ICES)* recently also published a roadmap to renew their approach to research on offshore wind [65].

*4.4.2. Co-existence with the fisheries.* The fishing industry is concerned about the resilience of the marine ecosystem and the sustainable yield of key species. It is therefore important to consider vulnerable spawning and nursing areas in MSP. Figure 7 indicates that this has been well-accounted for in the Norwegian Vestavind areas of the North Sea [46]. Still, early involvement is key for co-existence, and strongly emphasized in the recently published best practice guidance published by *Offshore Norway*, the *Norwegian Fishermen's Association* and their subsidiaries [66].

Some measures have been explored in different countries towards co-existence between offshore wind and fisheries [67]: financial disruption or displacement compensation in Denmark and passive fishing inside wind farms in the UK are examples. Negotiations during stakeholder consultations in Denmark include compensation for financial disruption and displacement of the fisheries sector. The amounts are estimated based on the impact assessment of the project [68]. Passive fishing and navigation are allowed inside OWFs in the UK. Studies have shown that turbine foundations can have a strong positive effect on crab abundance [69] and may provide a new opportunity for passive fisheries. However, uncertainties related to safety and liability are hindrances to the practice of fishing in OWF [70].

Journal of Physics: Conference Series **2745** (2024) 012004 doi:10.1088/1742-6596/2745/1/012004

*4.4.3. Co-existence and multi-use with maritime and offshore industry.* Shipping lanes are largely taken into considerations within the planning of areas for offshore wind. Unlike fishing, maritime traffic is nearly completely covered by automatic identification systems (AIS), and in densely navigated areas like the North Sea follows well-defined corridors. Still, a scenario where 12% of the North Sea is reserved for OWFs implies higher navigational risk compared to a scenario with only 3.5% area for offshore wind. Opportunities for sharing of peripheral infrastructure and logistics services [16] can be relevant to reduce costs and emissions related to logistics, for instance by using the same vessels [71].

*4.4.4. Co-existence and multi-use with aquaculture.* Aquaculture is currently mostly a coastal activity, despite ongoing trials to grow low-trophic aquaculture within OWFs. Multi-use of area with farming of species like mussels and kelp are promising, since they have a small logistical footprint compared to salmon farming. Figure 2 shows one area proposed for offshore fish farming, which so far has received little attention among the companies trying to develop salmon farms at sea.

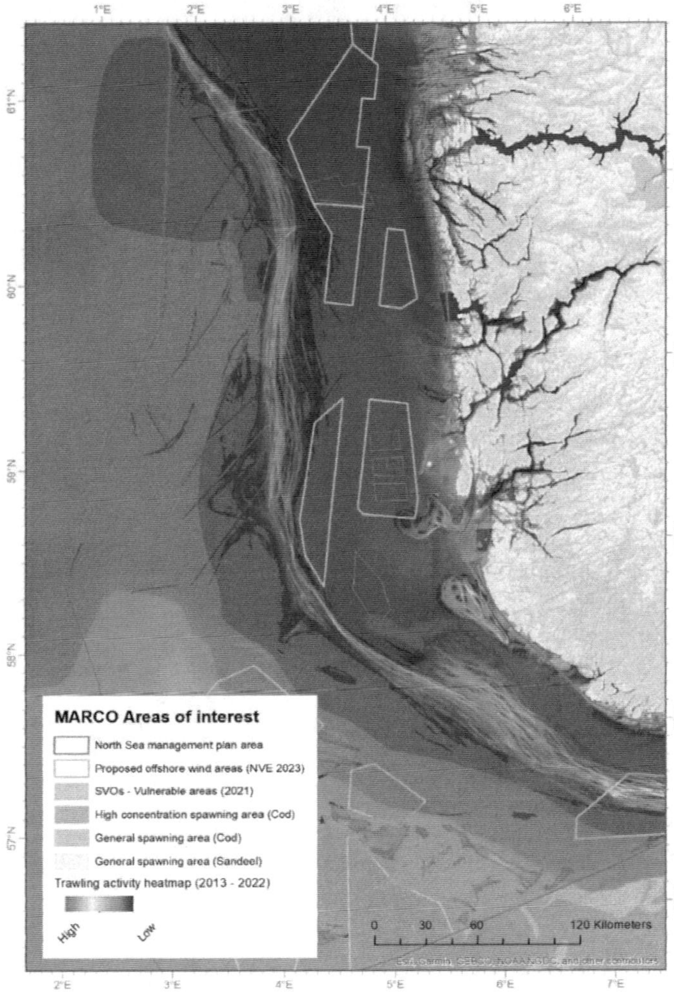


**Figure 7.** Vestavind proposed offshore wind areas [46], combined with the SVOs (particularly valuable and vulnerable areas) [48], high concentration spawning areas for cod, and general spawning areas for cod and sandeel overlaid [72], and historical trawling (based on vessel monitoring systems (VMS) and electronic reporting system (ERS)) activity heatmap from 2013 to 2022 [73].

## 5. Concluding remarks

This paper has briefly reviewed the decision support literature on spatial planning for offshore wind, highlighting particularly the need for considering scenarios to explore goal trade-offs during early project planning and in stakeholder dialogues. We have presented example scenarios for offshore wind development in the North Sea and the implications for co-existence. In the MARCO project, we will continue to explore such scenarios, including by further modelling of the implications discussed in Section 4. These will be explored with a wide range of ocean stakeholders, aiming to find common ground and good solutions for co-existence.

Our findings elucidate the need for more research on environmental impacts, as well as on the potential for co-existence among different ocean industries and their cumulative co-existence with nature. Scenarios can be a path forward for exploring alternative outcomes for nature, but when it comes to epistemic uncertainties like impacts of noise and electromagnetic fields on fish stocks, more basic physical and physiological investigations are necessary.

**References**

[1] *Ostend Declaration on the North Seas as Europe's Green Power Plant* 2023 https://www.regjeringen.no/contentassets/78bfc87bb04044c0933002ad7dd6e0f1/erklaring-energiministere.pdf
[2] Wind Europe 2023 *Offshore Renewable Industry Declaration to the North Sea Summit of 24th April 2023* https://windeurope.org/wp-content/uploads/files/policy/position-papers/20230424-Offshore-Renewable-Industry-Declaration.pdf
[3] United Nations 2023 *Summary of Global Climate Action at COP28* https://unfccc.int/sites/default/files/resource/Summary_GCA_COP28.pdf
[4] DNV 2023 *Energy Transition Outlook* https://www.dnv.com/energy-transition-outlook/index.html
[5] DNV 2023 *Ocean's Future to 2050: Spatial Competition Forecast* https://www.dnv.com/Publications/spatial-competition-forecast-237261
[6] Pettersen S S, Aarnes Ø, Arnesen B, Pretlove B, Ervik A K, and Rusten M 2023 Offshore wind in the race for ocean space: A forecast to 2050 *J. Phys.: Conf. Ser.* **2507** 012005
[7] United Nations 2023 *Convention of Biological Diversity* https://www.cbd.int/convention/
[8] Directive 2014/89/EU of the European Parliament and of the Council of 23 July 2014 establishing a framework for maritime spatial planning *Official Journal of the European Union* 2014
[9] Nordic Energy Research 2023 *Co-existence and nature inclusive design in Nordic offshore wind farms* https://www.nordicenergy.org/publications/coexistence-and-nature-inclusive-design-in-nordic-offshore-wind-farms/
[10] Pardo J C F, Aune M, Harman C, Walday M and Skjellum S F 2023 A synthesis review of nature positive approaches and coexistence in the offshore wind industry *ICES J. Mar. Sci.* 1-17
[11] Ehler C and Douvere F 2007 *Visions for a sea change: Report of the First International Workshop on Marine Spatial Planning* (Paris, France: UNESCO)
[12] Ehler C N 2021 Two decades of progress in Marine Spatial Planning *Mar. Policy* **132** 104134
[13] Santos C F, Agardy T, Andrade F, Crowder L B, Ehler C N and Orbach M K 2021 Major challenges in developing marine spatial planning *Mar. Policy* **132** 103248
[14] Crona B, Wassénius E, Lillepold K, Watson R A, Selig E R, Hicks C, Österblom H, Folke C, Jouffray J B and Blasiak R 2021 Sharing the seas: A review and analysis of ocean sector interactions *Environ. Res. Lett.* **16** 063005
[15] Gissi E, Fraschetti S and Micheli F 2019 Incorporating change in marine spatial planning: A review *Environ. Sci. Pol.* **92** 191–200
[16] Schupp M F, Bocci M, Depellegrin D, Kafas A, Kyriazi Z, Lukic I, Schultz-Zehden A, Krause G, Onyango V and Buck B H 2019 Towards a common understanding of ocean multi-use *Front. Mar. Sci.* **6** 165
[17] Gill A B, Degraer S, Lipsky A, Mavraki N, Methratta E and Brabant R 2020 Setting the context for offshore wind development effects on fish and fisheries *Oceanography* **33** 118–27
[18] Mooney T A, Andersson M H and Stanley J 2020 Acoustic impacts of offshore wind energy on

fishery resources: An evolving source and varied effects across a wind farm's lifetime *Oceanography* **33** 82–95

[19] van Berkel J, Burchard H, Christensen A, Mortensen L O, Petersen O S and Thomsen F 2020 The effects of offshore wind farms on hydrodynamics and implications for fishes *Oceanography* **33** 108–17

[20] Daewel U, Akhtar N, Christiansen N and Schrum C 2022 Offshore wind farms are projected to impact primary production and bottom water deoxygenation in the North Sea *Commun. Earth Environ.* **3** 1-8

[21] Degraer S, Carey D, Coolen J, Hutchison Z, Kerckhof F, Rumes B and Vanaverbeke J 2020 Offshore wind farm artificial reefs affect ecosystem structure and functioning: A synthesis *Oceanography* **33** 48–57

[22] De Mesel I, Kerckhof F, Norro A, Rumes B and Degraer S 2015 Succession and seasonal dynamics of the epifauna community on offshore wind farm foundations and their role as stepping stones for non-indigenous species *Hydrobiologia* **756** 37–50

[23] Öhman M C, Sigray P and Westerberg H 2007 Offshore windmills and the effects of electromagnetic fields on fish *AMBIO* **36** 630–3

[24] Maar M, Holbach A, Boderskov T, Thomsen M, Buck B H, Kotta J and Bruhn A 2023 Multi-use of offshore wind farms with low-trophic aquaculture can help achieve global sustainability goals *Commun. Earth Environ.* **4** 447

[25] Van den Burg S W K, Rockmann C, Banach J L and van Hoof L 2020 Governing risks of multi-use: seaweed aquaculture at offshore wind farms *Front. Mar. Sci.* **7** 60

[26] Pinarbasi K, Galparsoro I, Borja A, Stelzenmüller V, Ehler C N and Gimpel A 2017 Decision support tools in marine spatial planning: Present applications, gaps and future perspectives *Mar. Policy* **83** 83–91

[27] Stelzenmüller V, Lee J, South A and Rogers S I 2013 Practical tools to support marine spatial planning: A review and some prototype tools *Mar. Policy* **38** 214–27

[28] Bergh Ø, Beck A C, Tassetti A N, Olsen E, Thangstad T H, Gonzales-Mirelis G, Grati F, Bolognini L and Søvik G 2023 Analysis of spatial conflicts of large-scale salmonid aquaculture with coastal fisheries and other interests in a Norwegian fjord environment, using the novel GIS-tool SEAGRID and stakeholder surveys *Aquaculture* **574** 739643

[29] Hammar L, Molander S, Pålsson J, Crona J S, Carneiro G, Johansson T, Hume D, Kågesten G, Mattson D, Törnqvist O, Zillen L, Mattson M, Bergström U, Perry D, Caldow C and Andersen J H 2020 Cumulative impact assessment for ecosystem-based marine spatial planning *Sci. Tot. Environ.* **734** 139024

[30] Bonnevie I M, Hanse H S and Schrøder L 2020 SEANERGY – A spatial tool to facilitate the increase of synergies and to minimize conflicts between human uses at sea *Environ. Modell. Softw.* **132** 104808

[31] Hansen C, Hjøllo S S, Ottersen G and Skern-Mauritzen M 2022 *Miljøverdiers sårbarhet i norske havområder - En gjennomgang av sårbarhet til ulike typer påvirkninger i foreslåtte særlig verdifulle og sårbare områder i norske havområder* Rapport fra Havforskningen **2022-33** (Bergen, Norway: Institute of Marine Research)

[32] Hansen C, Aarflot J M, Eriksen E, Husson B, Fauchald P, Johansen G O, Jørgensen L L, van der Meeren G I, Mikkelsen N, Ottersen G, von Quillfeldt C and Skern-Mauritzen M 2022 *Samlet påvirkning i foreslåtte særlig verdifulle og sårbare områder i norske havområder* Rapport fra Havforskningen **2022-46** (Bergen, Norway: Institute of Marine Research)

[33] Hjelmeland M and Nøland J K 2023 Correlation challenges for North Sea offshore wind power: a Norwegian case study *Sci. Rep.* **13** 18670

[34] Solbrekke I M, Kvamstø N G and Sorteberg, A 2020 Mitigation of offshore wind power intermittency by interconnection of production sites *Wind Energy Sci.* **5** 1663–78

[35] Cazzaro D, Koza D F and Pisinger D 2023 Combined layout and cable optimization of offshore wind farms *Eur. J. Oper. Res.* **311** 301–15

[36] Göke C, Dahl K and Mohn C 2018 Maritime spatial planning supported by systematic site selection: Applying Marxan for offshore wind power in the western Baltic Sea *PLoS ONE* **13** 0194362

[37] Basirati M, Billot R, Meyer P and Bocher E 2021 Exact zoning optimization model for Marine Spatial Planning (MSP) *Front. Mar. Sci.* **8** 726187

[38] Virtanen E A, Lappalainen J, Nurmi M, Viitasalo M, Tikanmäki M, Heinonen J, Atlaskin E, Kallasvuo M, Tikkanen H and Moilanen, A 2022 Balancing profitability of energy production, societal impacts and biodiversity in offshore wind farm design *Renew. Sust. Energ. Rev.* **158** 112087

[39] Solbrekke I M, and Sorteberg A 2024 Norwegian offshore wind power-Spatial planning using multi-criteria decision analysis *Wind Energy* **27** 5-32

[40] Kyriazi Z, Lejano R, Maes F and Degraer S 2017 A cooperative game-theoretic framework for negotiating marine spatial allocation agreements among heterogenous players *J. Environ. Manage.* **187** 444–55

[41] Pinarbasi K, Galparsoro I, Depellegrin D, Bald J, Perez-Moran G and Borja A 2019 A modelling approach for offshore wind farm feasibility with respect to ecosystem-based marine spatial planning *Sci. Total Environ.* **667** 306–17

[42] Vermeulen-Miltz E, Clifford-Holmes J K, Scharler U M and Lombard A T 2023 A system dynamics model to support marine spatial planning in Algoa Bay, South Africa *Environ. Modell. Softw.* **160** 105601

[43] Abspoel L, Mayer I, Keijser X, Warmelink H, Fairgrieve R, Ripken M, Abramic A, Kannen A, Cormier R and Kidd S 2021 Communicating maritime spatial planning: The MSP challenge approach *Mar. Policy* **132** 103486

[44] Steins N A, Veraart J A, Klostermann J E M and Poelman M 2021 Combining offshore wind farms, nature conservation and seafood: Lessons from a Dutch community of practice *Mar. Policy* **126** 104371

[45] Couture J L, Froehlich H E, Buck B H, Jeffery K R, Krause G, Morris J A, Perez M, Stentiford, G D, Vehviläinen H and Halpern B S 2021 Scenario analysis can guide aquaculture planning to meet sustainable future production goals *ICES J. Mar. Sci.* **78** 821–31

[46] Norwegian Water Resources and Energy Directorate 2023 *Identifisering av utredningsområder for havvind* https://veiledere.nve.no/havvind/identifisering-av-utredningsomrader-for-havvind

[47] Norwegian Directorate of Fisheries 2022 *Havbruk til havs* https://portal.fiskeridir.no/portal/apps/webappviewer/index.html?id=74d9be72eb0e41d4b15f1d85490106e0

[48] Norwegian Environmental Agency 2021 *Særlige verdifulle områder i havområdene* https://kartkatalog.miljodirektoratet.no/dataset/Details/702?lang=en-us

[49] Institute of Marine Research 2024 *Fish spawning areas (seasonally closed for surveys)*

[50] MAREANO 2024 http://mareano.no/kart/mareano.html

[51] Norwegian Directorate of Fisheries 2022 *Fishing activity Norwegian vessels (all gears)*

[52] Norwegian Coastal Administration 2023 *Tilgang på AIS-data* https://www.kystverket.no/navigasjonstjenester/ais/tilgang-pa-ais-data/

[53] Royal Norwegian Ministry of Petroleum and Energy 2023 *Announcement of competition for project areas in Utsira Nord for offshore renewable energy production* https://www.regjeringen.no/contentassets/94b9f178d05849a1a5852ce129693f27/announcement-of-competition-for-project-areas-in-utsira-nord.pdf

[54] Day J, Dudley N, Hockings M, Holmes G, Laffoley D, Stolton S, Wells S and Wenzel L (eds) 2019 *Guidelines for Applying the IUCN Protected Area Management Categories to Marine Protected Areas* 2nd edn (Gland, Switzerland: IUCN)

[55] Defingou M, Bils F, Horchler B, Liesenjohann, T, and Nehls G (2019) *PHAROS4MPAs- a review of solutions to avoid and mitigate environmental impacts of offshore windfarms.*

*BioConsult SH on behalf of WWF France*

[56] Hammar L, Perry D and Gullström M 2015 Offshore wind power for marine conservation *Open J. Mar. Sci.* **6** 66–78

[57] Ashley M C, Mangi S C and Rodwell L D 2014 The potential of offshore windfarms to act as marine protected areas – A systematic review of current evidence *Mar. Policy* **45** 301–9

[58] Reubens J T, Braeckman U, Vanaverbeke J, Van Colen C, Degraer S and Vincx M 2013 Aggregation at windmill artificial reefs: CPUE of Atlantic cod (*Gadus morhua*) and pouting (*Trisopterus luscus*) at different habitats in the Belgian part of the North Sea. *Fish. Res.* **139** 28–34

[59] Wilson J C and Elliott M 2009 The habitat-creation potential of offshore wind farms *Wind Energy* **12** 203–12

[60] Hermans A, Bos O G and Prusina I 2020 *Nature-Inclusive Design: A catalogue for offshore wind infrastructure: Technical report* (Den Haag, Netherlands: Witteveen+ Bos)

[61] Kaiser M J, Clarke K R, Hinz H, Austen M C, Somerfield P J and Karakassis I 2006 Global analysis of response and recovery of benthic biota to fishing. *Mar. Ecol. Prog. Ser.* **311** 1–14

[62] Gerber L R, Botsford L W, Hastings A, Possingham H P, Gaines S D, Palumbi S R and Andelman S 2003 Population models for marine reserve design: a retrospective and prospective synthesis *Ecol. Appl* **13** 47–64

[63] McClanahan T R, and Mangi S 2000 Spillover of exploitable fishes from a marine park and its effect on the adjacent fishery *Ecol. Appl.* **10**1792–805

[64] De Jong K, Steen H, Forland T N, Wehde H, Nyqvist D, Palm A C U, Nilssen K T, Albretsen J, Falkenhaug T, Biuw M, Buhl-Mortensen L and Sivle L D 2020 *Potensielle effekter av havvindanlegg på havmiljøet* Rapport fra havforskningen **2020-42** (Bergen, Norway: Institute of Marine Research)

[65] ICES 2024 *ICES Roadmap for Offshore Renewable Energy: ICES Convention, policies, and strategy* https://doi.org/10.17895/ices.pub.24990198

[66] Fiskebåt, Offshore Norge, Norges Fiskarlag and Sør-Norges Fiskarlag 2023 *Dreiebok Anbefalt praksis for sameksistens mellom fiskeri og havvind* https://offshorenorge.no/contentassets/2a9710414f0349b99647e685bc431cd7/dreiebok---anbefalt-praksis-for-sameksistens-mellom-fiskeri-og-havvind---05.12.23.pdf

[67] Stelzenmüller V, Gimpel A, Haslob H, Letschert J, Berkenhagen J and Brüning, S 2021 Sustainable co-location solutions for offshore wind farms and fisheries need to account for socio-ecological trade-offs *Sci Total Environ.* **776** 145918

[68] Danish Energy Agency 2018 *Offshore Wind and Fisheries in Denmark.* https://ens.dk/sites/ens.dk/files/Globalcooperation/offshore_wind_and_fisheries_in_dk.pdf

[69] Krone R, Dederer G, Kanstinger P, Kramer P, Schneider C and Schmalenbach I 2017 Mobile demersal megafauna at common offshore wind turbine foundations in the German Bight (North Sea) two years after deployment - increased production rate of *Cancer pagurus Mar. Environ. Res.* **123** 53–61

[70] Hooper T, Ashley M and Austen M 2015 Perceptions of fishers and developers on the co-location of offshore wind farms and decapod fisheries in the UK *Mar. Policy* **61** 16–22

[71] Omrani P S, Poort J, Swamy S K, Uritsky V, Dick R, Peet L, Egbertsen J and Winters D 2022 *The Potential of Shared Offshore Logistics. North Sea Energy* https://north-sea-energy.eu/static/9890eefabe9a327a1cbad29d455a2f01/NSE-2020-2022-5.1-Logistics.pdf

[72] Institute of Marine Research 2024 *Spawning of cod and sandeel data*

[73] Norwegian Directorate of Fisheries 2017 *Electronic Reporting Systems data* https://www.fiskeridir.no/English/Fisheries/Electronic-Reporting-Systems

# SUSTAINABLE DESIGN OF ONSHORE WIND TURBINE FOUNDATIONS

**Claudio Masanet[a,c,*], Ignacio Navarro[b,c], Maria Collado[a], Víctor Yepes[d]**

[a] Energy Division, TYPSA, 46980, Paterna, València, Spain.

[b] Department of Structures, TYPSA, 46980, Paterna, València, Spain.

[c] Department of Construction Engineering, Universitat Politècnica de València, 46022, València, Spain.

[d] Institute of Concrete Science and Technology (ICITECH), Universitat Politècnica de València, 46022, València, Spain.

* Correspondence/Presenting author: cmasanet@typsa.es

**Abstract**. In recent times, wind power has emerged as a prominent contributor to electricity production. Minimizing the costs and maximizing sustainability of wind energy is required to improve its competitiveness against other non-renewable energy sources. This communication offers a practical approach to assess the sustainability of wind turbine generator foundations from a 3-dimensional holistic point of view. Specifically, the main goal of this study is to analyse the life cycle impacts of one shallow foundation design comparing three different concrete alternatives: conventional concrete, concrete with 66-80% of blast furnace slags and concrete with 20% fly ash, and then to apply a Multi-Criteria Decision-Making model based on TOPSIS method to evaluate and compare the resulting sustainability of each alternative considered. The study results in a methodology for quantifying sustainability rather than simply qualifying it. Therefore, this methodology can be employed for design optimization, such as geometry and materials, with a sustainable perspective in mind. Specifically in this study, concrete with blast furnace slags emerges as the top-ranked sustainable alternative, followed by conventional concrete in second place, and fly ash option in third position.

## 1. Introduction

Wind power is a major player in electricity generation, contributing significantly to Europe's power needs with a 15% share. The EU Comission anticipates an even greater role for wind energy, targeting it to make up half of Europe's electricity supply by 2050[1]. On global scale, the World Wind Energy Association (WWEA) reported a robust 13% growth in 2022, resulting in a total installed capacity of 874 Gigawatts [2]. The majority of this capacity increase comes from onshore wind [3], favored for its ease of installation compared to offshore alternatives. To compete with non-renewable sources and sustain the rapid market growth and energy development trend, it is crucial to minimize costs and maximize sustainability.

In 1987, the Brundland Commission coined the term sustainable development, defining it as a way to satisfying the actual needs of the society without compromising the ability of future generations to meet their own needs [4]. Since then, sustainability has guided the scientific community's efforts to assess the impacts of products on society, the economy and the environment. The construction sector, encompassing both building and civil engineering, serves as a notable example of the study and application of sustainability [5-9].

Sustainability has not been consistently applied to all aspects of wind energy. Given the significant capital investment and large-scale production involved in Wind Turbine Generators (WTG) [10], the scientific community has focused its efforts on applying sustainability criteria to the design and location of wind farms [11-14], as well as the positioning and design of WTGs [15-17]. Over the past decades, sustainability has conventionally been evaluated in terms of the environmental impacts derived from human actions. However, current trends in the construction sector indicate a shift towards incorporating economic and social dimensions in sustainability assessments.[18-24]. In the context of foundations, the prioritization of sustainability implementation seems to be diminished, potentially due to their economic impact representing 5-7% of the overall investment in a wind farm [25].

In recent decades, foundation research has focused on the material reduction to save costs and reduce environmental impacts. As a result, progress has been made mainly in the development of tools for design optimization [26] and development of new materials [27]. Moreover, investigations focused on decreasing the carbon footprint [28-29] or analyzing the combined environmental impact of the WTG-foundation [30-31] have been recently conducted.

Despite this partial development, high emphasis shall be put on the assessment of the social dimension of sustainability. Initiatives, such as the Social European Taxonomy, are currently under development aiming to effectively incorporate social and economic aspects in sustainability assessments. Sustainability certification systems for infrastructures, exemplified by ENVISION [32], recognize the economic and social dimensions as integral components for any infrastructure aspiring to promote sustainability. In light of these considerations, it is imperative to develop efficient methodologies that simultaneously incorporate the three dimensions of sustainability in life cycle assessments.

To cope with the actual trends and demands in sustainability assessments, the present paper proposes the application of the recognized methodology defined in ISO 14040 [33] combined with a Multi-Criteria Decision-Making model (MCDM) to quantify and compare the resulting sustainability of three different WTG foundation. Each foundation is designed with a different concrete type, specifically, conventional concrete, concrete incorporating blast furnace slags and fly ash concrete.

The paper is composed of three further sections: materials and methods, results and discussion and finally, the conclusions. The materials and methods section outlines the main objective and details, all the applied methodologies, including the definition of the functional unit, system boundaries, inventory analysis and impact assessment. The results and discussion section presents all the data obtained from the analysis along with corresponding comments. The final section emphasizes key conclusions.

## 2. Materials and methods

The main objective of this paper is to analyse the life cycle impacts of a WTG foundation built with different concretes from an economic, environmental and social perspective individually. Then, a Multi-Criteria Decision-Making model (MCDM) is applied to evaluate and compare the resulting sustainability of each WTG foundation alternatives considered from a holistic perspective.

For that purpose, the calculation methodology to evaluate the sustainability of civil constructions defined in UNE-EN 17472 [34] is applied. This Standard allows to use the life cycle assessment methodology developed in ISO 14040 [33] and ISO 14044 [35] for all the dimensions of the sustainability. According to ISO 14040 [33], a rigorous LCA assessment must consist of four steps 1) the definition of the objective and scope of the study, 2) an analysis of the inventory to be accounted for, 3) a description of the methods and assumptions used for the impact assessment and, 4) the presentation and discussion of the results obtained.

### *2.1. Goal and scope*

WTG foundations contribute to huge material consumption, primarily concrete. Given that the production of cement is responsible of 8% of global $CO_2$ emissions [36-37], it becomes imperative to examine the influence of concrete types on the sustainability of foundation design.

For shallow foundations, the most common geometry used around the world is the circular foundation. As it is shown in Figure 1, the circular foundation presents a pedestal area in the center of the concrete mass where the anchor cage (AC) is installed. Then, the height of the slab decreases with constant slope up to the border. Usually, the main designs drivers are 0% ground gap condition or overturning, dimensioning total diameter of the foundation. Total height of the slab and pedestal are defined considering the internal bending moment. Generally three different concrete grades can appear in the concrete volume. Usually, concrete grade C35/45 is used for the slab, C45/50 for the pit (area near the bottom flange of the AC) and C50/60 for the pedestal (area near the top flange of the AC). The main reason is the excessive compression forces which appear concentrated near the anchor templates.

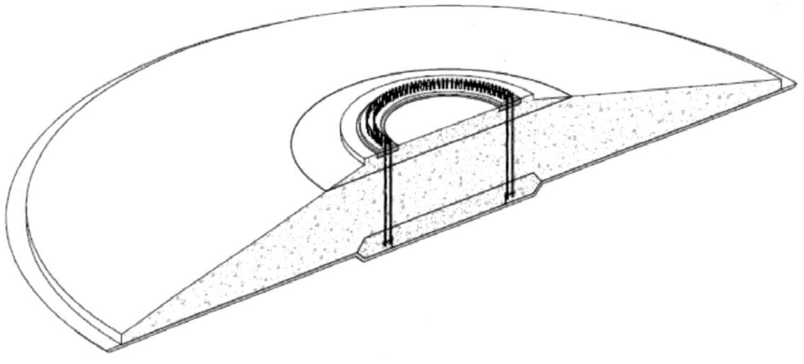

**Figure 1.** Isometric view of shallow circular WTG foundation

This study aims to assess the life cycle impacts of a particular shallow foundation design, specifically by comparing three alternative concrete options within the slab area (Alternative 1: conventional concrete (CONV, hereafter) Alternative 2: concrete with 66-80% of blast furnace slags (GBFS, hereafter) and Alternative 3: concrete with 20% fly ash (FA, hereafter), and then to apply a Multi-Criteria Decision-Making model for evaluating and comparing the sustainability of each considered alternative. All these three types of concretes are selected taking into consideration the recommendations and required cement type for big reinforced concrete volumes according to Spanish Standards [38-39]. For the sustainability life cycle performance, economic, environmental and social dimensions are simultaneously considered.

### *2.2. Functional unit*

According to ISO 14040[33], a life cycle analysis must be based on the same functional unit for the results of the assessments to be comparable.

In this study, the functional unit is a circular shallow foundation designed for a WTG 5.9MW located in Requena (Valencia, Spain). The design of the functional unit contemplates two different stages: stability and structural integrity. Initially, an analytical calculation is made to define the geometry of the foundation.

In these calculations, the backfill presents a specific weigh of 18.5 $kN/m^3$ and a soil resistance of 200 kPa. Additionally, the base of the tower presents a diameter of 4.70m connected to the foundation with an anchor cage composed 2x112 M42 10.9 bolts. The WTG loads are defined in Table 1 for 25 years of service life.

**Table 1.** WTG loads at tower base (Characteristic values)

| Load case | Fz [kN] | Fxy [kN] | Mz [kNm] | Mxy [kNm] | Safety Factor |
|---|---|---|---|---|---|
| **Characteristic extreme normal loads** | 7500.00 | 1200.00 | 1450.00 | 155000.00 | 1.35 |
| **Characteristic extreme abnormal loads** | 7530.00 | 1143.00 | 990.00 | 168500.00 | 1.10 |
| **Quasi-permanent loads** | 7550.00 | 900.00 | 3000.00 | 123400.00 | 1.00 |

According to IEC 64001-6[40] verifications of overturning, sliding, ground gapping, soil stresses and differential settlements are conducted. Specifically, no ground gapping condition for quasi-permanent load is the design driver for this functional unit. The results of no ground gapping verification can be checked in Table 2.

**Table 2.** Ground gapping verification

| Parameter | Results |
|---|---|
| **Eccentricity of WTG foundation** | 3.14 m |
| **Limit of eccentricity for 0% ground gapping** | 3.19 m |
| **% of foundation base in contact with soil** | 100.00 % |

The geometry of the foundation is defined in Figure 2 below. The foundation exhibits a total diameter of 25.50 m (Øtot), with a pedestal of 6.5 m (Øped). The height of the foundation is variable, measuring 0.35 m at the border (Hbor) and reaching a total height of 3.10 m at the center (Hslab + Hped), with 0.20 m attributed to the pedestal (Hped). Additionally, to properly allocate the AC, a 0.30m pit is defined in the center (Hpit). The diameter of the pit is the same than pedestal diameter (Øpit).

The backfill is engineered as a structural element. Therefore, its geometry and weight (18.50 $kN/m^3$) shall be preserved during all the service life. The geometry of the backfill is determined based on a pedestal protrusion of 0.10m and 2% slope.

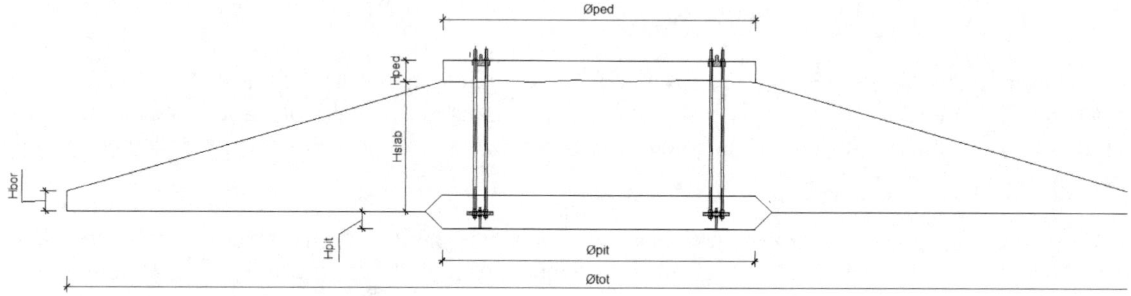


**Figure 2.** Geometry of the foundation

Once the geometry is established, a Finite Element Model (FEM) is implemented in SAP2000v.24 to analyse internal forces and define the reinforcement layout. Specifically, the FEM is based on shell and frame elements. The concrete material is modelled as linear elastic material type C35/45 according to Eurocode 2[41].

In the last step, boundary conditions are simulated by incorporating soil elasticity through non-linear springs. These springs only experiments force in compression according to the modulus of subgrade reaction (ks,stat=5MPa/m). The outcomes are then extracted as sectional forces and moments along a path aligned with the principal wind direction. The design exclusively accounts for dry conditions, excluding any buoyancy effects. The reinforcement bars have a characteristic yield strength of 500MPa and are arranged in a radial and circumferential layout.

Finally, local stresses calculations are implemented in the anchor cage area to define the concrete grade required in pedestal, pit and slab. The results require a C45/55 concrete grade in pit area, C50/60 in the pedestal area and a slab formed by concrete C35/45. A concise overview of the quantity of each material considered for the three alternatives can be reviewed in Table 3.

### *2.3. System boundaries*
After establishing the functional unit, the system boundaries needs to be delineated in accordance with ISO 14040[33] guidelines to ensure a comprehensive life cycle assessment. For this purpose, the boundary system is constructed using a "cradle to site" approach covering from production of the different construction materials in their respective production centers up to the end of the construction of the WTG foundation.

As a cut-off criterion and considering the comparison-oriented scope of the present assessment, identical processes and common to every alternative have been omitted from the system definition. Consequently, all processes associated with the anchor cage and its assembling, as well as those related to the grout pocket, are excluded. Furthermore, as the foundation is a buried structure inaccessible for maintenance, this particular stage is not factored into the analysis. Figure 3 shows a detail of the system boundaries considered for the analysis presented in this communication.

Finally, the end-of-life stage is not incorporated into the design. Two potential scenarios for the end-of-life stages may arise; a wind farm life-extension or dismantling it. Both situations are right now uncertain[42].

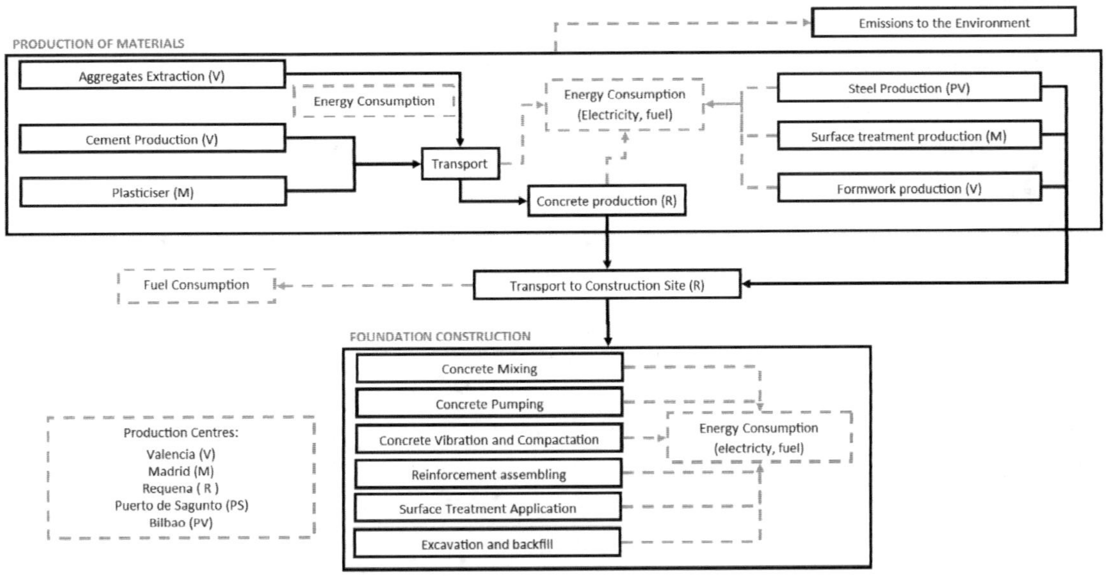


**Figure 3.** Boundaries of the product system considered for the functional unit

## 2.4. *Inventory analysis*

All the data required to properly quantify the relevant inputs and outputs of the product system under study are defined in this section. As recommended in ISO 14040 [33], all data used for this purpose have come from accepted sources.

Table 3 presents the different materials used to define the functional unit. These products are the concrete and reinforcing steel used to build the foundation, but also the formwork required for the construction, the backfill to cover the foundation and guarantee its stability and the bituminous paint required to protect the buried structure.

**Table 3.** Materials used to build the functional unit. Quantities and properties.

| Product | Quantity | Properties [39] |
|---|---|---|
| Slab concrete C35/45 | 716.88 $m^3$ | 2500 kg/$m^3$ |
| Pit concrete C45/50 | 25.44 $m^3$ | 2500 kg/$m^3$ |
| Pedestal concrete C50/60 | 25.07 $m^3$ | 2500 kg/$m^3$ |
| Lean concrete C20/25 | 55.84 $m^3$ | 2400 kg/$m^3$ |
| Reinforcing steel B500SD | 96696 kg | 7850 kg/$m^3$ |
| Formwork | 32.12 $m^2$ | 3 uses |

| | | |
|---|---|---|
| Excavation | 1362.69 m³ | 1850 kg/m³ |
| Backfill | 1268.91 m³ | 1850 kg/m³ |
| Bituminous paint | 1070.44 m² | 0.8 kg/m² |

In Table 4, the concrete components used for each concrete considered in the study are defined. The different types of cement considered fulfil the requirements of Spanish Standards [38-39].

**Table 4.** Concrete components for 1m³ of concrete.

| Concrete mix component | C50/60 | C45/55 | Blinding concrete | C35/45 CONV | C35/45 GBFS | C35/45 FA |
|---|---|---|---|---|---|---|
| Cement I (kg/m³) | 390 | 370 | 220 | 360 | - | - |
| Cement II/A-V (kg/m³) | - | - | - | - | - | 500 |
| Cement III/B (kg/m³) | - | - | - | - | 330 | - |
| Gravel (kg/m³) | 1020 | 1020 | 990 | 990 | 990 | 940 |
| Sand (kg/m³) | 860 | 864 | 1035 | 850 | 880 | 750 |
| Plasticiser (kg/m³) | 4.37 | 4.44 | 1.76 | 3.7 | 3.3 | 4 |
| Water (l/m³) | 97.50 | 74 | 129.80 | 162 | 158.4 | 185 |

Table 5 specifies the km required for transport of each material used in the concrete mixtures. Transport with truck is considered if distance is lower than 500 km. For higher distances, then train is selected as the main transport.

**Table 5.** Transport distances in km for each material considered in this study.

| Production process | Transport with lorry (km) | Transport with train (km) | Total (km) |
|---|---|---|---|
| Aggregates | 58.4 | - | 58.4 |
| Cement | 67.6 | - | 67.6 |
| Plasticiser | 418.5 | - | 418.5 |
| Concrete | 30.2 | - | 30.2 |
| Steel | 30 | 932.7 | 962.7 |

| | | | |
|---|---|---|---|
| Reinforcing steel | 107.5 | - | 107.5 |
| Bituminous paint | 448.7 | - | 448.7 |

The inventory data to perform the environmental assessment of each alternative have been collected from the Ecoinvent 3.8 environmental database [43]. Ecoinvent database is a diverse repository with over 20.000 datasets modeling human activities, covering global and regional sectors. The datasets include information on processes, resources used, emissions and resulting products, co-products and wastes.

The economic data to evaluate costs of each of the alternatives throughout the life cycle have been collected from the Spanish construction cost database [44]. The costs of each construction element include the proportional part of the machinery and labour involved in the manufacture of the material and its installation on site. The unit costs of each economic concept are indicated in Euros (€) updated to 2023. Table 6 shows the costs considered for each material.

**Table 6.** Unit cost of each material.

| Material | Cost |
|---|---|
| Excavation | 12.82 €/m$^3$ |
| Blinding concrete | 17.40 €/m$^2$ |
| Concrete C45/55 | 116.45 €/m$^3$ |
| Concrete C35/45 CONV | 109.45 €/m$^3$ |
| Concrete C35/45 GBFS | 101.20 €/m$^3$ |
| Concrete C35/45 FA | 129.85 €/m$^3$ |
| Concrete C50/60 | 120.45 €/m$^3$ |
| Reinforcing steel ($f_{yk}$=500MPa) | 1.70 €/kg |
| Backfill (18.5 kN/ m3) | 9.63 €/m$^3$ |
| Formwork | 22.74 €/m$^2$ |
| Bituminous paint | 7.79 €/m$^2$ |

Finally, social dimension is focused on workers from different production sites. Special emphasis is put on the problems related to gender discrimination and high unemployment. Local economy is also important since benefits from the economic inflows due to the production and construction helps the region.

Inventory data are gathered through web research from the Spanish National Institute of Statistics [45], Spanish Tax Office database [46] and the official OECD (Organization for Economic Co-operation and Development) databases [47]. To understand the meaning of the social context of the regions involved in the present study in relation to the rest of the regions in the Spanish territory, information has been collected as well on the minima and maxima values to be found in the Spanish regions for each of the social indicators. It is noted that this information does not allow the evaluation of the social impact of a specific activity per se, but does contextualize it. Performance values regarding material production and those related to worker activities are obtained also from [48] and shown in Table 8.

**Table 7.** Background data on Unemployment and gender discrimination

| Data | Requena | Buñol | Sagunto |
|---|---|---|---|
| Unemployment rate (%) | 12.87 | 12.74 | 15.35 |
| Maximum and minimum national unemployment (%) | 7.13-17.05 | 7.13-17.05 | 7.13-17.05 |
| Male unemployment (%) | 12.30 | 12.00 | 11.50 |
| Female unemployment (%) | 14.80 | 14.50 | 14.30 |

**Table 8.** Performance values considered for production and construction activities

| Activies | Performance values |
|---|---|
| Cement | 0.165 h/tn |
| Aggregate extraction | 0.1925 h/tn |
| Concrete production | 0.18 h/tn |
| Concreting | 0.35 h/$m^3$ |

### *2.5. Impact assessment*

The assessment of the environmental life cycle impacts is based on the international ReCiPe 2016 methodology [49]. This method integrates the environmental problem-oriented and damage-oriented approaches by transforming life cycle inventory data into two sets of impact categories: 18 midpoints indicators and 3 end points indicators. In both cases, emissions and resource extractions are translated into corresponding environmental impact scores using characterization factors based on three cultural perspectives, namely individualistic, hierarchical and egalitarian. This analysis adopts hierarchical perspective as the most consensus model and the endpoint approach that considers damage to ecosystem diversity (ED), damage to resource availability (RA) and damage to human health (HH).

In evaluating the economic dimension, the economic resources used in each phase of the life cycle considered are quantified in Euros (€)[50].

Finally, regarding the social dimension, the methodology described by [48] is used which is aligned with the Guidelines for social life cycle assessment of products and organizations 2020 reported by the United Nations[51]. Local employment is the indicator selected to quantify the social dimension as indicated following equation (1):

$$x_{local\ employment} = \frac{y_{local} - Y_{min}}{Y_{max} - Y_{min}} \quad (1)$$

Where:

- $y_{local}$ is the unemployment rate at the activity location.
- $Y_{max}$ is maximum national unemployment rate.
- $Y_{min}$ is minimum national unemployment rate.

A summary of the indicators considered for each sustainability dimension considered is exposed in Table 9.

**Table 9.** Sustainability indicators considered in this study.

| Sustainability Field | Criterion Id. | Criterion description | Impact assessment |
|---|---|---|---|
| Environment | 1 | Damage to human health | ReCiPe methodology. |
| | 2 | Damage to ecosystem | ReCiPe methodology. |
| | 3 | Damage to resource availability | ReCiPe methodology. |
| Economy | 4 | Construction costs | BoQ – Spanish construction data base. |
| Social | 5 | Local employment | [48] |

## 3. Results and discussion

The results obtained in this study are exposed in this section. Section 3.1, 3.2 and 3.3 present the results of the analysis for each considered dimension, including the outcomes for each indicator. Finally, section 3.4 shows the result of the multicriteria analysis that provides a quantitative assessment of the sustainability for each studied foundation alternative.

### *3.1. LCA results*

The environmental impact resulting from each design analysed in this study are shown in Figure 4. The higher the score obtained, the greater the impact. Based on the indicators, there is a noticeable disparity between the impact on the ecosystem and resources. Almost all the alternatives affect the ecosystem twice as much as they do resources. As for human health, all the alternatives impact it to a third compared to their impact on the ecosystem.

Regarding the comparison among the analysed alternatives, the results exhibit a similar trend the different concrete types analyzed. GBFS emerges as the optimal choice, displaying the lowest score in each impact category. The second alternative, CONV, ranks as the second-best in terms of environmental impacts, while FA option is considered the least favourable.

Nevertheless, it is worth noting that the difference between alternatives for each indicator is very small. When comparing the results of both GBFS and FA with CONV, it is observed that, in regard to ecosystem quality, they show a 2.46% decrease and a 1% increase in impact, respectively. For the resources, the percentages are 0.2% decrease and 1.4% increase, respectively. Finally, the impact on human health for GBFS compared to CONV is 6% lower, whereas for FA compared to CONV, it is 1.4% higher.

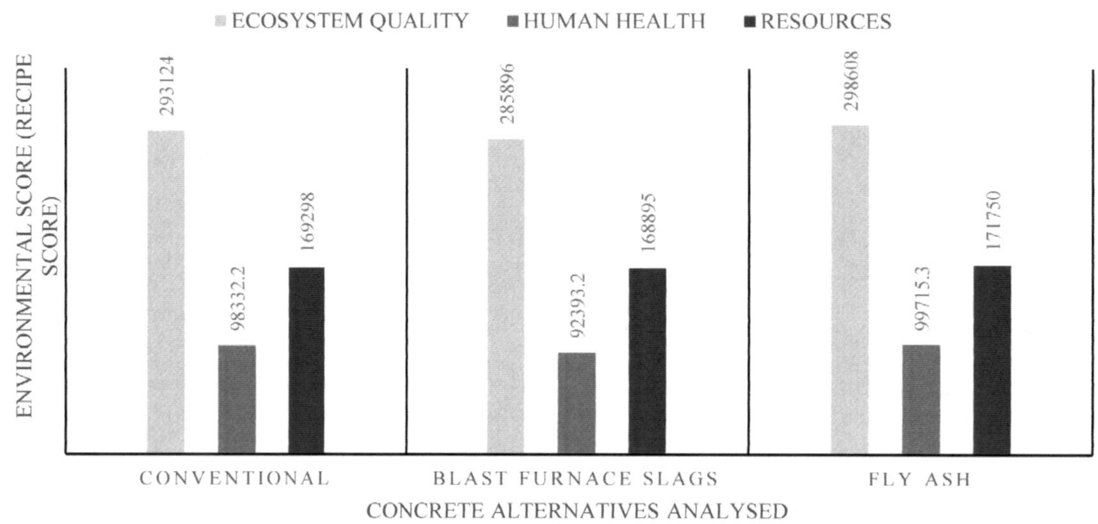


**Figure 4.** Environmental results obtained for each alternative and indicator.

### *3.2. Economic results*

Figure 5 exposes the economic results obtained for each alternative analysed. These results clearly show that the alternatives to conventional concrete incur higher costs. The economic comparison reveals CONV is the most cost-effective solution, with a total budget of approximately 283.000€. Following this, GBFS proves to be 2.09% more expensive than CONV. Finally, FA emerges as the last economically favorable option, exhibiting a higher cost difference of 7.26% compared to CONV.

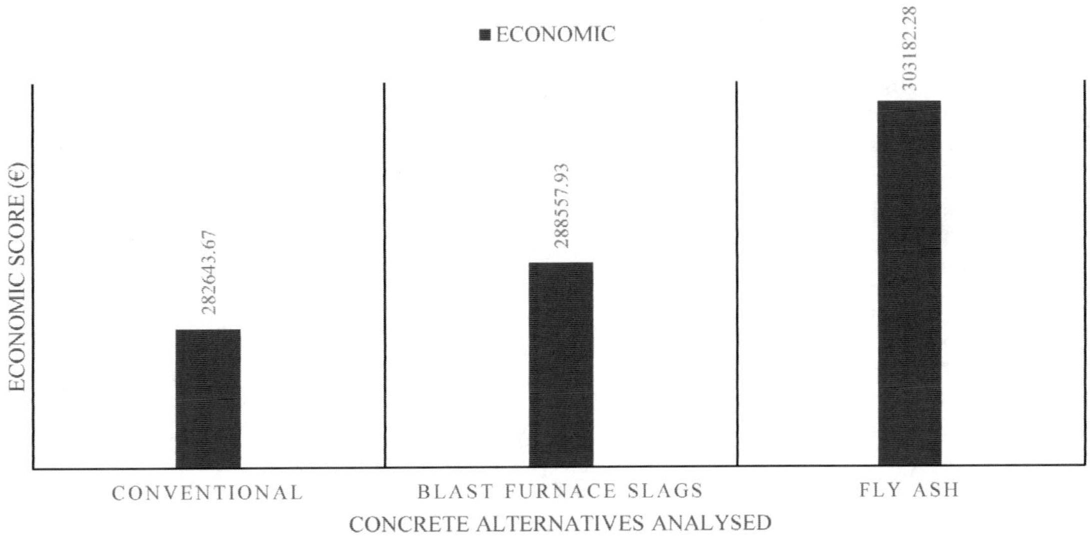


**Figure 5.** Results obtained for economic indicator per functional unit.

### *3.3. Social results*

The employment generated through building a foundation with each of the concrete types considered in this study is shown in Figure 6. The social impact of each alternative is comparable between CONV and FA. However, it is more pronounced for GBFS. Despite these differences, the results indicate that using a concrete different from the conventional generates more employment.

Comparing the results shown in Figure 6, FA shows a 0.595% higher employment generation than CONV, whereas GBFS demonstrates significantly higher results of 43.19%. It is important to consider that these results are strongly dependent on the production centre as they rely on regional unemployment rates.

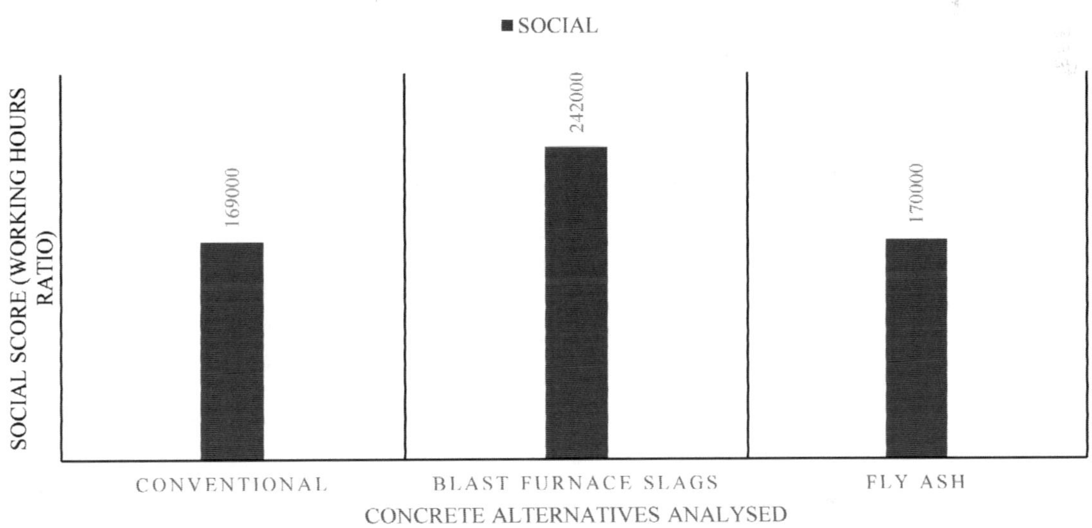


**Figure 6.** Results obtained for social indicator in each alternative analysed in this study.

### *3.4. Multicriteria Decision-Making Procedure*

The last stage of the sustainability assessment consists of aggregating the life cycle performance results, for each sustainability dimension and each alternative, into a single three-dimensional index that allows ranking preferences. Therefore, the results obtained for the economic, environmental and social dimension are aggregated using a distance-based method called TOPSIS, which is the second most popular technique used to deal with MCDM problems [52].

The weightings of each criterion are determined through the application of the Analytical Hierarchy Process (AHP) [53], which has been widely used to derive the relative relevance of each criterion involved in MCDM assessments related to infrastructure projects based on induvial preferences of decision-makers [54][55]. In this study, three experts in wind energy and civil engineering have been involved in the AHP. These experts have 5, 10 and 20 years of experience in the respective fields. The voting power coefficient for each expert is calculated based on the years of experience, reflecting their expertise in the field. The weights considered for the analysis are exposed in Table 10 and the results for the sustainability analysis in Figure 7.

**Table 10.** Weights considered for each indicator.

| Indicator | Weight |
|---|---|
| Ecosystem quality | 0.11 |
| Human health | 0.25 |
| Resources | 0.22 |
| Economy | 0.17 |
| Social | 0.25 |

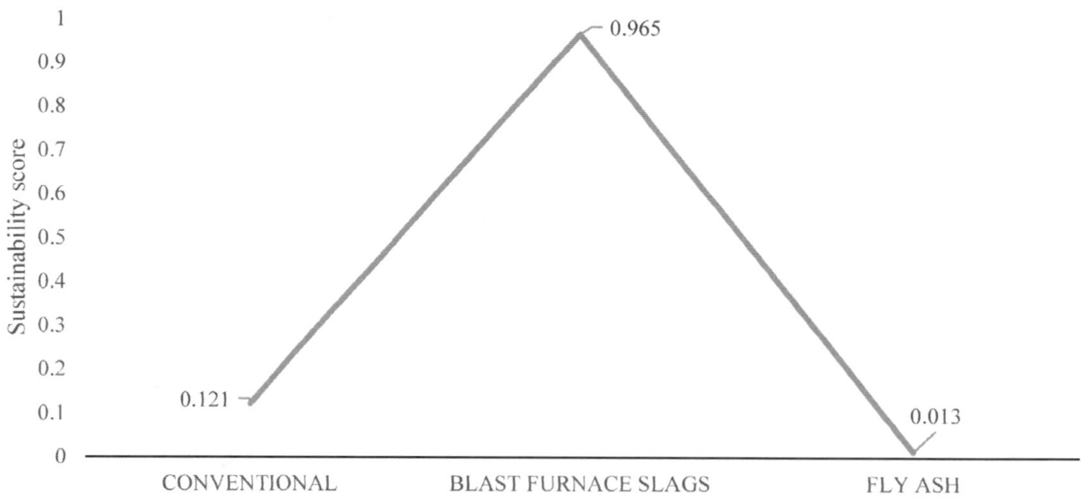


**Figure 7.** TOPSIS results.

In view of the results, it can be affirmed GBFS is the option that has proven to be the most sustainable of the three analysed for this particularly study, followed by CONV and FA. According to TOPSIS, Positive Ideal Solution (PIS) is defined as the sum of all the best value that can be achieved for each indicator. It can be observed that GBFS is positioned very close to the Positive Ideal Solution, as the obtained result is nearly unity.

## 4. Conclusion

The present communication proposes a methodology to assess the sustainability of WTG foundation designs according to the recognized international guidelines and ISO 14040[33]. In view of the results, the main conclusion is that GBFS material presents higher sustainability ratios than CONV and FA for this specific case.

- The GBFS alternative outperforms both FA and CONV in terms of environmental impacts regarding the three primary indicators (Human health, ecosystem quality and resources).
- Regarding costs, CONV proves to be 2.09% more economical than GBFS and 7.26% more cost-effective than FA.
- Considering social indicators based on workers indicator, the GBFS is more prominent than CONV and FA. Both, CONV and FA exhibit comparable social impact. These results are strongly related to the location of the concrete production centres.
- Regarding TOPSIS outcomes, GBFS emerges as the top-ranked sustainable alternative, followed by CONV in second place, and FA in third position.

To sum up, this communication offers a practical approach to assess the sustainability of WTG foundations from a 3-dimensional holistic point of view. In particular, following objectives may be drawn:

- The communication presents a methodology for quantifying sustainability rather than simply qualifying it. Therefore, this methodology can be employed for design optimization, such as geometry and materials, etc, with a sustainable perspective in mind.
- The presented methodology can be easily automated.
- As results shown, the ideal economic design point may not align with the optimal environmental design point. Both dimensions need to be considered simultaneously.
- The social dimension is essential and has the potential to influence the design when it is integrated with both the environmental and economic dimensions.

## References

[1] Wind Europe, "Wind Energy Today. Wind delivers the energy society wants," https://windeurope.org/about-wind/wind-energy-today/.

[2] WWEA, "WWEA Half-year Report 2022: Worldwide Windpower Boom Continues in 2022," https://wwindea.org/worldwide-windpower-boom-continues-in-2022/.

[3] C. Maienza, A. M. Avossa, F. Ricciardelli, D. Coiro, G. Troise, and C. T. Georgakis, "A life cycle cost model for floating offshore wind farms," *Appl Energy*, vol. **266**, 2020, doi: 10.1016/j.apenergy.2020.114716.

[4] United Nations, *Report of the World Commission on Environment and Development. Our Commmon Future*. 1987.

[5] I. Negrin, M. Kripka, and V. Yepes, "Multi-criteria optimization for sustainability-based design of reinforced concrete frame buildings," *J Clean Prod*, vol. **425**, 2023, doi: 10.1016/j.jclepro.2023.139115.

[6] Z. W. Zhou, J. Alcalà, and V. Yepes, "CARBON IMPACT ASSESSMENT OF BRIDGE CONSTRUCTION BASED ON RESILIENCE THEORY," *Journal of Civil Engineering and Management*, vol. **29**, no. 6, pp. 561–576, 2023, doi: 10.3846/jcem.2023.19565.

[7] M. Hadizadeh-Bazaz, I. J. Navarro, and V. Yepes, “Life-Cycle Cost Assessment Using the Power Spectral Density Function in a Coastal Concrete Bridge,” *J Mar Sci Eng*, vol. **11**, no. 2, 2023, doi: 10.3390/jmse11020433.

[8] I. J. Navarro, V. Yepes, and J. V. Martí, “Role of the social dimension in the sustainability-oriented maintenance optimization of bridges in coastal environments,” in *WIT Transactions on the Built Environment*, 2020, pp. 205–215. doi: 10.2495/HPSM200211.

[9] A. Allahyarzadeh-Bidgoli and J. I. Yanagihara, “Energy efficiency, sustainability, and operating cost optimization of an FPSO with CCUS: An innovation in CO<inf>2</inf> compression and injection systems,” *Energy*, vol. **267**, 2023, doi: 10.1016/j.energy.2022.126493.

[10] N. Stavridou, E. Koltsakis, and C. C. Baniotopoulos, “A comparative life-cycle analysis of tall onshore steel wind-turbine towers,” *Clean Energy*, vol. **4**, no. 1, pp. 48–57, 2020, doi: 10.1093/ce/zkz028.

[11] R. Lehneis and D. Thrän, “Temporally and Spatially Resolved Simulation of the Wind Power Generation in Germany,” *Energies (Basel)*, vol. **16**, no. 7, 2023, doi: 10.3390/en16073239.

[12] T. Zuo *et al.*, “A Review of Optimization Technologies for Large-Scale Wind Farm Planning With Practical and Prospective Concerns,” *IEEE Trans Industr Inform*, vol. 19, no. 7, pp. 7862–7875, 2023, doi: 10.1109/TII.2022.3217282.

[13] Z. Lian, K. Liu, and T. Yang, “Potential Influence of Offshore Wind Farms on the Marine Stratification in the Waters Adjacent to China,” *J Mar Sci Eng*, vol. **10**, no. 12, 2022, doi: 10.3390/jmse10121872.

[14] B. Josimović, D. Srnić, B. Manić, and I. Knežević, “Multi-Criteria Evaluation of Spatial Aspects in the Selection of Wind Farm Locations: Integrating the GIS and PROMETHEE Methods,” *Applied Sciences (Switzerland)*, vol. **13**, no. 9, 2023, doi: 10.3390/app13095332.

[15] H. R. E. H. Bouchekara, Y. A. Sha’aban, M. S. Shahriar, M. A. M. Ramli, and A. A. Mas’ud, “Wind Farm Layout Optimization/Expansion with Real Wind Turbines Using a Multi-Objective EA Based on an Enhanced Inverted Generational Distance Metric Combined with the Two-Archive Algorithm 2,” *Sustainability (Switzerland)*, vol. **15**, no. 3, 2023, doi: 10.3390/su15032525.

[16] J. J. Wimhurst, J. S. Greene, and J. Koch, “Predicting commercial wind farm site suitability in the conterminous United States using a logistic regression model,” *Appl Energy*, vol. **352**, 2023, doi: 10.1016/j.apenergy.2023.121880.

[17] X. Huang, K. Hayashi, and M. Fujii, “Resources time footprint analysis of onshore wind turbines combined with GIS-based site selection: A case study in Fujian Province, China,” *Energy for Sustainable Development*, vol. **74**, pp. 102–114, 2023, doi: 10.1016/j.esd.2023.03.012.

[18] M. Delogu, L. Berzi, C. A. Dattilo, and F. Del Pero, “Definition and sustainability assessment of recycling processes for bonded rare earths permanent magnets used on wind generators,” *Advances in Materials and Processing Technologies*, vol. **9**, no. 2, pp. 608–654, 2023, doi: 10.1080/2374068X.2022.2095142.

[19] R. Kasner, “The environmental efficiency of materials used in the lifecycle of a wind farm,” *Sustainable Materials and Technologies*, vol. 34, 2022, doi: 10.1016/j.susmat.2022.e00512.

[20] M. Asadi, M. Ramezanzade, and K. Pourhossein, “A global evaluation model applied to wind power plant site selection,” *Appl Energy*, vol. 336, 2023, doi: 10.1016/j.apenergy.2023.120840.

[21] S. Karamountzou and D. G. Vagiona, “Suitability and Sustainability Assessment of Existing Onshore Wind Farms in Greece,” *Sustainability (Switzerland)*, vol. **15**, no. 3, 2023, doi: 10.3390/su15032095.

[22] K. E. S. Jones and M. Li, “Life cycle assessment of ultra-tall wind turbine towers comparing concrete additive manufacturing to conventional manufacturing,” *J Clean Prod*, vol. **417**, 2023, doi: 10.1016/j.jclepro.2023.137709.

[23] Z.-Y. Zhuo, M.-J. Chen, and X.-Y. Li, "A comparative analysis of carbon reduction potential for directly driven permanent magnet and doubly fed asynchronous wind turbines," *Energy Sci Eng*, vol. **11**, no. 3, pp. 978–988, 2023, doi: 10.1002/ese3.1425.

[24] J. G. Rueda-Bayona, J. J. Cabello Eras, and T. R. Chaparro, "Impacts generated by the materials used in offshore wind technology on Human Health, Natural Environment and Resources," *Energy*, vol. **261**, 2022, doi: 10.1016/j.energy.2022.125223.

[25] A. Mathern, V. Penadés-Plà, J. Armesto Barros, and V. Yepes, "Practical metamodel-assisted multi-objective design optimization for improved sustainability and buildability of wind turbine foundations," *Structural and Multidisciplinary Optimization*, vol. **65**, no. 2, 2022, doi: 10.1007/s00158-021-03154-0.

[26] B. D. Lago *et al.*, "Experimental tests on shallow foundations of onshore wind turbine towers," *Structural Concrete*, vol. **23**, no. 5, pp. 2986–3006, 2022, doi: 10.1002/suco.202100655.

[27] Q. Shen, F. Vahdatikhaki, H. Voordijk, J. van der Gucht, and L. van der Meer, "Metamodel-based generative design of wind turbine foundations," *Autom Constr*, vol. **138**, 2022, doi: 10.1016/j.autcon.2022.104233.

[28] M. L. Berndt, "Influence of concrete mix design on CO<inf>2</inf> emissions for large wind turbine foundations," *Renew Energy*, vol. **83**, pp. 608–614, 2015, doi: 10.1016/j.renene.2015.05.002.

[29] M. , C. V. N. Van Tran, "Mass Concrete Placement of the Offshore Wind Turbine Foundation: A Statistical Approach to Optimize the Use of Fly Ash and Silica Fume," *Int J Concr Struct Mater*, vol. **50**, 2021.

[30] E. Martínez, F. Sanz, S. Pellegrini, E. Jiménez, and J. Blanco, "Life cycle assessment of a multi-megawatt wind turbine," *Renew Energy*, vol. **34**, no. 3, pp. 667–673, 2009, doi: 10.1016/j.renene.2008.05.020.

[31] E. Martínez, E. Jiménez, J. Blanco, and F. Sanz, "LCA sensitivity analysis of a multi-megawatt wind turbine," *Appl Energy*, vol. **87**, no. 7, pp. 2293–2303, 2010, doi: 10.1016/j.apenergy.2009.11.025.

[32] DC. Institute for Sustainable Infrastructure. Washington, *Envision: Sustainable Infrastructure Framework Guidance Manual*. 2018.

[33] ISO 14040:2006, "Environmental management - Life cycle assessment - Principles and framework," 2006.

[34] UNE-EN 17472:2022, "Sustainability of construction works - Sustainability assessment of civil engineering works - Calculation methods," 2022.

[35] ISO 14044:2006, "Environmental management — Life cycle assessment — Requirements and guidelines," 2006.

[36] SACYR, "Reducing the carbon footprint of concrete," https://www.sacyr.com/en/-/hormigon-para-reducir-la-huella-de-carbono.

[37] X. Li, H. Grassl, C. Hesse, and J. Dengler, "Unlocking the potential of ordinary Portland cement with hydration control additive enabling low-carbon building materials," *Commun Mater*, vol. **5**, no. 1, 2024, doi: 10.1038/s43246-023-00441-9.

[38] RC-16, "Real Decreto 256/2016, de 10 de junio, por el que se aprueba la Instrucción para la recepción de cementos (RC-16).," *Ministerio de la Presidencia. Gobierno de España.*, 2016.

[39] Código Estructural., "Real Decreto 470/2021, de 29 de junio, por el que se aprueba el Código Estructural.," *Ministerio de la Presidencia, Relaciones con las Cortes y Memoria Democrática. Gobierno de España*, 2021.

[40] IEC 61400-6:2020, "Wind energy generation systems - Part 6: Tower and foundation design requirements," 2020.

[41] EN 1992-1-1:2004, "Eurocode 2: Design of concrete structures - Part 1-1: General rules and rules for buildings," 2004.

[42] A. Bonou, A. Laurent, and S. I. Olsen, "Life cycle assessment of onshore and offshore wind energy-from theory to application," *Appl Energy*, vol. **180**, pp. 327–337, 2016, doi: 10.1016/j.apenergy.2016.07.058.

[43] "Ecoinvent, 2021. Ecoinvent. Ecoinvent Database v3.8, Swiss Centre for Life Cycle Inventories: St Gallen, Switzerland".

[44] ITeC BEDEC, "Base de Datos Construcción.," *Fundación Instituto de Tecnología de la Construcción de Cataluña – ITeC* , 2023.

[45] Spanish National Institute of Statistics, "https://www.ine.es/."

[46] Ministerio de Hacienda, "Agencia tributaria," https://sede.agenciatributaria.gob.es.

[47] OECD, "Organization for Economic Co-operation and Development," https://data.oecd.org/.

[48] I. J. Navarro, V. Yepes, and J. V. Martí, "Social life cycle assessment of concrete bridge decks exposed to aggressive environments," *Environ Impact Assess Rev*, vol. **72**, pp. 50–63, 2018, doi: 10.1016/j.eiar.2018.05.003.

[49] S. Z. E. P. S. G. V. F. V. M. H. A. Z. M. van Z. R. Huijbregts MAJ, *ReCiPe 2016 : A harmonized life cycle impact assessment method at midpoint and endpoint level Report I: Characterization*. 2017.

[50] T. E. Swarr *et al.*, "Environmental life-cycle costing: A code of practice," *International Journal of Life Cycle Assessment*, vol. **16**, no. 5, pp. 389–391, 2011, doi: 10.1007/s11367-011-0287-5.

[51] C. Benoît Norris *et al.*, "UNEP, 2020. Guidelines for Social Life Cycle Assessment of Products and Organizations 2020. ," https://www.lifecycleinitiative.org/wp-content/uploads/2021/01/Guidelines-for-Social-Life-Cycle-Assessment-of-Products-and-Organizations-2020-22.1.21sml.pdf .

[52] E. K. Zavadskas, A. Mardani, Z. Turskis, A. Jusoh, and K. M. Nor, "Development of TOPSIS Method to Solve Complicated Decision-Making Problems - An Overview on Developments from 2000 to 2015," *Int J Inf Technol Decis Mak*, vol. **15**, no. 3, pp. 645–682, 2016, doi: 10.1142/S0219622016300019.

[53] I. J. Navarro, J. V. Martí, and V. Yepes, "GROUP ANALYTIC NETWORK PROCESS FOR THE SUSTAINABILITY ASSESSMENT OF BRIDGES NEAR SHORE," in *WIT Transactions on the Built Environment*, 2022, pp. 143–154. doi: 10.2495/HPSU220131.

[54] M. S. Ali, M. S. Aslam, and M. S. Mirza, "A sustainability assessment framework for bridges – a case study: Victoria and Champlain Bridges, Montreal," *Structure and Infrastructure Engineering*, vol. **12**, no. 11, pp. 1381–1394, 2016, doi: 10.1080/15732479.2015.1120754.

[55] M. R. Pryn, Y. Cornet, and K. B. Salling, "Applying sustainability theory to transport infrastructure assessment using a multiplicative ahp decision support model," *Transport*, vol. **30**, no. 3, pp. 330–341, 2015, doi: 10.3846/16484142.2015.1081281.

Journal of Physics: Conference Series **2745** (2024) 012006 doi:10.1088/1742-6596/2745/1/012006

# An efficient frequency-domain based methodology for the preliminary design of FOWT substructures

**M A Reig[1], I Mendikoa[1] and V Petuya[2]**

E-mail: maria.alonso@tecnalia.com
[1] Tecnalia, Basque Research and Technology Alliance (BRTA), Edificio 700, 48160, Derio, Bizkaia, Spain
[2] Department of Mechanical Engineering, University of the Basque Country - UPV/EHU, Spain

**Abstract.** An efficient frequency domain numerical tool for the preliminary design of the Floating Offshore Wind (FOW) substructures is developed, and validated against one of the leading state-of-the-art time domain based numerical tools: OpenFAST. It has been proven that the present tool enables to reach to similar conclusions as the state-of-the-art tools with a substantial reduction of the computational effort. This is achieved through two novel methods that obtain the hydrodynamic loads acting on the floating substructures. One estimates the linear hydrodynamic coefficients, and the other predicts the second-order wave loads. When combined with a simplified frequency domain response model, a quick assessment of the dynamics of numerous platform designs can be performed, accelerating the floater preliminary design process. The proposed method has been validated by carrying out a comparative analysis on a semi-submersible platform design space, characterised by the column diameter, the separation between the columns and the platform draft. The response motions have been compared against OpenFAST. Furthermore, the high efficiency of the proposed frequency domain tool makes it suitable to perform several sensitivity analyses that will allow for a better understanding of the dynamics and fine tune the foundations design accordingly from the early stages of design.

## 1. Introduction

Global efforts to combat climate change focus on increasing the use of renewable energy, particularly emphasising wind power [1]. As onshore wind sites reach their capacity, attention is shifting to offshore wind, especially Floating Offshore Wind (FOW). However, the material and construction costs of the platform make up about 17% of the total cost [2], making it one of the main cost drivers. Optimising the substructure from the early design stages can reduce overall costs [3].

The lack of efficient tools specific to FOW hinders quick analysis of structure behavior under different offshore conditions. With the rapid growth of this technology and the expectation of larger turbines in the future, understanding structure dynamics and analysing the impact of design parameters through sensitivity analyses are crucial. This knowledge empowers designers to make informed decisions, fine-tuning platform designs for optimal performance and stability.

The level of modelling and analysis varies based on the design stage. Advanced stages employ fully coupled aero-hydro-servo-elastic models with numerous Degrees of Freedom (DoF) and time-consuming time-domain numerical analyses like Fast or OpenFAST [4] and HAWC2

[5]. Initial stages use frequency domain analysis with simplified models, featuring only a few DoFs to represent primary structure motions. Wind turbine control system non-linearities add complexity, often requiring integration of aerodynamic loads from advanced models. Hydrodynamic loads are computed using commercial software like ANSYS-AQWA [6] or WAMIT [7] based on potential flow theory. Optimising a semi-submersible platform involves evaluating numerous solutions, requiring hydrodynamic analysis for each design variation, a potentially time-consuming process.

This study aims to streamline the preliminary design phase of semi-submersible platforms and enhance understanding of dynamic behavior and design parameter influence. A novel design tool has been developed, integrating two innovative modules dedicated to hydrodynamic analysis of FOW platforms [8, 9]. These modules offer a faster alternative to traditional radiation-diffraction analyses, accelerating the floating platform design process. The tool, referred to as *FD Tool*, efficiently narrows down potential solutions meeting design requirements in the floater preliminary design stage. Validation against results from the leading time-domain numerical tool, OpenFAST [4], has been conducted.

This paper is structured as follows. In section 2 the case study is described, providing the information regarding the Floating Offshore Wind Turbines (FOWT) assessed. In section 3 the *FD Tool* implemented for the previously explained case study is explained, including a brief description of the novel methods for the calculation of the hydrodynamic loads. In section 4 validation of the proposed design tool is exposed, and the most relevant results of the maximum dynamic response from both *FD Tool* and state-of-the-art tool are shown. Finally, the conclusions are outlined.

## 2. Case Study

In this work, the DTU 10 MW WT supported by the semi-submersible Nautilus platform used in the LIFES50+ [10] project has been employed (see figure 1), since the model and results are publicly available.

The Nautilus platform is a steel platform composed of four columns and a pontoon base. Following the procedure of the platform conceptual design, a wide design space was defined, as shown in figure 2. To simplify the design process and streamline the number of parameters involved, several key assumptions and relationships were implemented. Thus, the platform dimensions were characterised by three parameters: the diameter of the columns, the separation between the columns and the floater's draft. According to the common semi-submersible

**Figure 1.** DTU 10 MW WT supported on the Nautilus semi-submersible platform.

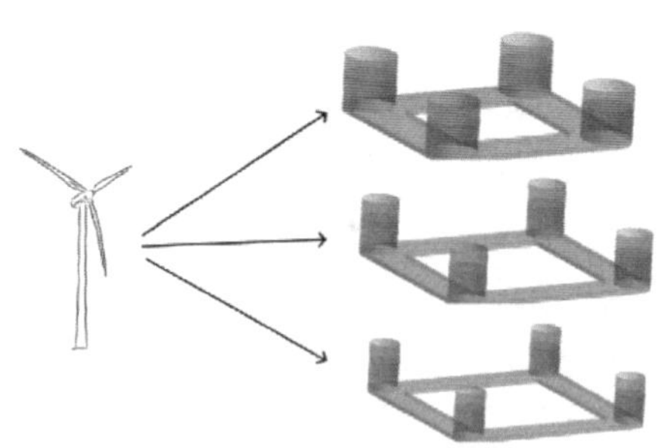

**Figure 2.** Design process of FOW substructures.

platform dimensions, these values ranged from 10 to 15 m of columns diameter, separations between 50 to 70 m and drafts between 11 to 20 m. A total of thirty-three platform configurations were generated combining different aspect ratios.

The site chosen for the study is the Gulf of Maine, the same as the one used in the LIFES50+ project. The thirty-three designs were subjected to a wide range of Design Load CaseS (DLCs), covering operational and extreme conditions, with mean wind speeds from $U_{ref} = 7 - 22\ m/s$, and sea states of $H_s = 1.67 - 10.90$ m and $T_p = 8 - 16$ s.

## 3. Methodology

In this section the methodology followed for the development of the *FD Tool* for the design of FOW substructures is described.

### 3.1. Simplified Frequency Domain Response Model

FOWT are composed of four main components: the wind turbine, the tower, the platform and the mooring system which is anchored to the seabed. They are complex systems since they are subject to many load interactions, such as wind, wave, current or ice. In the initial stages of design, the dynamic analysis of the FOWT is performed in the frequency domain. The equation of motion written in the matrix form is given in equation (1).

$$\hat{\ddot{\xi}}(\mathbf{M} + \mathbf{A}) + \hat{\dot{\xi}}\mathbf{B} + \hat{\xi}\mathbf{C} = \hat{F}_{aero} + \hat{F}_{hyd} \tag{1}$$

Where **M** is the system structural mass matrix, **A** is the hydrodynamic added mass matrix, **B** is the system damping matrix and **C** the system stiffness matrix. $\hat{F}_{aero}$ is the Fourier Transform amplitude of the wind excitation load and $\hat{F}_{hyd}$ is that of the wave loads to which the FOWT is subjected. The total load acting on the FOWT is the sum of the aerodynamic and hydrodynamic loads. $\hat{\xi}(\omega)$ is the frequency domain response vector.

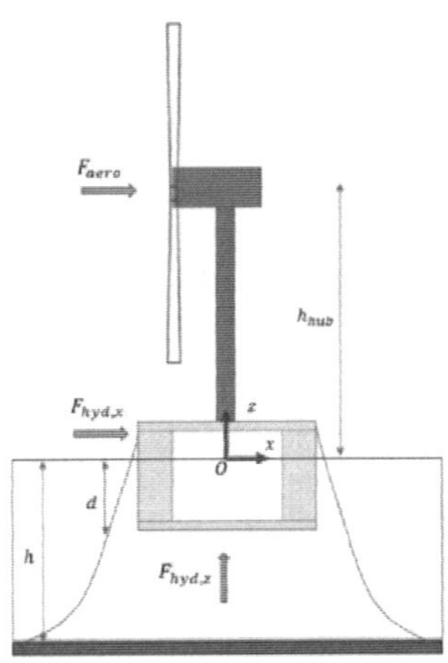


**Figure 3.** Simplified FOWT model.

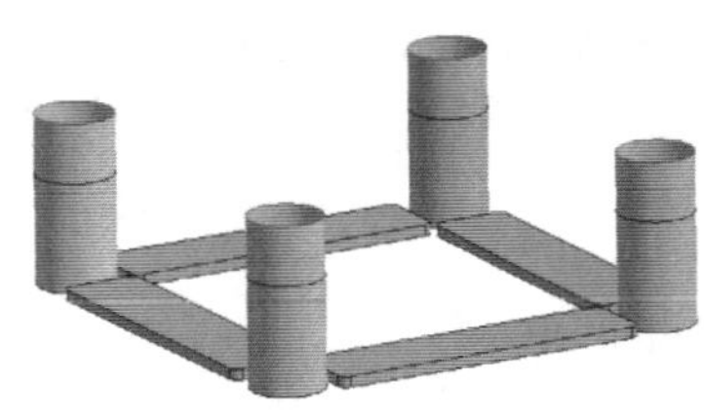

**Figure 4.** Decomposition of the Nautilus floater.

For the present work, the QuLAF model developed by Pegalajar et al [11] was used as a reference, which considers the Rotor and Nacelle Assembly (RNA) as a lumped mass, the tower is the only flexible member, the platform is considered a rigid body and the mooring system is represented with a stiffness matrix that changes with the platform position. It is a planar model represented by four DoFs: the surge, heave and pitch of the floater, and the tower first fore-aft modal deflection, as shown in figure 3.

The structure is subject to the action of turbulent wind, modelled with the Kaimal spectrum, and irregular waves, with Jonswap spectrum. The global reference frame is depicted with the red axes, where the origin is located at the Mean Sea Level (MSL), the $x$-axis aligned with the mean wind speed direction, $z$-axis pointing upwards and the $y$-axis resulted from the right hand rule.

Similarly to the QuLAF, aerodynamic loads were pre-computed using advanced tools and then loaded to the simplified model. For the given wind turbine, the wind speed time series were generated in Turbsim [12], and then, loaded to OpenFAST to extract the aerodynamic loads and damping. The tower mode shapes were obtained through the NREL tool, BModes [13], and the mooring stiffness matrices associated to each mean wind speed were obtained using OrcaFlex software, performing static analyses. Drag loads were also considered, represented by a linear drag damping coefficient derived by the Morison equation, as described in [11]. Contrary to the QuLAF, that pre-computes the hydrodynamic loads in the radiation-diffraction code WAMIT, herein we have integrated the two novel methods developed for the main purpose of reducing the computational cost of this process. These methods are explained in the following section.

### *3.2. Hydrodynamic Analysis. Novel Efficient Methods*

Radiation-diffraction analysis integrated in floater design process can be time-consuming and inefficient. To address this, we developed two standalone methods that rapidly calculate first and second-order hydrodynamic loads. These were denoted as *HAAC* [8] and *Accelerated Method* [9], respectively, and both will be further briefly explained. When combined with the frequency domain response model, these methods enable quick assessment of the dynamics of numerous platform designs, accelerating the floater pre-design process.

Both of the developed methods share common characteristics: they decompose the floater into simple members (see figure 4) and estimate the loads acting on each of them. Superposition is then applied, disregarding cross-interactions. The methods were validated against radiation-diffraction analysis and have been integrated into the frequency domain simplified response model.

*3.2.1. Hydrodynamic Analysis on Aggregated Components (HAAC)* This method efficiently estimates the hydrodynamic coefficients of the platform: added mass, radiation damping and excitation wave coefficients. The first step is to decompose the floater into simple members. Then, the process involves three main steps applied to the different geometries (isolated column and pontoon):

(i) **Build a database**: radiation-diffraction analyses are run offline to create a database of hydrodynamic coefficients of bodies with different aspect ratios.

(ii) **Interpolation**: hydrodynamic coefficients are interpolated for the target aspect ratio. This is done for the isolated cylinder and for the pontoon base.

(iii) **Scaling**: hydrodynamic coefficients are scaled back to the dimensions of the target platform members.

Two additional steps were introduced to address specific issues. Firstly, acknowledging the non-negligible interaction between members, a semi-empirical formulation was applied to consider this effect. This adjustment proved to significantly enhance results, particularly in the context of radiation damping for an array of columns. Secondly, when coefficients are scaled, they correspond to a pontoon submerged at a scaled depth, not that of the target platform. To rectify this, novel semi-empirical formulations were employed, demonstrating notable improvements compared to results from radiation-diffraction computations. Further details on these semi-empirical formulations and their impact on results can be found in [8].

The last step is to apply superposition. For that purpose, the hydrodynamic coefficients of the isolated cylinders are referred to the PoF of the platform, taking into account the locations of each of the columns and the phase change.

*HAAC* shown to significantly reduce the computational cost of the hydrodynamic analysis with respect to the commonly used radiation-diffraction analysis [8]. The most expensive part, in computational terms, is dedicated to build the database, although it is built just once. It was concluded that if more than five floaters are evaluated, the *HAAC* methodology would save computational cost compared to radiation diffraction analysis.

*3.2.2. Accelerated Method on Semi-submersible Platforms* This method efficiently estimates the second-order hydrodynamic loads, and it is denoted *Accelerated Method.* It was originally developed by Bredmose et al. [14] for the second-order hydrodynamic surge force and pitch moment on a slender vertical cylinder. It was validated for a spar and a monopile against the results from radiation-diffraction analysis computed in WAMIT, and here we have extended it to a semi-submersible platform [9]. For that purpose, the floater is decomposed into simple members, and finally superposition is applied. This time, the platform is split into simple members that can be considered, to certain extent, slender bodies, such as four isolated columns and four isolated rectangular prisms (see figure 4). New load formulation was required, such as the heave force on the cylinder and all the components on the pontoon base. This can be done provided that the interaction between the members can be assumed negligible.

This method is based on the slender body theory, using the Morison and Rainey force model, thus, it obtains the second-order wave loads by means of analytical integration from the first and second order wave velocity potential. For instance, in equation (2) the second-order surge force acting on a vertical cylinder is given by

$$F^{(2)} = (C_m + 1)i \sum_{m}^{N} \sum_{n}^{N} \hat{B}_m \hat{B}_n e^{i((\omega_m+\omega_n)t-(k_m+k_n)x)} \mathscr{F}, \tag{2}$$

where $N$ is the size of the frequency array, $\hat{B}$ is the wave amplitude, $\omega$ the angular wave frequency, $k$ the wave number, and $\mathscr{F}$ is the QTF derived from the Morison and Rainey force model. This QTF is symmetric and real, thus it can be decomposed in eigenvalues, $\lambda_j$, and eigenvectors, $V_j$, where $j$ corresponds to each mode. The equation above then can be rewritten as in (3).

The evaluation of equation (2) involves a double sum over the number of frequencies, which can become time consuming. Therefore, rewriting this equation into matrix form equation (3), and then decomposing the QTF into eigenvalues and eigenvectors, it enables to obtain the load time series in a much quicker way. Thus, equation (4) results in the product of two identical pseudo time series that can be reformulated considering symmetry and that provide the load time series via the iFFT.

$$F^{(2)} = i \sum_{j}^{2N} \left[ \cdots \quad \hat{B}_m e^{i(\omega_m t - k_m x)} \quad \cdots \right] \mathbf{V}_j \lambda_j \mathbf{V}_j \begin{bmatrix} \vdots \\ \hat{B}_n e^{i(\omega_n t - k_n x)} \\ \vdots \end{bmatrix} \tag{3}$$

$$F^{(2)} = i \sum_{q=1}^{2N} \lambda_q \left[ \sum_{m=-N}^{N} \mathbf{V}_{qm} \hat{B}_m e^{i(\omega_m t - k_m x)} \right]^2. \tag{4}$$

The core principle of this method is bypassing the time-consuming double sum over frequencies by utilizing the eigenvalue decomposition of the QTF, followed by modal truncation.

This approach ensures that only a few modes are needed to accurately represent second-order wave loads. The same idea is applied to compute other load components on the semi-submersible platform, with subsequent superposition to obtain total second-order loads.

The *Accelerated Method*'s primary advantage lies in its significant reduction of computational cost compared to conventional methods. Whilst the radiation-diffraction analysis of a full Nautilus platform, plus the double sum over frequencies for a given wave spectrum, took nearly 20,000 seconds, the *Accelerated method*, using 128 modes, required less than 7 seconds. This efficiency makes it well-suited for the preliminary design of floating platforms for Floating Offshore Wind (FOW), where multiple platform designs are evaluated.

### *3.3. FD Tool Assembly*

In figure 5 a scheme of the efficient tool presented in this paper is given, which gives a quick view of the integration of the two novel methods that were developed for the calculation of the hydrodynamic loads into the simplified frequency domain based response model. Additionally, the aerodynamic model is pre-computed in OpenFAST and loaded into the frequency-domain simplified model.

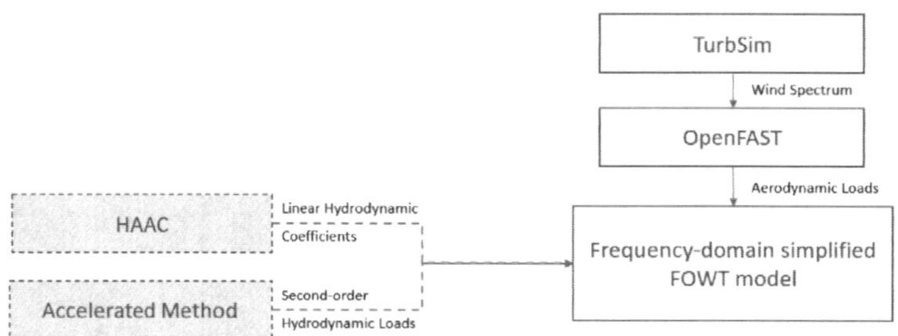


**Figure 5.** Scheme of the *FD Tool.*

## 4. Results and Discussion

### *4.1. FD Tool Validation Procedure*

In order to validate the *FD Tool*, the results obtained have been compared against the results from the time domain simulations run in OpenFAST fed with the hydrodynamic loads, first- and second-order, from the radiation-diffraction panel code ANSYS-AQWA.

Both the frequency and time domain model use the same mooring stiffness and the viscous damping, in order to eliminate other potential sources of error. From each model, the static component and standard deviation are extracted, and the maximum response has been defined as a function of these two and a dynamic factor, which was defined and calibrated with the time domain simulations. A value of 3.6 was found to agree very well for all response signals, as given in equation (5).

$$\xi_{max} = \xi_{st} + 3.6\sigma_{\xi} \tag{5}$$

The error measurement defined is given in equation (6). It is the ratio between the differences of maxima response and the threshold value for that motion. This way the accuracy of the prediction is pondered as the maximum response is closer to the maximum value allowed. The thresholds are based on the reference standards for FOW structures design [15, 16, 17].

$$\epsilon = \frac{\xi_{max}^{FD} - \xi_{max}^{TD}}{threshold} \tag{6}$$

*4.2. Response Comparative Analysis*

In figure 6 and 7 the maximum platform pitch angle and nacelle acceleration of the different platforms are shown, respectively. The surge and heave are not here discussed, since they are not critical parameters for the platform design, although the former would be useful for the mooring design. The threshold values are depicted with the red dashed lines, considering a tolerance of 10%. Herein, only one DLC is shown, which corresponds to a mean wind speed of 12 $m/s$, that produces maximum thrust, and a severe sea state with a significant wave height of 10.9 m and a wave peak period of 16 s.

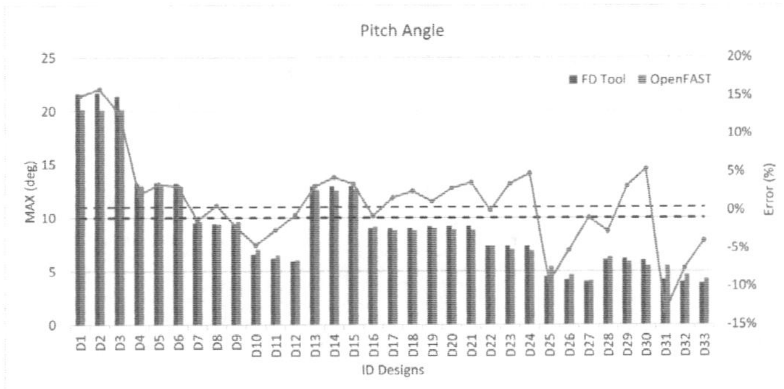


**Figure 6.** Pitch maximum angle for the different platform designs supporting the DTU 10 MW WT.

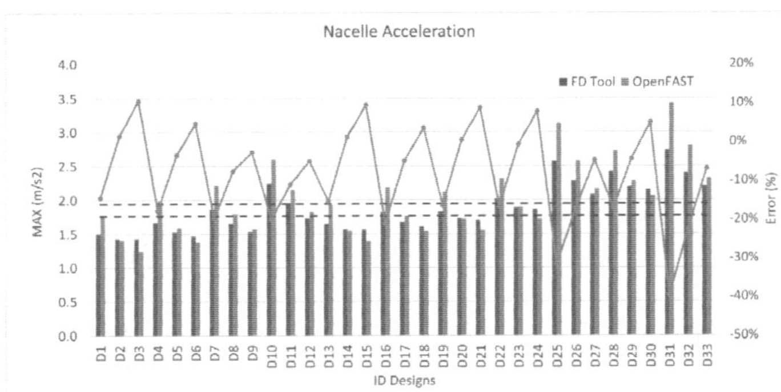


**Figure 7.** Nacelle maximum acceleration for the different platform designs supporting the DTU 10 MW WT.

The pitch maximum angle was found to be very dependent on the thrust force, achieving better agreement with the SoA results when aerodynamic loads were significant. This is explained because the main difference on both methods resides on the hydrodynamic loads, thus, if these are high compared to the aerodynamic loads, the error would increase. The most remarkable aspect of these results is that the same platforms discarded by the SoA tool are discarded by the *FD Tool*.

Regarding the nacelle acceleration, the maximum response was more influenced by the sea state, rather than by the aerodynamic loads. The *FD Tool* resulted to underestimate the maximum nacelle accelerations compared to the SoA. This was related to the double derivative of the deflection signal, since each derivative increases the error. Despite this, the tendency followed across the designs using the *FD Tool* is similar to the SoA tools' predicted response. This time, some of the platforms that would be discarded by SoA tool are accepted by the *FD Tool*, but just a few of them have this behaviour and it is preferable and conservative to accept a platform design and discard it in the next design stage rather than eliminate it in this stage.

Moreover, the tool's inclusion of second-order wave loads in the preliminary design phase has proven to lead to different, and more accurate conclusions compared to analyses that do not consider these loads.

The efficiency gained is thanks to the developed hydrodynamic modules, which it even improves with the number of designs to evaluate. This fact relates to two key factors. Firstly, the SoA tool necessitates performing a radiation-diffraction analysis for each floater design to obtain hydrodynamic loads, which are subsequently fed into OpenFAST. In contrast, the *FD Tool* streamlines this process. Secondly, the developed *HAAC* method consistently takes a constant time to build the database, regardless of the number of platform designs being evaluated.

To put this into perspective, employing the conventional SoA tool for dynamic response analysis, which involves radiation-diffraction analysis in ANSYS-AQWA followed by time-domain simulation in the fully coupled OpenFAST tool, for each of the thirty different platforms supporting the DTU 10 MW WT would take 480 hours. However, the *FD Tool* can complete the same analysis in just 83 hours while providing reasonably accurate results.

### *4.3. Potential Applications of the FD Tool*

Providing an efficient design tool for FOWT offers numerous advantages beyond substructure design. The *FD Tool* holds great potential for conducting multiple sensitivity studies, gaining insights into the influence of various components, design parameters, and metocean conditions on FOWT design.

One avenue for exploration involves investigating how the continuously increasing size of wind turbines affects platform required dimensions. The scaling of the platform may not be proportional, and the tool allows designers to define a broad platform design space and swiftly assess dynamic responses under specific wind and wave conditions, providing insights into potential main dimensions for the most suitable platform.

Additionally, the tool facilitates concurrent development of the mooring system and platform design. While mooring line definitions were not evaluated in this work, the tool represents the mooring system with a stiffness matrix that varies with mean wind speed. This enables assessing the minimum mooring stiffness to predict the minimum pretension required for a given floater solution. Considering mooring pretension as a design parameter, the tool allows for concurrent development of the mooring system and platform design. Especially, the significant impact of the second-order loads in the mooring design can be assessed.

## 5. Conclusions and Future Work

This study introduces a novel frequency domain-based design tool, referred to as *FD Tool*, specifically designed for the preliminary design of FOW substructures. The uniqueness of this tool lies in the efficiency of its hydrodynamic analysis, thanks to two standalone methods that significantly reduce computational costs. Notably, the inclusion of second-order wave loads in the initial design stages is now feasible with the developed *Accelerated Method*, which demonstrated reasonable accuracy in estimating these loads.

The *FD Tool* allows for a rapid assessment of the dynamic behavior of floating offshore wind turbines, reaching conclusions similar to state-of-the-art tools in the time domain. This capability aids in narrowing down potential platform solutions for further evaluation in the next stage of design, based on specific design thresholds. Leveraging the efficiency of the *FD Tool* enables the efficient execution of multiple sensitivity analyses, thereby enhancing the understanding of FOWT behavior and the most influential parameters related to substructure design. Additionally, it supports the co-design of FOWT components by facilitating interaction among them.

Regarding the steps for future work, several topics have not been covered in this study, and some have not undergone an in-depth analysis. Exploring these areas could greatly enhance the precision of the results, leading to a more reliable tool. For instance, in relation to the proposed methods for the hydrodynamic analysis of semi-submersible platforms, to account for the interaction between members would improve the accuracy of the results, especially under DLCs where hydrodynamic loads dominate over aerodynamics. Moreover, it could be studied the possibility to wider to misalignment of the excitation, since this can become critical for some structures. Finally, the logical next step for this tool, taking into account the important reduction of the computational cost achieved, it to integrate it into optimisation loops, where just the range of platform dimensions would be identified and the optimizer itself would determine the characteristics of the most cost effective platform.

## Acknowledgments

This work was funded by the Basque Government through the BIKAINTEK PhD support program [grant number 48-AF-W2-2019-00011].
Financial support received from the Spanish Government (Project PID2020-116176GB-I00) financed by MCIN/AEI/10.13039/ 501100011033 and the support through Project IT1480-22 provided the Regional Basque Government.
The authors acknowledge TECNALIA and SENER for their support to the PhD thesis of María Alonso Reig.
The authors also thank the Wind Energy team from the Technical University of Denmark for the support and learning received during the international stay.

## References

[1] WindEurope 2021 *Technical Report. Getting fit for 55 and set for 2050*
[2] WindEurope 2023 *Offshore wind energy 2023 mid-year statistics.*
[3] Leimeister M, Kolios A and Collu M 2018 **1104** 012007
[4] J Jonkman M S *OpenFAST. An open source wind turbine simulation tool. NREL*
[5] Larsen T J and Hansen A M 2007 *How 2 HAWC2, the user's manual* (Risø National Laboratory)
[6] Ansys A 2016 *USA. Canonsburg (PA): ANSYS Inc*
[7] Lee C H 1995 *WAMIT theory manual* (Massachusetts Institute of Technology, Department of Ocean Engineering)
[8] Reig M A, Mendikoa I, Touzon I and Petuya V 2023 *Ocean Engineering* **285** 115318
[9] Reig M A, Pegalajar-Jurado A, Mendikoa I, Petuya V and Bredmose H 2023 *Marine Structures* **90** 103430
[10] Yu W, Müller K, Lemmer F, Bredmose H, Borg M, Sanchez G and Landbo T 2017 *LIFES50+ Deliverable* **4**
[11] Pegalajar-Jurado A, Borg M and Bredmose H 2018 *Wind Energy Science* **3** 693–712
[12] Jonkman B 2014 *Natl. Renew. Energy Lab*
[13] Modes C 2007 User's guide to bmodes Tech. rep. Colorado: National Renewable Energy Laboratory, 2007: 1- 24
[14] Bredmose H and Pegalajar-Jurado A 2021 *Coastal Engineering* **170** 103952
[15] DNV 2019 *Recommended practice, DNV*
[16] Ghigo A, Cottura L, Caradonna R, Bracco G and Mattiazzo G 2020 *Journal of Marine Science and Engineering* **8** 835
[17] Collu M and Borg M 2016 *Offshore wind farms* (Elsevier) pp 359–385

Journal of Physics: Conference Series **2745** (2024) 012007 doi:10.1088/1742-6596/2745/1/012007

# Influence of platform dynamic yaw motions on wake, and subsequent impacts on downwind turbine power and loads.

**Marit I Kvittem[1], Lene Vien Eliassen[1], Balram Panjwani[2], Tom Levick[3], and Mark Thomas[3]**

[1] SINTEF Ocean, Trondheim, Norway, [2] SINTEF Industry, Trondheim, Norway, [3] DNV, Bristol, United Kingdom

E-mail: marit.kvittem@sintef.no

**Abstract.** The floating foundations of FOWTs cause turbine motions that introduce new unknowns to the field of wind turbine wake estimation, that again leads to uncertainties in the load and power estimation of wake affected turbines. Recent studies have shown significant effects of floater motions on power production and emphasizes the need for the effect to be integrated into existing wind farm energy yield tools. Also, field measurements in installed FOWT wind farms have shown that there are measurable effects of the wake on floater yaw motions and mooring line loads.

This paper describes a study that aims to quantify the effect of time-varying yaw misalignment, from floater motion and/or turbine yaw control, on the power production and turbine loads. 2 IEA 15 MW wind turbines supported by the UMaine VolturnUS-S semi-submersible, serves as the case study. The wind field and power production was simulated including the effect of steady and time-varying yawed inflow. The response of the downwind turbine in the waked wind field was simulated.

Power production and loads for the downwind turbine in yaw cases are benchmarked against results for the downwind turbine without time varying turbine yaw and show that including the upwind turbine yaw motion resulted in higher mean power (up to 4%), but also larger power fluctuations. It also lead to larger standard deviations in the mooring line tension. These effects seem to decrease with increasing wind speeds and turbine spacing, and increase with the amplitude of yaw motion.

## 1. Introduction

The floating foundations of floating offshore wind turbines (FOWTs) cause turbine motions that introduce new unknowns to the field of wind turbine wake estimation, that again leads to uncertainties in the load and power estimation in waked wind conditions. The DNV Energy Transition Outlook predicts 270 GW of installed wind power on floating foundations by 2050. The largest floating wind farm installed today is the 88 MW Hywind Tampen [1], which is much smaller than commercial scale wind farms with bottom fixed foundations. Thus, the experience with wind fields within floating wind farms is limited.

Recent studies have shown significant effects of floater motions on power production [2], which emphasizes the need for the effect to be integrated into existing wind energy assessment tools. Also, field measurements from the Hywind Scotland wind farm showed that there are measurable effects of the wake on floater yaw motions and mooring line loads [3]. It is suspected, but not confirmed, that wake meandering can excite sideways platform motion (sway and yaw), because of partial wake immersion.

Advanced supervisory farm control uses wake steering by yawing or tilting the rotor to mitigate power loss and to alleviate turbine loads in the wake [4, 5]. For wind turbines on a floating foundation, the total yaw or tilt that influences wake deflection, will depend on the floater motions as well. Innovative mooring designs even utilize these effects on wake steering to increase energy yield [6]. Therefore, it is important to consider floater motions when implementing farm control strategies for floating wind farms. This paper describes a study that aims to quantify the effect of time-varying yaw misalignment, from floater motion and/or turbine yaw control, on the power production and turbine loads.

## 2. Wake

### 2.1. *Waked Wind Field*

It is well known that the wind field in a wind farm is strongly influenced by the presence of the turbines. For some distance downstream of a turbine in operation, the mean wind speed is reduced and the turbulence intensity is increased [7], often referred to as the wake of a turbine. The drop in wind speed causes reduced power production and the increased turbulence can contribute to increased fatigue loading [8, 9] for downwind turbines. Even with a considerable distance between wind turbines, there is a notable increase in fatigue load within the wake, amounting to approximately 80% [9]. Today, wind farm layout optimization and use of supervisory farm control is used to deal with wake effects, both to optimize power production and to reduce turbine loads. In addition, there are large scale effects that cause these wakes to oscillate around the mean wind direction [10]. This is referred to as wake meandering, which can be modeled by the DWM method [11]. The strength of the wake, in the form of a velocity deficit, decays with distance from the turbine. There are also other large scale effects, such as blockage and farm-to-farm wake effects, but these will not be addressed in this study.

In the idealized case of steady wind in one direction, acting normal to the rotor plane, the wake travels in a straight line downwind of the turbine. In a more realistic case with turbulent wind, the incident wind direction changes and turbine yaw control and support structure rotations cause the wind inflow to deviate from the normal to the rotor plane.

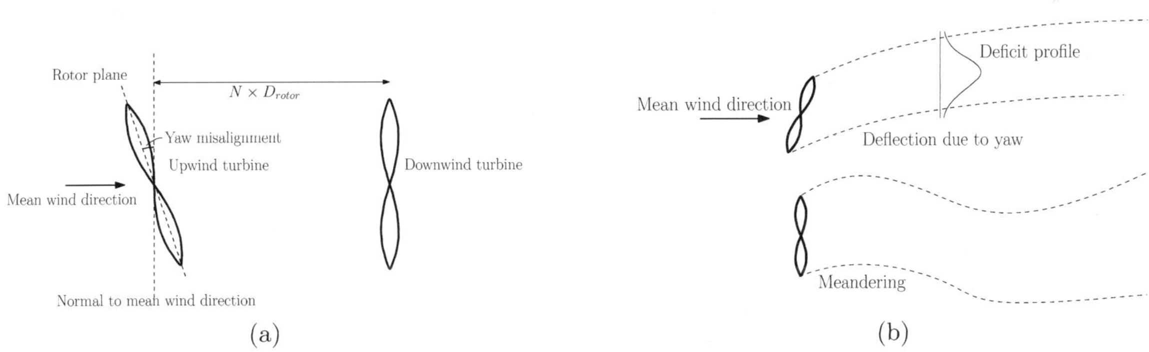


Figure 1: (a) Two-turbine case with definition of yaw misalignment. (b) Illustration of wake deflection and -meandering.

An inflow angle which is not 90°, which in this paper we call *yaw misalignment* (see Fig. 1a) causes the wake intensity to reduce and the trajectory to deflect from the wind direction. This effect can be used in farm control, to steer the wake away from downwind turbines. In the following sections, we divide the wake effect into a static- and a dynamic part.

A report by the Carbon Trust Floating Wind JIP [2] summarized conclusions from an internal study, where energy yield was estimated for a 3 × 3 array of turbines with floater motions from three types of floating foundations was studied. Their conclusion was that the effect of dynamic floater yaw motion on energy yield is negligible compared to static floater yaw, but that the general trend was that floater yaw lead to increased power production, because there is increased wake recovery when yaw is included. The impacts of yaw appeared larger for wider turbine spacing, and they argue that this may be due to the width of the wake being wider far downstream. They also found that the tilt of a floating turbine can be important to include in energy yield calculations. Since the details of the study behind this conclusion is not publicly available, the assumptions and limitations behind this conclusion are not known.

### *2.2. Yawed Inflow*

Site measurements show that there there is an additional wake displacement for yaw conditions [12]. Engineering models for steady-state deflection wakes in yawed conditions is implemented into wake models. We have in this study considered the methods developed by Jimenéz et al. [13] and Bastankhah and Porté-Agel [14]. Jimenéz et al. [13] used large eddy simulations (LES) and wind tunnel experiments to characterize the wake deflection based on inflow angle and thrust coefficient, and proposed a simple analytical model based on their results. Bastankhah and Porté-Agel [14] also used wind tunnel experiments and full scale LES, but proposed a different analytical model for the wake defelction.

Yaw misalignment is not necessarily only a static condition, but may vary over shorter time periods. The dynamic behaviour of the wake in such an event is not well known. One possible model is to assume quasi-steady behaviour and superimpose the quasi-steady deflection with the wake meandering. An illustration of these two effects can be found in Fig. 1b.

## 3. Case Study

### *3.1. Environment*

Four different mean hub height wind speeds are chosen for comparison: 5.5 m/s, 7.5 m/s, 9.5 m/s and 11.5 m/s. Wake effects are most dominant below rated wind speeds, thus, wind speeds above 11.5 m/s are not included in the study. The turbulence intensity is taken as 6%, which represents a generic offshore site with relatively high turbulence. This is lower than the recommended value for IEC Class 2C, but more in line with measured values for offshore sites, as shown by Wang et al. [15]. The Mann model [16] was used to generate turbulence boxes that are input to DIWA (see Sec. 3.2). Input parameters to the turbulence model are formulated as to fit a Kaimal model, using 3.9 for the shear distortion parameter ($\gamma$), and 42 m for the integral length scale parameter ($\Lambda_1$), according to the IEC standard [17]. Three random realizations of 1-hour long turbulence are used in the simulations. Neutral atmospheric conditions were assumed.

### *3.2. Disturbed Wind Field Generation*

The disturbed inflow wind field for the downstream turbine was generated using wind farm wake analysis software DIWA, which is developed by SINTEF and included in the SIMA Workbench. DIWA calculates the wake deficit, added turbulence and wake meandering for a wind farm based on farm layout, turbine characteristics and input turbulence boxes for the incoming wind field. DIWA can also return turbulence boxes including the disturbed wind field, which can be used in load and response simulations with aero-elastic software. A more detailed description of DIWA can be found in Panjwani et al. [18] and the DIWA User and Theory manual [19].

A recent implementation of steady and varying yawed inflow in wake simulation software DIWA was used to simulate the wind field and power production, including the effect of varying degree of time-varying yaw motion. Two yaw models were implemented, Jimenéz [13] and

Bastankah [20], and in a validation study, the Bastankah model gave the closest comparison to CFD results [21]. Thus, the Bastankah yaw deflection model was applied in the current study.

The DIWA generated turbulence box, including wake effects, was subsequently applied in an aero-, hydro-, servo elastic model of the FOWT in Bladed [22] to simulate the response. The procedure can be summarized as follows:

(i) Generate disturbed windfield using DIWA. A bi-harmonic signal was used to invoke forced yaw motion of the upwind turbine.
(ii) Extract the disturbed windfield observed at the downwind turbine.
(iii) Apply this wind field in an aero-, hydro-, servo-elastic model and extract power and loads.
(iv) Compare the results to a simulation where no time-varying yaw was applied in the disturbed wind field generation.

The yaw motions of the upwind turbine are imposed using a bi-harmonic signal consisting of the yaw natural frequency (0.011 Hz), and a smaller amplitude (0.2 deg) signal representing the rotor frequency (Hz). Combinations of different of yaw motion amplitude (1°, 3° and 6°), wind speed (5.5 m/s, 7.5 m/s, 9.5 m/s and 11.5 m/s) and turbine spacing (4, 6, 8 and 10 x rotor diameters) are simulated. The benchmark cases are run for the same wind speeds and turbine spacing, but with no yaw motion of the upwind turbine.

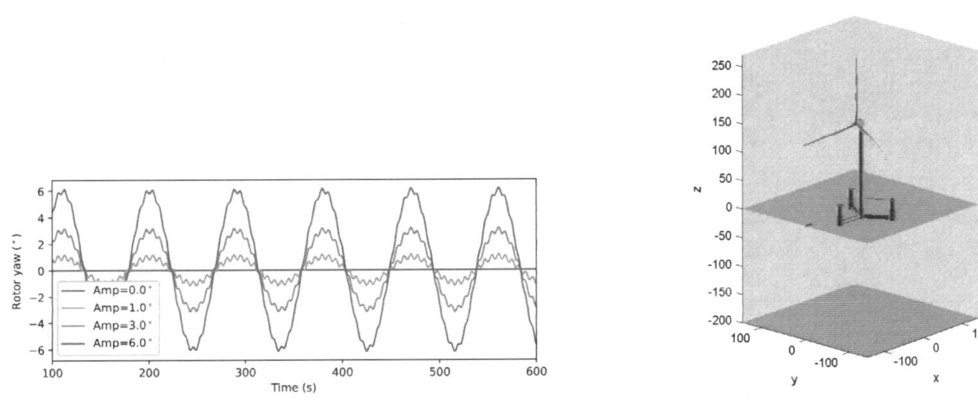


(a) (b)

Figure 2: (a) Sample of upwind turbine yaw motion signal. (b) IEA 15MW UMaine VolturnUS-S Bladed model vizualisation.

The yaw response of a FOWT in operation is mainly excited by aerodynamic loads on the rotor and wave or current loads on the hull. The response amplitudes and frequencies depends on the inertia of the platform and the stiffness and damping of the mooring system. The VolturnUS-S 15MW has a yaw resonant frequency of 0.011 Hz, and a rotor frequency at rated of 0.143 Hz [23], thus these were chosen as the two signal frequencies in the yaw motion signal, see Fig. 2a. Based on observations of yaw-behaviour in earlier studies of semi-submersible wind turbines (e.g. [3, 24]) with various mooring systems, yaw motion amplitudes ranging from 1° to 6° were chosen to represent relevant yaw-resonant behaviour for semi-submersible wind turbines.

Yaw misalignment is only prescribed for the upstream/wake generating turbine in DIWA, and a wind field is generated at the location of the downstream turbine. This wind field is then used as input for load and power analysis in Bladed, where the aeroelastic model is free to move in all degrees of freedom, which ensures that floater motions of are included in the power production and loads.

Journal of Physics: Conference Series **2745** (2024) 012007 doi:10.1088/1742-6596/2745/1/012007

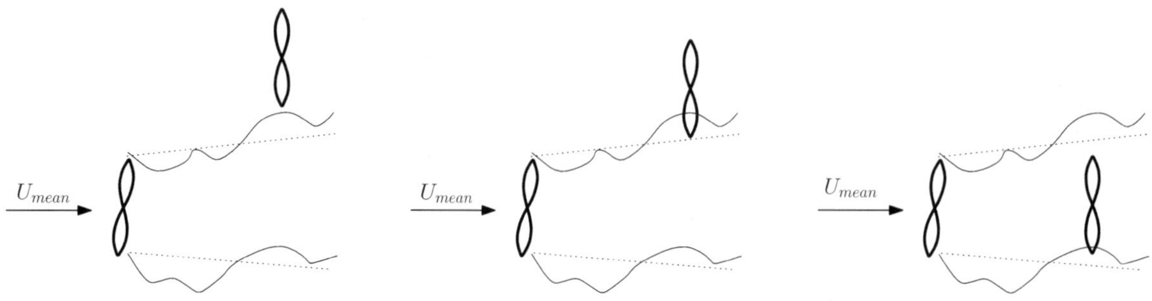


(a) No wake (b) Partial wake (c) Full wake

Figure 3: Illustration of different scenarios of wake immersion.

### *3.3. Wind Farm Layout and FOWT*

A simple two-turbine case, with two IEA 15 MW UMaine VolturnUS-S wind turbines, is used to quantify the effect of yaw motion. An illustration of the layout is shown in in Fig. 1a. This is a 15 MW wind turbine [25] supported by a three-column semi-submersible foundation [23]. The ROSCO controller is applied for blade pitch- and torque control and the inflow angle dependent performance curves that served as input to DIWA are generated with Bladed.

The support structure is anchored with a catenary chain system of three lines. The platform Bladed model (Fig. 2b) applies first order wave force transfer functions in addition to accounting for frequency dependent added mass and damping through the retardation function. The mooring line forces formulation is non-linear and quasi-static. Since the objective of this study is to investigate wind induced effects, the same irregular wave sea state is applied in all simulations. A benign sea state with a significant wave height of 1.5 m and peak period of 7 seconds is chosen to represent operational conditions at rated and below. A Jonswap spectrum with shape parameter of 3.3 is applied.

### *3.4. Farm Level Effects of Yawed Inflow*

In our study, we only consider the effects of yaw deflection on a turbine directly downwind of the yawing turbine (illustrated in Fig. 3c). In an actual wind farm, with a layout likely optimized to reduce wake losses, situations where turbines operate in fully waked condition will depend on wind direction distribution. Wake meandering and time varying yaw of upstream turbines, may create situation where a turbine which is normally out of the wake experiences partial wake immersion (see Fig. 3b). Partial wake immersion is suspected to lead to increase in power- and load fluctuation. Whether a turbine experiences partial wake immersion, depends on its position relative to other turbines in the farm and the amplitude of wake path of upstream turbines.

## 4. Results

The objective of this study is to investigate the influence of time-varying yaw motion on the downwind wind field, and the subsequent effects on power production and loads on the downwind turbine. Thus, the results in Figures 4, 6 and 7, are shown as percentage deviations from the case where the upwind turbine has zero yaw. Three different seeds are used when generating the ambient turbulence, and the deviations are calculated by comparing results using the same seed. The error-bars in Figures 4, 6 and 7 show mean values and standard deviations of these three seeds.

### *4.1. Wind Field*

Results from the wind field analysis of the case study with two turbines are shown in Fig. 4, for varying distance between the turbines, ambient mean wind speed and yaw amplitude of the upwind turbine. The general trend is that the mean incoming wind speed at the hub of the downwind turbine increases with increasing yaw amplitude, if the upwind turbine has time varying yaw. The effect is largest for low wind speeds and low turbine spacing. The maximum observed difference is 1.1% increase in mean inflow wind for a wind speed of 5.5 m/s, turbine spacing of 4 rotor diameters and a yaw amplitude of 6°. At 11.5 m/s, the wake effects are not so prominent, and the trends are a bit more unclear. Note that this is at a controller regime where the turbine is not extracting all the available energy.

The standard deviations of the horizontal wake path show significant increase for yaw amplitudes 3° and 6°, up to 30% increase for the lowest wind speeds at 4 rotor diameter spacing. It is inherent in the model that wake deflection due to yaw misalignment adds to the wake meandering, thus the increase in horizontal wake motion is as expected.

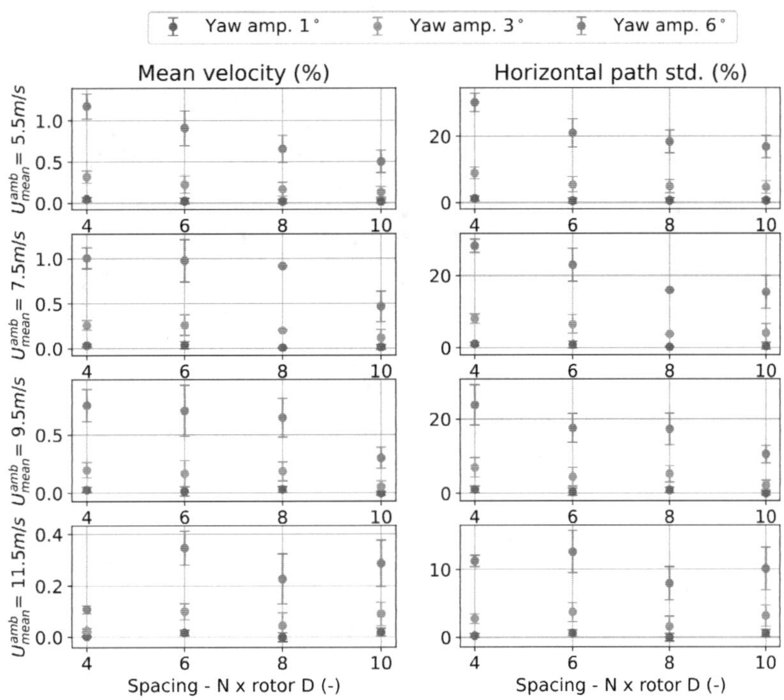


Figure 4: Relative change in hub mean inflow hub wind velocity and horizontal wake path standard deviation at the downwind turbine, for varying yaw of the upwind turbine.

Figure 5a shows an example of the u-component of the turbulent wind field, seen from above. The wind field implicitly contains the wake deficit, which explains the non-zero negative mean. Fig. 5a shows a clear meandering tendency, which is present both with and without yaw motion of the upwind turbine. By comparing the difference between them, a clear side-to-side pattern can be observed. The sine-curve like shape has a frequency of around 0.011 Hz, which matches the imposed yaw frequency. When looking at the crossflow v-component in Fig. 5b, differences in the order of 0.001 m/s are observed, which are probably insignificant for power and loads.

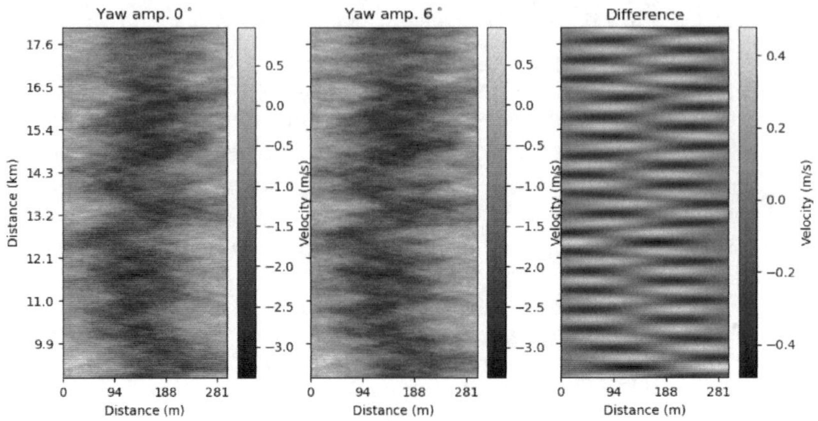


(a) u-component

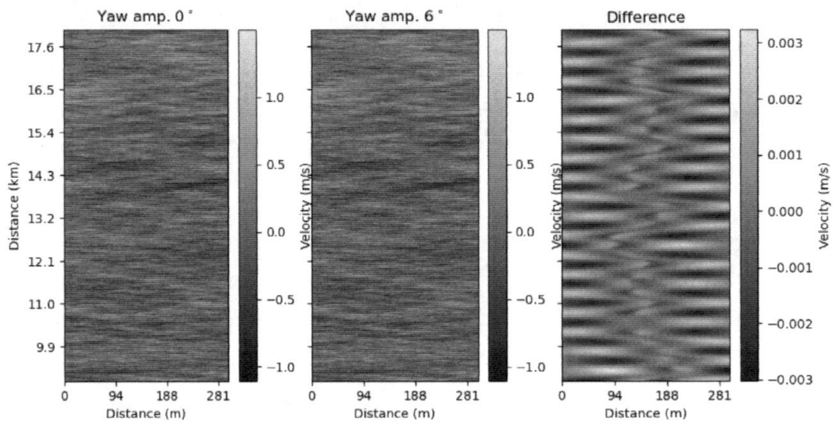


(b) v-component

Figure 5: Top view of lateral components of turbulence at hub height for 5.5 m/s with yaw amplitudes 0°, 6° and the difference between them.

### 4.2. Power Production

When one turbine is directly downwind of another turbine, increased oscillations of the wake path from the upwind turbine means that the downwind turbine spends less time in the wake, and it is expected that this is beneficial for the power production. Figure 7 shows that the mean power production (error-bars) is larger for a turbine downwind of a yawing turbine. The sum of the produced power is shown as dotted lines for one single seed, to show that the produced power follows the trend of the mean power. The standard deviations of power also show an increase, which means that power fluctuations are larger with a wind field influenced by time-varying yaw.

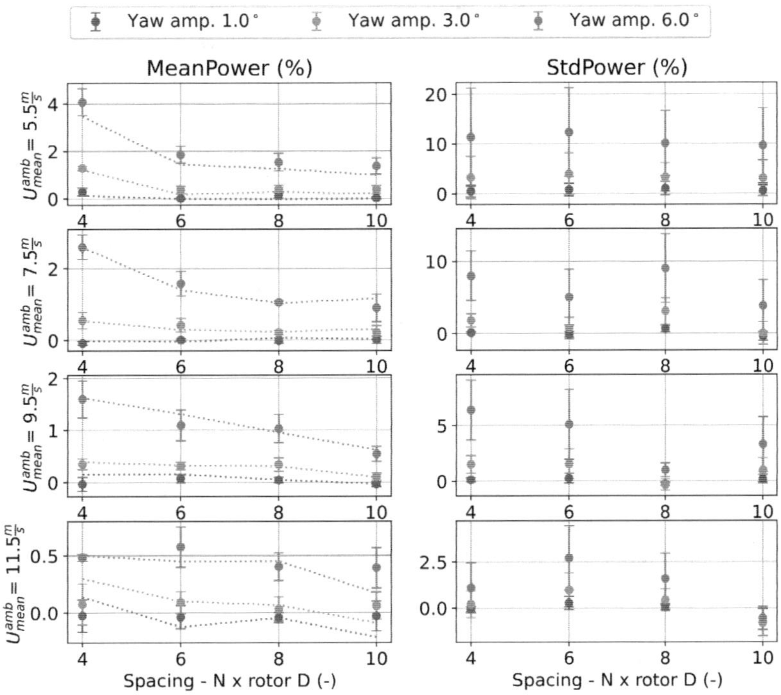


Figure 6: Relative change in mean and standard deviation of power for the downwind turbine, for varying yaw of the upwind turbine. Dotted lines show the change in total power production.

### *4.3. Turbine and Mooring Loads*

The increased oscillation of the wake is expected to give higher excitation of platform motions in sway and yaw, which will lead to higher mooring line loads. Figure 7 shows an insignificant increase in mean mooring line tension, but an increase of up to 17% in the standard deviation. A possible consequence of this is increased fatigue damage compared to the case where the yawing of the upwind turbine is not considered. Figure 7 also shows a small increase in blade root flapwise moment. Other responses, not included here due to space limitations, display similar increases in the standard deviation as the blade flapwise moment (tower fore-aft bending moment) or no siginificant increase (blade edgewise moment). There is some uncertainty in the side-to-side tower moments due to undamped oscillations from the initialisation phase of the simulation, thus they were not included.

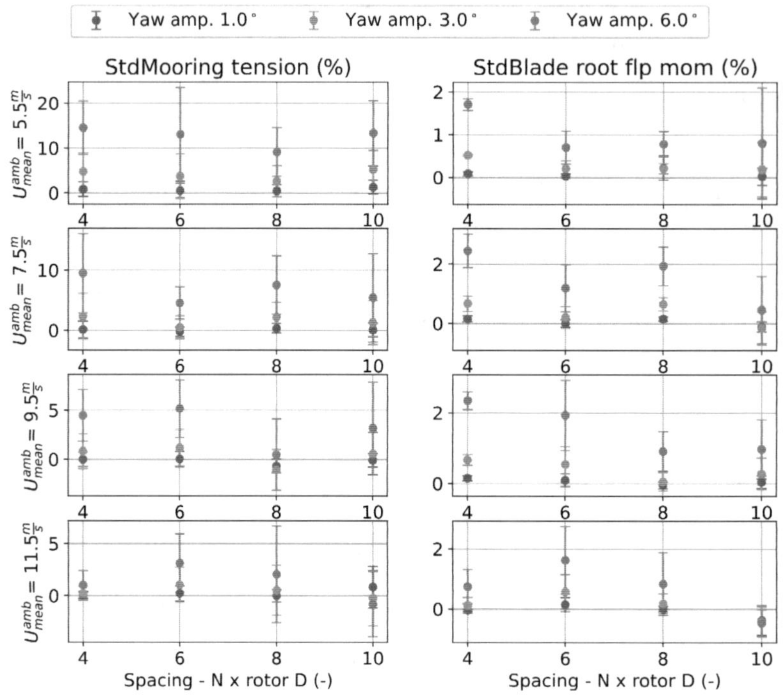


Figure 7: Relative change in standard deviation of side-to-side tower base bending moment and mooring line tension for the downwind turbine, for varying yaw of the upwind turbine.

### *4.4. Farm Level Effects*

This study only considers the case where one turbine is located directly downwind of the yawing turbine. The results from this study are also relevant more generally, since yaw is found to increase the total horizontal deflection of the wake, which may have implications also for turbines not directly in line with an upstream turbine. Figure 8 shows the standard deviation of the horizontal wake path for yaw amplitudes 0-6°. 1 and 3° yaw implies quite modest increases of the wake path oscillation (less than 10%, also see Fig. 4), whereas for 6° the additional sideways fluctuation amplitudes is more pronounced.

In practice, this means that if a wake model does not account for time varying yaw of the wake generating turbine, all partial wake immersion situations (Fig. 3b) at a downstream location may not be accounted for. The total effect of this would have to be considered for a specific farm layout.

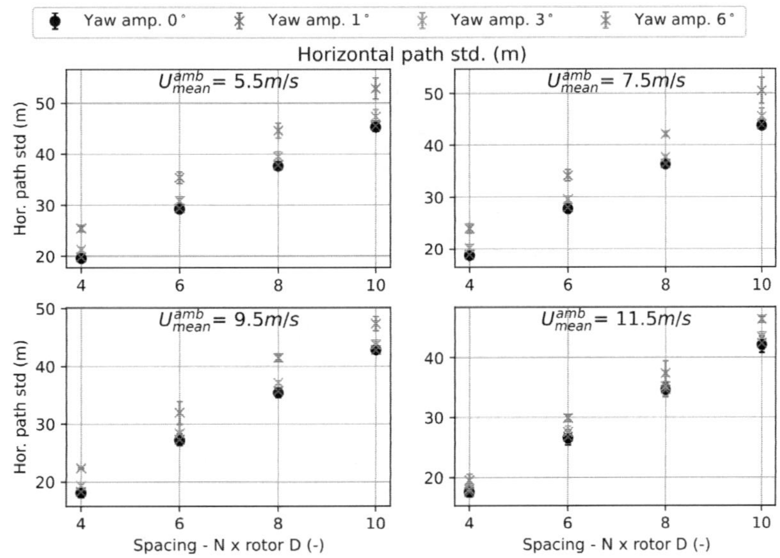


Figure 8: Standard deviation of lateral wake path. Error-bars show mean values and standard deviation of three seeds.

## 5. Conclusions

In this study, a two-turbine wind farm case was used to investigate the effect of dynamic yaw motions of the upstream wind turbine, on the incoming wind field, power production and turbine loads of the downwind turbine, using the wind field simulation tool DIWA, and the aero-, hydro-, servo-elastic wind turbine simulation tool Bladed. Bi-harmonic yaw signals representing yaw of a FOWT with different magnitudes (1° to 6°) was applied on the upwind turbine. Mean wind speeds from 5.5 m/s to 11.5 m/s and turbine spacing from 4 rotor diameters to 10 rotor diameters are considered.

Resulting wind speeds, power production and structural- and mooring line loads were compared to a case where the upwind turbine had no yaw motion. The comparison showed that including yaw in wake simulations gave higher mean power (up to 4%), but also higher power fluctuations. It also lead to larger standard deviations in the mooring line tension and a small increase in the blade root flapwise moment. These effects seem to decrease with increasing wind speeds and turbine spacing, and increase with the amplitude of yaw motion.

This study used a simple case where one turbine is directly downwind of the upwind turbine. In a real, modern wind farm, wind directions vary, and the layout is more optimized to avoid wake losses. Thus it is not possible, at this stage, to comment on the total effect of yaw motions on the annual energy production. In the present study, the influence of yaw motion of an upstream turbine on a second turbine not directly downwind of the first, was not considered.

Nevertheless, these results indicate a not insignificant effect of dynamic yaw motions on power production. The increased fluctuations in power could be important to consider in the design of turbine controllers and farm control. The total effect needs to be considered using a realistic farm layout, and may be different from farm to farm. Also, the increased mooring line tension variations could lead to increased fatigue, and design calculations not taking yaw motions into account could end up non-conservative. It should be mentioned that quasi-static mooring lines, and no quadratic damping of the platform motions were assumed in this study, which could influence the mooring line response. Further work should assess the importance of this.

## Acknowledgments
The work is part of the ImproveFLOW R&D project, funded by the Research Council of Norway through the ENERGIX programme (grant 321593), and industry partners.

## References
[1] Hywind Tampen URL `https://www.equinor.com/no/energi/hywind-tampen`
[2] Harvey M, Strivens S, Evans H and Morris A 2022 Floating Wind Joint Industry Programme - Phase IV Summary Report Tech. rep. Carbon Trust
[3] Jacobsen A and Godvik M 2021 *Wind Energy* **24** 149–161 ISSN 1099-1824
[4] Fleming P, Gebraad P M, Lee S, van Wingerden J W, Johnson K, Churchfield M, Michalakes J, Spalart P and Moriarty P 2015 *Wind Energy* **18** 2135–2143 ISSN 1099-1824
[5] Liew J, Juhl Andersen S, Troldborg N and Göçmen T 2022 *Journal of Physics: Conference Series* **2265** 022069 ISSN 1742-6588, 1742-6596
[6] Mahfouz M Y, Hall M and Cheng P W 2022 *Journal of Physics: Conference Series* **2265** 042004 ISSN 1742-6596
[7] Burton T, Jenkins N, Sharpe D and Bossanyi E 2011 *Wind Energy Handbook* 2nd ed (Wiley) ISBN 978-1-119-99272-1
[8] Larsen T J, Larsen G C, Pedersen M M, Enevoldsen K and Madsen H A 2017 *Journal of Physics: Conference Series* **854** 012027 ISSN 1742-6596
[9] Vermeer L J, Sørensen J N and Crespo A 2003 *Progress in Aerospace Sciences* **39** 467–510 ISSN 0376-0421
[10] Ainslie J F 1988 *Journal of Wind Engineering and Industrial Aerodynamics* **27** 213–224 ISSN 0167-6105
[11] Larsen G C, Madsen H A, Thomsen K and Larsen T J 2008 *Wind Energy: An International Journal for Progress and Applications in Wind Power Conversion Technology* **11** 377–395 ISSN 1095-4244
[12] Bromm M, Rott A, Beck H, Vollmer L, Steinfeld G and Kühn M 2018 *Wind Energy* **21** 1011–1028 ISSN 1099-1824
[13] Jiménez Á, Crespo A and Migoya E 2010 *Wind Energy* **13** 559–572 ISSN 1099-1824
[14] Bastankhah M and Porté-Agel F 2016 *Journal of Fluid Mechanics* **806** 506–541 ISSN 0022-1120, 1469-7645
[15] Wang H, Barthelmie R J, Pryor S C and Kim H G 2014 *Wind Energy* **17** 1587–1604 ISSN 1099-1824
[16] Mann J 1998 *Probabilistic Engineering Mechanics* **13** 269–282 ISSN 0266-8920
[17] International Electrotechnical Commission 2019 IEC61400-3-2: Wind energy generation systems - part 1: Design requirements
[18] Panjwani B, Kvittem M, Eliassen L, Ormberg H and Godvik M 2019 *Journal of Physics: Conference Series* **1356** 012039 ISSN 1742-6596
[19] SINTEF Ocean 2023 *DIWA User and Theory Manual, v. 2.9.1*
[20] Bastankhah M and Porté-Agel F 2015 *Journal of Physics: Conference Series* **625** 012014 ISSN 1742-6596
[21] Panjwani B, Vien Eliassen L, Fevang-Gunn L and Løvøy Alvestad B P 2023 Effect of Wind Turbine Yaw Misalignment on Wake Meandering. EERA DeepWind, Trondheim
[22] DNV 2023 *Bladed User Manual*
[23] Allen C, Viscelli A, Dagher H, Goupee A, Gaertner E, Abbas N, Hall M and Barter G 2020 Definition of the UMaine VolturnUS-S Reference Platform Developed for the IEA Wind 15-Megawatt Offshore Reference Wind Turbine Tech. Rep. NREL/TP-5000-76773 National Renewable Energy Lab. (NREL)
[24] Kvittem M I, Fredriksen A G, de Souza C S and Hanssen F C W 2023 *Journal of Physics: Conference Series* **2626** 012049 ISSN 1742-6596
[25] Gaertner E, Rinker J, Sethuraman L, Zahle F, Anderson B and Barter G 2020 IEA Wind TCP Task 37 Definition of the IEA Wind 15-Megawatt Offshore Reference Wind Turbine Technical Report NREL/TP-5000-75698 National Renewable Energy Lab. (NREL)

Journal of Physics: Conference Series **2745** (2024) 012008 doi:10.1088/1742-6596/2745/1/012008

# Development of a floating Vertical Axis Wind Turbine for the Mediterranean Sea

**Alberto Ghigo***, **Ermando Petracca**, **Gabriele Mangia**, **Giuseppe Giorgi**, **Giovanni Bracco**

Marine Offshore Renewable Energy (MOREnergy) Lab,
Department of Mechanical and Aerospace Engineering (DIMEAS),
Politecnico di Torino, Turin, Italy

*alberto.ghigo@polito.it, ermando.petracca@polito.it, gabriele.mangia@polito.it, giuseppe.giorgi@polito.it, giovanni.bracco@polito.it

**Abstract**. Differently from Horizontal Axis Wind Turbines (HAWTs), which are the reference technologies in the wind market for their reliability and maturity, Vertical Axis Wind Turbine (VAWT) applications are related to small-scale contexts, such as providing electricity in isolated areas or urban settings. Consequently, the capacity of VAWTs results significantly lower than the order of megawatts and does not exceed a few tens of kilowatts. A promising field of application for VAWTs is the floating offshore: among the main advantages there are an increased static stability, by placing the rotor-nacelle assembly (RNA) at the base of the VAWT and reduced operational and maintenance (O&M) costs. Moreover, the different wake dynamics allows to reduce the aerodynamic losses, allowing closer turbine installations. However, to be competitive with floating HAWTs, it is necessary to have numerical models for the analysis and simulation of multi-megawatt VAWTs. This paper aims to introduce a time domain model of a floating vertical axis wind turbine, developed within the Matlab-Simscape environment. The model comprises an aerodynamics module, based on Double Multiple Stream Tube theory while hydrodynamics is modelled using WEC-Sim. A case study, involving a Darrieus H-rotor VAWT tested in the Mediterranean Sea and supported by a semi-sub foundation, the OC4-DeepCwind, is introduced. The results obtained are compared with those from QBlade, an software developed by TU Berlin, demonstrating a good agreement between the two codes.

## 1. Introduction

One of the primary applications of Vertical Axis Wind Turbines (VAWTs) is in small-scale energy production, with installations in both urban and remote areas. In urban settings, these turbines can be placed on building rooftops, while in remote areas without access to the electricity grid, they can satisfying the energy needs of small-scale users, such as water pumping and desalination. In recent years, the application of VAWTs for offshore wind generation has been explored. The **DeepWind** technology, introduced in 2012, featuring a 5 MW $\phi$-Darrieus turbine supported by a spar-buoy foundation, was one of the most studied concepts. The floating system incorporates a generator located at the bottom of the structure. Experimental verification was conducted in both a test tank and a wind tunnel, and a 1 kW prototype tested at sea in Roskilde Fjord near Risø campus (Denmark).

Another attempt at a pre-commercial scale involved the Savonius Keel and Wind Turbine Darrieus (**Skwid**), developed by Modec. This hybrid wind-currents platform comprised a 500 kW H-Darrieus turbine supported by a floater, with a 60 kW Savonius turbine arranged on top to harness currents. However, in 2014 the installation of the demonstrator encountered difficulties due to buoyancy issues with the structure and the technology has been abandoned after huge financial investments.

Among the more recent technologies is **SeaTwirl**, featuring a Darrieus H-shape turbine supported by a spar-buoy foundation. The floater, rotating along with the turbine, is anchored to the seabed with multiple catenary moorings, while a static, non-rotating generator is enclosed just above sea level on the substructure. This device offers scalability for higher powers and reduced maintenance. A 30 kW prototype was installed in Norway in 2015, and a 1 MW device is scheduled for installation in Norway by the end of 2023.

Compared to floating HAWTs, VAWTs boast some interesting advantages [1]. By placing the rotor-nacelle assembly (RNA) at the base of the VAWT, the system gains increased static stability, thanks to a lower moment of inertia. Moreover, this allows greater accessibility to the RNA, especially in more intense metocean conditions, and reduces operational and maintenance (O&M) costs. Furthermore, the different wake dynamics, which dissipates in a more confined space compared to HAWTs, allows for reducing aerodynamic losses, increasing Annual Energy Production (AEP), and placing the VAWTs closer.

To evaluate the performances and to quantify the advantages previously introduced, it is essential to use numerical models with different levels of accuracy and computational time required. Several review works on numerical models are available in the literature [2], [3]. However, many state-of-the-art software, only allow to simulate HAWTs due to greater diffusion and technological maturity. The main ones are listed below.

**FloVAWT** (Floating Vertical Axis Wind Turbine), developed by Cranfield and Strathclyde Universities, is an aero-hydro-servo-elastic coupled model based on Paraschivoiu's Double-Multiple Streamtube model for aerodynamics and a time-domain model for hydrodynamics [4],[5].

**HAWC2**, created by DTU, is a versatile multibody time-domain program suitable for both floating and bottom-fixed offshore wind turbines, incorporating blade element momentum theory for aerodynamics and Morrison's equation for hydrodynamic loads [6].

**QBlade**, an open-source code from TU Berlin, allows the design and simulation of wind turbines, with the latest version, QBlade Ocean, enabling the simulation of complete systems of floating offshore wind turbines, both horizontal and vertical axis [7]. Qblade utilizes a multi-body formulation for structural dynamics and employs the Lifting Line Free Vortex Wake method for aerodynamics.

**OpenFast**, is a state-of-the-art software for simulating floating offshore wind turbines, primarily designed for HAWTs. Despite this limitation, there have been attempts to incorporate VAWTs by coupling OpenFast with the Offshore Wind ENergy Simulator (OWENS) [8], [9].

**Simo-Riflex-DMS** is a aero-hydro-servo-elastic simulator operating in the time domain that couples three codes [10], [11]. Simo computes hydrodynamic loads based on the actual displacement of the floater, while the DMS code calculates aerodynamic loads on the blades. Riflex conducts full equilibrium iterations at each time step.

Another obstacle is the lack of publicly available turbines, whose data can be accessed for analysis and simulations, such as the NREL 5 MW or the IEA 15 MW for HAWTs. Among the most studied, there is the DeepWind concept: it entails a two-bladed Darrieus-Φ rotor with a modified Troposkien shape to accommodate the substantial blade weight [12], [13]. Designed for a nominal power of 5 MW, the turbine had a diameter of 60.5 m, a height of 130 m and a chord of 5 m. Different upscales were made, up to a nominal power of 20 MW.

The objective of this study is to present a numerical model of a floating vertical axis wind turbine. The model, created in Matlab-Simscape, includes a floating platform, a wind turbine and a mooring system. Aerodynamics is simulated using the Double Multiple Stream Tube method, while hydrodynamics is modeled using WEC-Sim. The VAWT used for the simulations was designed starting from a Darrieus H-rotor VAWT: the chord length and type of airfoil were analyzed, and then the power and thrust were compared with QBlade. Finally, a case study of the floating VAWT, with a semi-sub foundation, the OC4-DeepCwind, under operative and extreme conditions is presented.

Journal of Physics: Conference Series **2745** (2024) 012008 doi:10.1088/1742-6596/2745/1/012008

## 2. Material and methods

The numerical model introduced in this work is based on Most (Matlab for OFWT Simulations Tools), developed by MOREnergy Lab of Politecnico di Torino [14]. Most main application regards the design and analysis of a FOWT: due to its flexibility it is possible to simulate different floating foundations, wind turbines and mooring layouts. The use of look-up tables for the aerodynamics also allow a remarkable time reduction while the accuracy has been compared with OpenFast in [15].

To assess VAWT performances, the following implementations have been made:

- *Aerodynamics*: The simulation of aerodynamics utilizes the Double Multiple Stream Tube (DMST) method, which is based on the blade element momentum (BEM) theory. Aerodynamic forces are computed using pre-calculated look-up tables.
- *Hydrodynamics*: The hydrodynamics of the floating foundation is calculated using WEC-Sim, employing a time-domain approach to solve the dynamics of floating bodies based on a frequency-domain boundary element method. Hydrodynamic characteristics, including linear hydrostatics, added mass, radiation damping, and wave excitation coefficients, can be assessed using external software tools such as Wamit, Ansys Aqwa, Nemoh, or Capytaine.
- *Moorings*: Two options have been incorporated: quasi-static mooring lookup tables, obtained by solving catenary equations, or Moordyn,

Figure 1 shows a representation of the model.

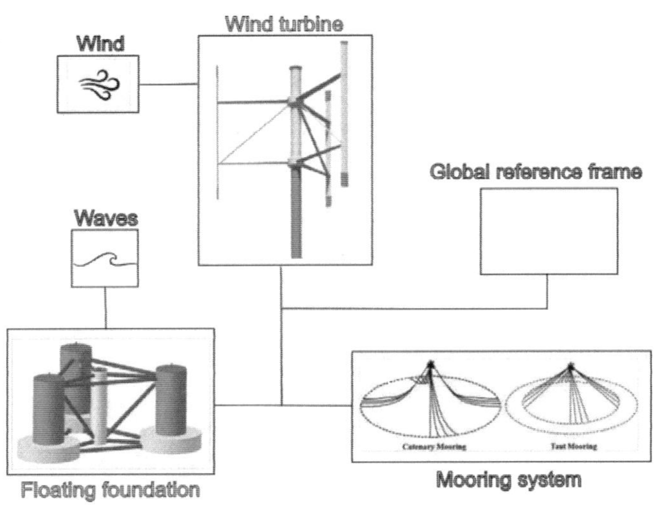


**Figure 1.** The numerical model representation.

### *2.1. Aerodynamics*

The approach used in this work for the estimation of aerodynamic forces acting on the wind turbine is based on the Double Multiple Stream Tube (DMST) theory, developed by Parashivoiu [16]. In this model, several aerodynamically independent streamtubes are employed, wherein the upwind and downwind sections of the rotor are substituted by distinct actuator disks with equivalent loading.

This methodology enables the model to calculate different induction factors for each streamtube and separately for the upwind and downwind components. Similar to the actuator disk model utilized for Horizontal Axis Wind Turbines (HAWTs), the solution is attained through an iterative process. The induction factor from the momentum model serves as input for the blade element model, and conversely, the load from the blade element model is incorporated into the momentum model.

After dividing the rotor into N streamtubes, the induction factor is assumed to be zero. For each streamtube, assuming a specific airfoil type and obtaining tabulated values of the lift and drag coefficients as functions of the angle of attack, the calculation proceeds for main aerodynamic parameters such as lift, drag, relative velocity, angle of attack, and normal and tangential coefficients. To determine the induction factor value, the convergence criterion is set as the thrust coefficient, calculated using both the momentum theory and the blade-element equation. In the model, for each streamtube the cycle is iterated up to 500 iterations, allowing for an error in the thrust coefficient of up to 1%. The loop is iterated for all streamtubes, both upwind and downwind. Finally, by summing the upstream and downstream contributions, it is possible to determine the value of the torque, power and thrust. The methodology used is summarised in Figure 2.

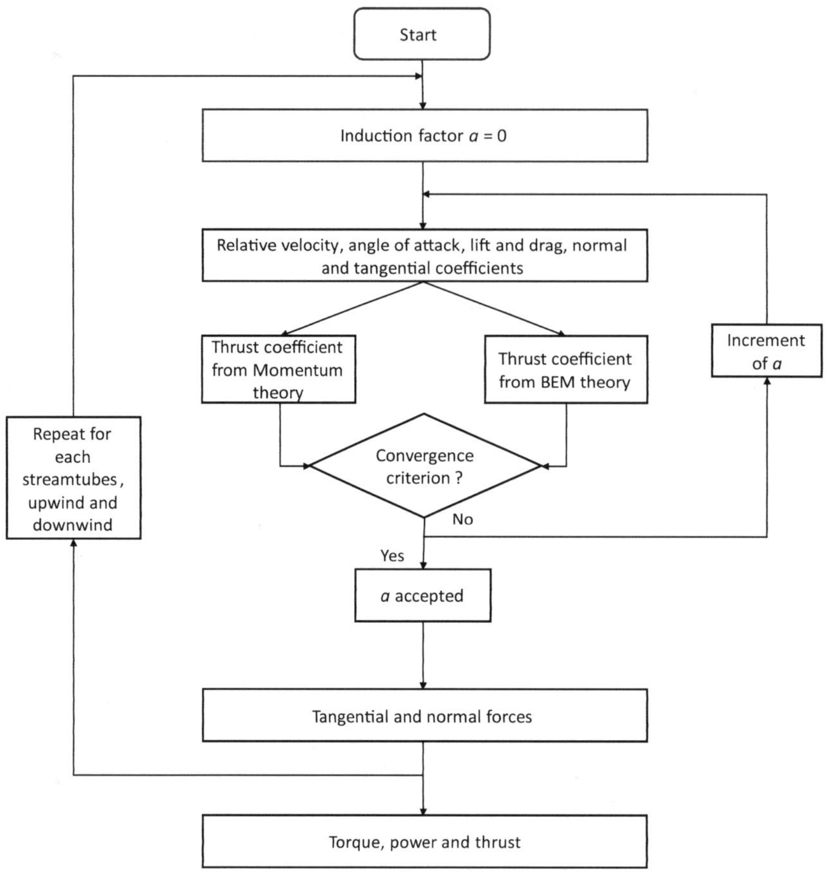


**Figure 2.** The DMST methodology used for the calculation of torque, power and thrust.

Some corrections in particular have been applied to the proposed methodology, including Glauert correction and the dynamic stall, as well as the blade pitch mechanism to allow the blades to rotate along the pitch angle to reduce the thrust force and to keep a constant power.

## 3. VAWT design

The wind turbine considered in this study is a Darrieus H-rotor VAWT with straight blades. The adoption of this type of turbine lies in the simplicity of the system, where the blades are easily manufacturable compared to the curved blades of the Darrieus-Φ turbine. For the input data, turbines developed for two European projects were considered: the H2OCEAN (2012) [17] and S4VAWT (2016) [18] projects. The final VAWT design is the result of several analysis, in which the influence of the chord size and the type of airfoil are investigated, as detailed in the following sections.

*3.1. Blade chord influence*

In this section, the influence of the chord is investigated by considering five different values of chord lengths, from 2 up to 10 m. In Figure 3 the power and thrust coefficient are reported by considering the different chord lengths, while in Figure 4 the torque coefficient is shown.

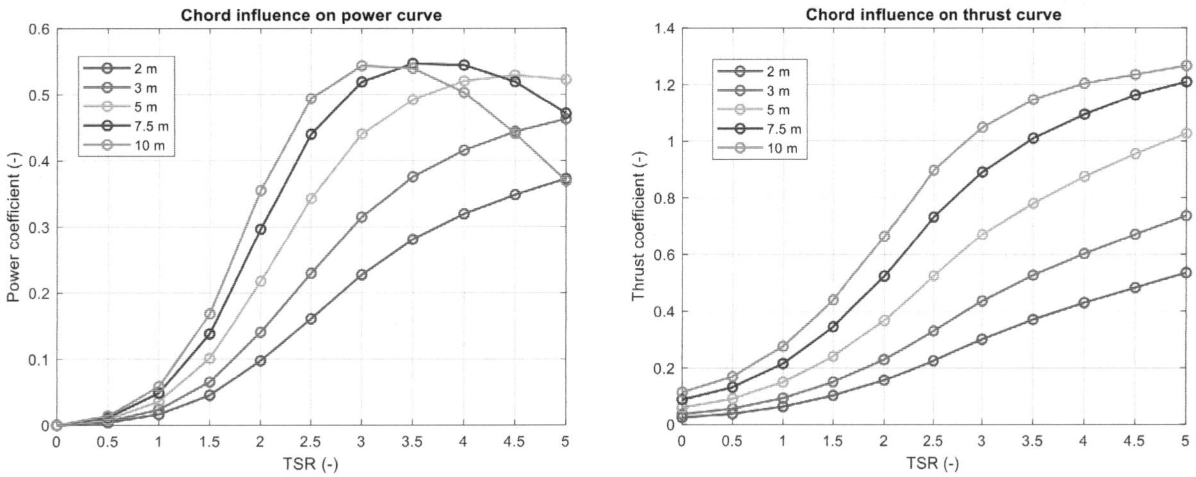


**Figure 3.** Power and thrust coefficient for different chord lengths.

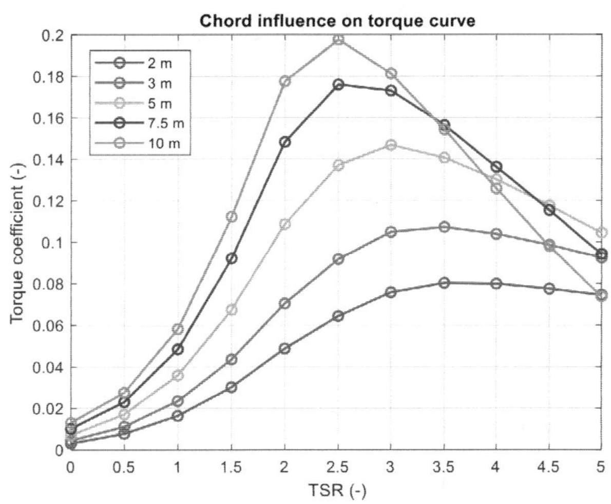


**Figure 4.** Torque coefficient for different chord lengths.

From the previous figures, it is observed that as the chord length increases, there is an increase in the thrust force acting on the blade. Regarding the power curve, for longer chords, the peak of the curve is reached earlier, to decrease rapidly for higher TSR values. Consequently, for the subsequent analyses, the intermediate chord value of 5 m is chosen as it is a good compromise in terms of power extracted and thrust force acting on the blades.

*3.2. Airfoil influence*

In this section, the influence of the airfoil is investigated by considering five types of airfoil profiles: NACA 0012, NACA 0015, NACA 0018, NACA 0024 and NACA 0027. The choice to consider NACA airfoils, in addition to data availability, is linked to their adoption in other studies about multi-megawatt VAWTs for offshore applications.

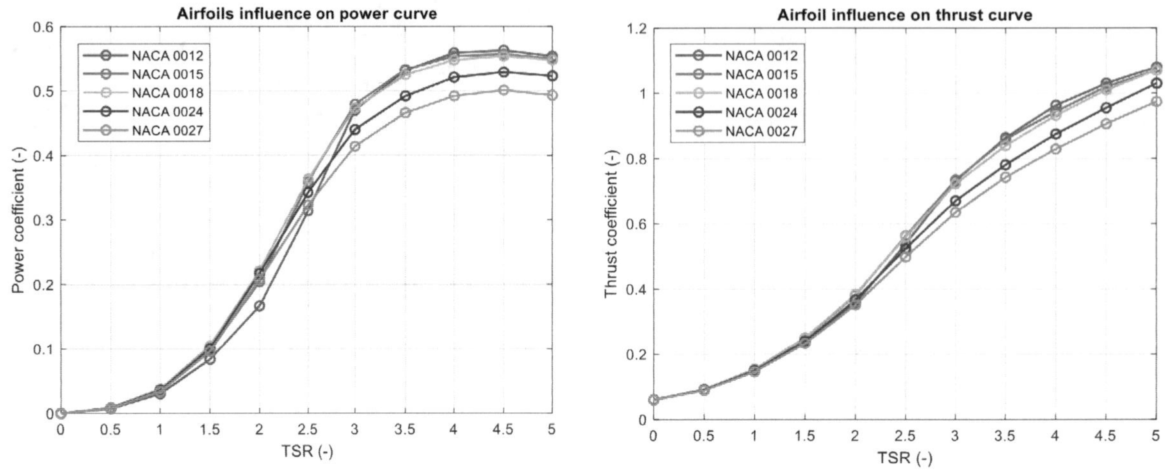


**Figure 5.** Power and thrust coefficient for different NACA airfoils.

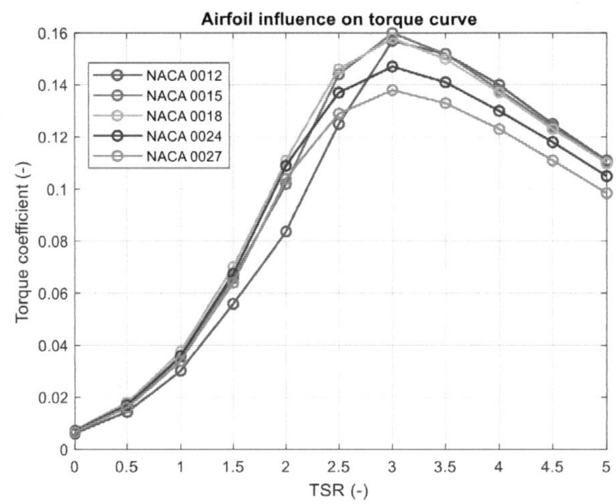


**Figure 6.** Torque coefficient for different NACA airfoils.

From Figures 5 and 6, it is evident that lower NACA airfoils exhibit higher maximum values for power and torque coefficients, as well as thrust. Conversely, larger NACA airfoils provide lower performance in terms of power and torque but result in lower thrust values acting on the blade. Consequently, since the blade height for a multi-MW VAWT is significant, the NACA 0024 has been chosen.

### *3.3. VAWT 6 MW final design*

From the previous analyses, assumptions have been made about the characteristics of the turbine used in the study. The criteria used was to obtain a compromise in terms of productivity and reducing the thrust forces acting on the turbine. Table 1 shows the main VAWT data.

**Table 1.** Wind turbine data.

| **Wind turbine** | ***VAWT*** |
|---|---|
| Nominal power | 6 MW |
| V_cut_in, V_rated, V_cut_off | 3, 10, 25 m/s |
| Ω_min, Ω_rated | 2.1, 6.6 rpm |
| Wind turbine radius | 63 m |
| Wind turbine height | 140 m |
| Chord length | 5 m |
| Airfoil | NACA 0024 |

In Figure 7, the power and torque curves are shown as a function of the rotor speed and the wind speed. The minimum rotor speed is 2.1 rpm, while over nominal conditions, i.e., for wind speeds exceeding 10 m/s, is constant and equal to 6.6 rpm.

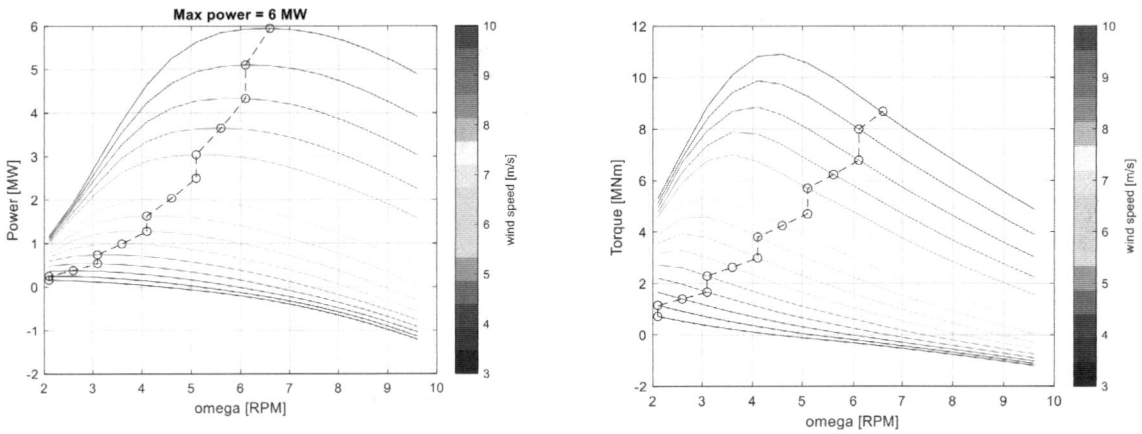


**Figure 7.** Power and torque curve.

In Figure 8, the thrust and power curves are presented. The blade pitch mechanism causes a decrease in the peak of the thrust curve once the nominal speed is reached, while the power curve follows a constant trend, equal to the nominal power. However, the power value is not instantaneous but is averaged over the entire rotation of the turbine.

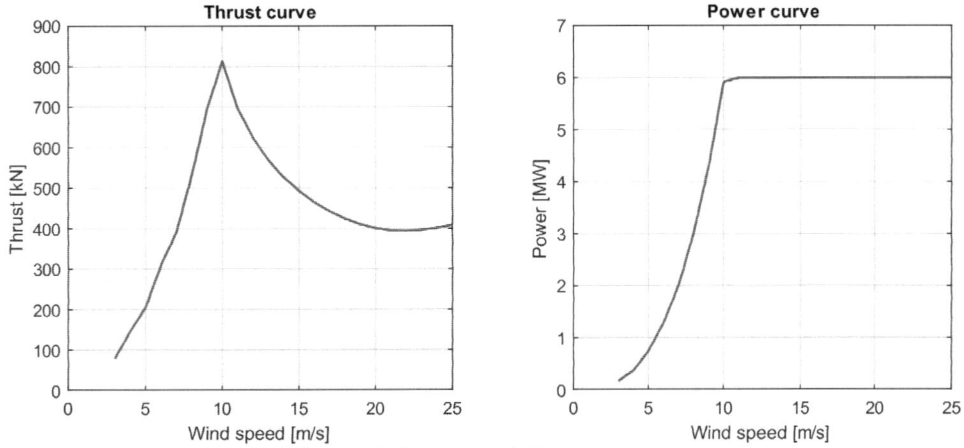


**Figure 8.** Power and thrust curve.

### *3.4. Comparison with QBlade*

In this paragraph, a comparisons of the aerodynamic performances between the numerical model and QBlade is reported. In Figure 9, the power and thrust curved by varying the wind speed are shown.

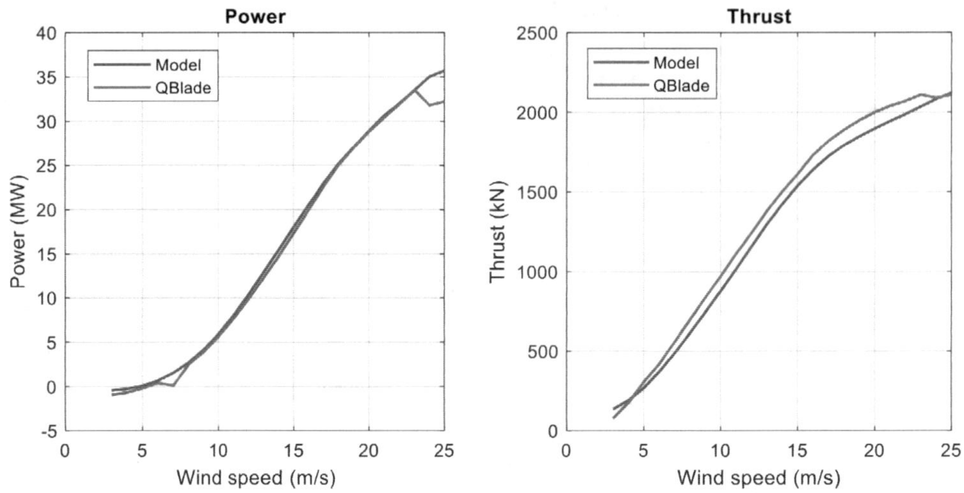


**Figure 9.** Power and thrust comparison between the numerical model and QBlade.

While QBlade slightly overestimate the thrust, there is a good agreement between the two codes for both the power and the thrust. In this simulation, to make the results comparable, the blade pitch mechanism for wind speed higher than the nominal speed was not implemented.

### *3.5. Floating VAWT simulations*

The floating foundation selected for the case study is a semi-submersible platform, the OC4 [19]. This platform features a triangular shape with three columns positioned at its vertices, along with a central column that provides support for the wind turbine. The complete system, consisting of the 6 MW VAWT installed on the OC4 floating foundation, is simulated in an operational environment. The site is located in the Sicily Streat, next to the island of Pantelleria (Italy). In particular, two sea-states have been considered, one for operational and one for near cut-off wind speed condition, as reported in Table 2.

**Table 2.** the two sea states considered for the floating VAWT.

| **Sea states** | ***Operational*** | ***Near cut-off*** |
|---|---|---|
| Average wind speed | 7 m/s | 21 m/s |
| Wave height | 0.65 m | 4.65 m |
| Wave period | 3.65 s | 8.13 s |

The two sea-states have been considered from the MORE-RES platform, a web tool developed by Politecnico di Torino for the analysis of offshore wind and waves [20]. Wind and wave data are provided by ECMWF ERA5 database, from 2010 to 2019. In Figure 10, on the left the most occurrent triplets of wave height, wave period and wind speed are reported, while on the right there is the wave scatter for the site of Pantelleria.

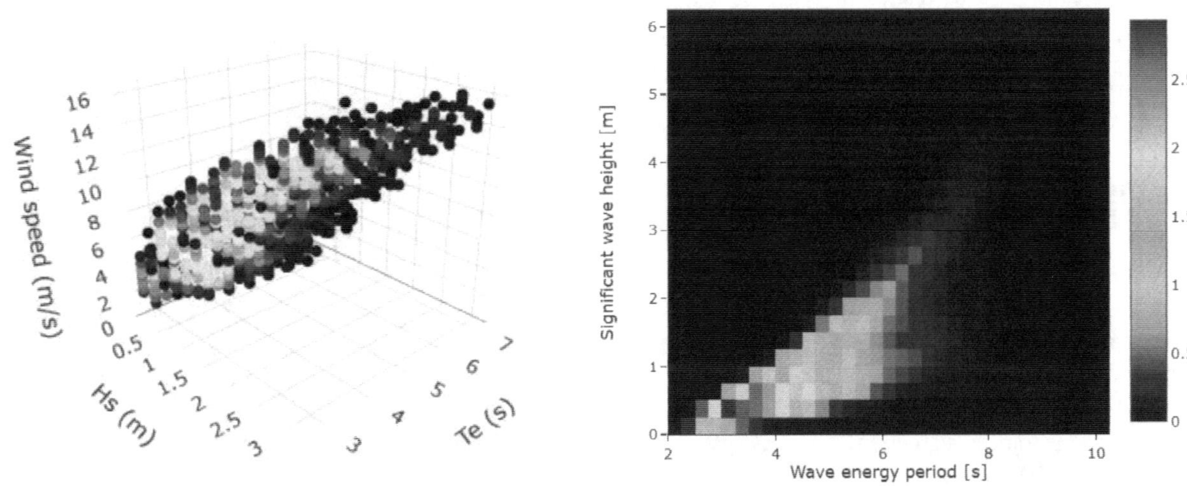


**Figure 10.** Wind speed, wave height and wave period triplets (on the left) and wave scatter (on the right) for the site near to Pantelleria, from MORE-EST platform.

The depth of the site is 200 m, while the mooring configuration includes three catenary lines arranged at 120° between them. The wind speed profile is generated by using the TurbSim software, developed by NREL. The turbulence model considered is a Kaimal, with a IEC turbulence characteristic B of 0.14. The power law exponent considered is equal to 0.11.

In Figure 11, on the left the VAWT performances are reported, while on the right the OC4 platform displacements.

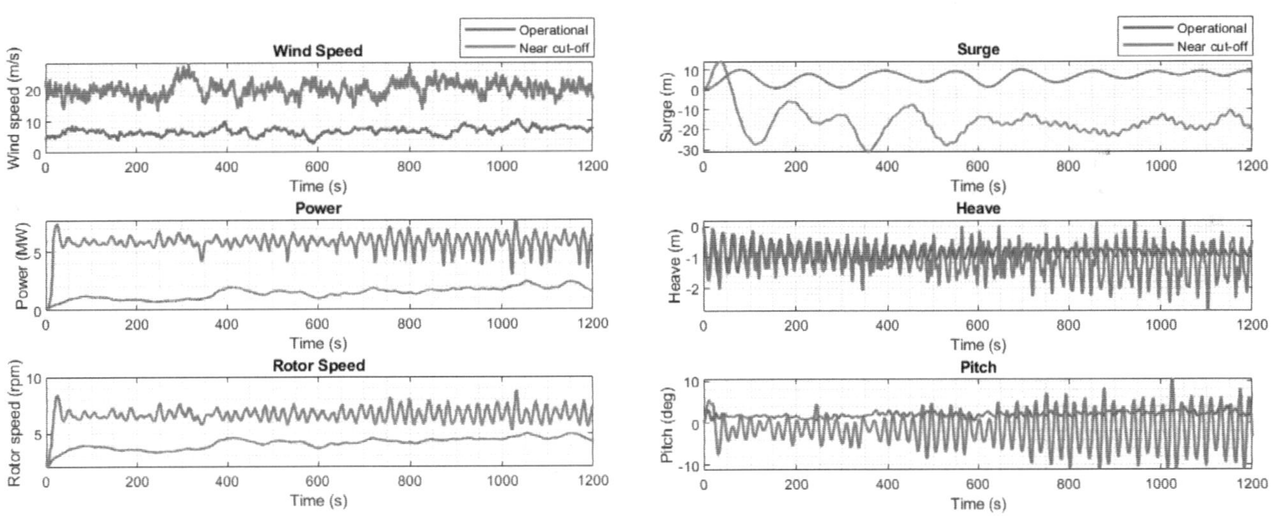


**Figure 11.** VAWT performances and displacements for the operational and extreme sea-states.

Regarding the operational conditions, with a wind speed of 7 m/s, lower than the nominal value, the turbine cannot achieve its nominal power and rotor speed. The initial transient appears slower, then settles at the value predicted by the power curve, with few oscillations. Conversely, in the extreme case, after a faster initial transient, the power and the rotor speed oscillate around the nominal values, with a greater amplitude than the operational case. As for displacements, there are significative differences between the two cases for the surge displacement. In the near cut-off conditions, the amplitude of surge is higher due to the significant wave height acting on the floating VAWT. Similarly, for the pitch, the system results more stressed, with oscillations up to 10°, while for operational conditions they are assessed to an average value of 3°.

Therefore, the design of an improved foundation and the study of an improved mooring configuration for the VAWT could help reduce these maximum values.

## 4. Results and discussion

The purpose of this work is twofold: firstly, to introduce the numerical model of a floating VAWT developed in the Matlab-Simscape environment; secondly, to exploit the model for the design of a multi-megawatt VAWT intended for the Mediterranean Sea. The numerical model, based on the Most tool, features aerodynamics grounded in the DMST model, with some corrections such as the Glauert correction and dynamic stall to improve the accuracy.

The VAWT design process has analyzed the airfoil type and the chord length, by considering the torque, power, and thrust as performance parameters. Once the VAWT design was completed, the power and thrust characteristics were compared with QBlade, demonstrating a good agreement. Finally, a case study near the island of Pantelleria (Italy) was presented, simulating a floating VAWT under both operational and extreme conditions.

The flexibility of the numerical model stands out as its primary advantage. Each Simulink block can be accessed and modified, including governing equations, wind turbine parameters, foundation configuration, and mooring layout. This flexibility ensures a balance between accuracy and computational efficiency, making the model an adaptable tool for preliminary assessments of floating VAWT projects.

## References

[1] A. Ghigo, E. Faraggiana, G. Giorgi, G. Mattiazzo, and G. Bracco, "Floating Vertical Axis Wind Turbines for offshore applications among potentialities and challenges: A review," *Renewable and Sustainable Energy Reviews*, vol. 193, p. 114302, Apr. 2024, doi: 10.1016/j.rser.2024.114302.

[2] E. Faraggiana, G. Giorgi, M. Sirigu, A. Ghigo, G. Bracco, and G. Mattiazzo, "A review of numerical modelling and optimisation of the floating support structure for offshore wind turbines," *Journal of Ocean Engineering and Marine Energy*, vol. 8, no. 3. Springer Science and Business Media Deutschland GmbH, pp. 433–456, Aug. 01, 2022. doi: 10.1007/s40722-022-00241-2.

[3] A. Otter, J. Murphy, V. Pakrashi, A. Robertson, and C. Desmond, "A review of modelling techniques for floating offshore wind turbines," *Wind Energy*, vol. 25, no. 5. John Wiley and Sons Ltd, pp. 831–857, May 01, 2022. doi: 10.1002/we.2701.

[4] M. Collu, M. Borg, A. Shires, F. Rizzo, and E. Lupi, "FLOVAWT: Further progresses on the development of a coupled model of dynamics for floating offshore VAWTS," in *Proceedings of the International Conference on Offshore Mechanics and Arctic Engineering - OMAE*, 2013. doi: 10.1115/OMAE2013-10717.

[5] M. Collu, M. Borg, A. Shires, and F. P. Brennan, "FLOVAWT: Progress on the development of a coupled model of dynamics for floating offshore vertical axis wind turbines," 2013. [Online]. Available: http://www.energytechnologies.co.uk

[6] F. Vorpahl, M. Strobel, J. M. Jonkman, T. J. Larsen, P. Passon, and J. Nichols, "Verification of aero-elastic offshore wind turbine design codes under IEA Wind Task XXIII," *Wind Energy*, vol. 17, no. 4, pp. 519–547, 2014, doi: 10.1002/we.1588.

[7] D. Marten *et al.*, "QBLADE: An Open Source Tool for Design and Simulation of Horizontal and Vertical Axis Wind Turbines," 2013. [Online]. Available: www.ijetae.com

[8] M. C. Devin, N. R. Mendoza, A. Platt, K. Moore, J. Jonkman, and B. L. Ennis, "Enabling Floating Offshore VAWT Design by Coupling OWENS and OpenFAST," *Energies (Basel)*, vol. 16, no. 5, Mar. 2023, doi: 10.3390/en16052462.

[9] B. C. Owens, T. W. Strganac, and R. Talreja, "Theoretical Developments and Practical Aspects of Dynamic Systems in Wind Energy Applications A Dissertation," 2013.

[10] K. Wang, N. / Cesos, N. Trondheim, M. Otto, and L. Hansen, "A method for modeling of floating vertical axis wind turbine," *Proceedings of the ASME 2013 32nd International Conference on Ocean, Offshore and Arctic Engineering.* , vol. Volume 8, Nov. 2013, doi: https://doi.org/10.1115/OMAE2013-10289.

[11] K. Wang, T. Moan, and M. O. L. Hansen, "Stochastic dynamic response analysis of a floating vertical-axis wind turbine with a semi-submersible floater," *Wind Energy*, vol. 19, no. 10, pp. 1853–1870, Oct. 2016, doi: 10.1002/we.1955.

[12] V. , Luca, *Offshore Vertical Axis Wind Turbine with Floating and Rotating Foundation*. APA, 2011. [Online]. Available: www.risoe.dtu.dk

[13] U. S. Paulsen *et al.*, "Outcomes of the DeepWind conceptual design," in *Energy Procedia*, Elsevier Ltd, 2015, pp. 329–341. doi: 10.1016/j.egypro.2015.11.437.

[14] M. Sirigu, E. Faraggiana, A. Ghigo, and G. Bracco, "Development of MOST, a fast simulation model for optimisation of floating offshore wind turbines in Simscape Multibody," in *Journal of Physics: Conference Series*, 2022. doi: 10.1088/1742-6596/2257/1/012003.

[15] L. Cottura, R. Caradonna, A. Ghigo, R. Novo, G. Bracco, and G. Mattiazzo, "Dynamic modeling of an offshore floating wind turbine for application in the mediterranean sea," *Energies (Basel)*, vol. 14, no. 1, 2021, doi: 10.3390/en14010248.

[16] I. Paraschivoiu, "Double-Multiple Stream Tube for Darrieus Wind Turbines."

[17] "H2OCEAN D4.2 Mechanical design of VAWT ad Power Take Off system FINAL".

[18] F. Huijs *et al.*, "Integrated design of a semi-submersible floating vertical axis wind turbine (VAWT) with active blade pitch control," in *Journal of Physics: Conference Series*, Institute of Physics Publishing, Nov. 2018. doi: 10.1088/1742-6596/1104/1/012022.

[19] A. Robertson *et al.*, "Definition of the Semisubmersible Floating System for Phase II of OC4," 2014. [Online]. Available: www.nrel.gov/publications.

[20] G. Giorgi, R. Novo, G. Cervelli, and G. Bracco, "Wave energy converters technology database for a web-based platform for evaluating wave energy resource and productivity potential," in *Trends in Renewable Energies Offshore - Proceedings of the 5th International Conference on Renewable Energies Offshore, RENEW 2022*, CRC Press/Balkema, 2023, pp. 33–41. doi: 10.1201/9781003360773-5.

# Offshore field experimentation for novel hybrid condition monitoring approaches

**Kayacan Kestel[1,2,3], Faras Jamil[1,2,4], Jens Jo Matthys[1,2], Konstantinos Vratsinis[1,2], Jonathan Sterckx[1,2], Rebeca Marini[1,2], Cédric Peeters[1,2], Jan Helsen[1,2]**

[1]Acoustics Vibration Research Group, Vrije Universiteit Brussel, Belgium
[2]Offshore Wind Infrastructure Lab, Vrije Universiteit Brussel, Belgium
[3]Univ Lyon, INSA Lyon, LVA, EA677, Villeurbanne, France
[4]Artificial Intelligence Lab, Vrije Universiteit Brussel, Belgium

E-mail: `kayacan.kestel@vub.be`

**Abstract.** This study details the development of a fully automated pipeline for the condition monitoring of wind turbine drive trains. Vibration data is collected using hardware designed and manufactured in-house and used directly to monitor the condition of the drive trains. The complex nature of wind turbine vibration signals, due to the large number of components and highly variable operating conditions, makes drive train condition monitoring a challenging task. This paper details the full data measurement and analysis flow from sensor to insights and proposes a hybrid automated pipeline with signal processing and data-driven techniques to address the complexity of dealing with wind turbine vibration data. The vibration signals are directly employed to estimate the wind turbine's instantaneous angular speed to compensate for any rotation speed fluctuations. Pre-processing is performed on the speed-independent signals to evaluate condition indicators in both the time and spectral domain for the vibration signals and their envelopes. Machine learning is then employed to distinguish the healthy state of the machine from a faulty one using the computed condition indicators. Besides the scalar indicators, also two-dimensional vibration decompositions such as the cyclic spectral correlation maps are used as inputs to the machine learning pipeline. This comprehensive and automated approach ensures both an early and reliable fault detection. Experimental results demonstrate that the fully automated hybrid pipeline can effectively be used for fleet-based health tracking of offshore wind turbine drivetrains.

## 1. Introduction

Over the past decade, wind energy capacity in Europe has experienced consistent growth, with an additional 3 GW of offshore capacity installed in 2020, bringing the total to 25 GW [1]. The growth rate of wind energy has increased in recent years, thanks in part to significant reductions in the levelized cost of energy (LCOE) for both onshore and offshore wind, making them more competitive with traditional energy sources [1, 2]. To further expand this growth, wind turbine manufacturers are striving to optimize their designs, minimizing the operational expenses of wind farms [3]. This necessitates identifying the machine components that have the most significant impact on downtime. Among these, the gearbox stands out as the component with the longest downtime per failure [4]. It was found that bearings are the most vulnerable to failure, followed by gears. Notably, high-speed stage (HSS) bearings exhibit the highest failure rate. Besides the gearbox, the data measurement and analysis workflow detailed in this paper is also employed for the monitoring of the main bearing and generator to enable comprehensive drivetrain monitoring. To reduce operational costs for wind farms, it is crucial to predict drivetrain failures beforehand, enabling preventive maintenance [5, 6]. Additionally, identifying the root causes of these failures is essential to prevent recurrence in future design iterations.

Vibration data serves as a predictive tool for anticipating failures through the application of advanced signal processing techniques. This enables the implementation of predictive maintenance, as opposed to reactive or periodic maintenance strategies. When it comes to monitoring rotating machinery components, vibration analysis is probably the most widely used approach. This prevalence stems from its capability to analyze high-frequency events typically linked with gears and bearings. Moreover, its straightforward installation, proven effectiveness, and ability to diagnose fault types and approximate locations contribute significantly to its widespread adoption.

Current practice in industrial condition monitoring systems typically involves tracking time-domain statistical indicators and more component-specific frequency-domain indicators [7, 8, 9]. Time-domain indicators offer the advantage of not requiring prior knowledge of characteristic fault frequencies. This simplifies the analysis process by eliminating the need to consider the number and complexity of components. However, this simplicity comes at the cost of reduced pinpointing capabilities. While time-domain indicators can identify abnormal behavior, they cannot pinpoint the exact component causing it. This limitation is not a major drawback at the initial stages of analysis, where a quick overview of faulty behavior across sensors is more valuable than immediate pinpointing. Subsequently, frequency-based analysis can be employed to delve deeper into the data and determine the exact location of the fault.

Conventional vibration monitoring frameworks often rely on rudimentary statistical indicators and spectral amplitudes, limiting their ability to detect and diagnose faults effectively. To address this limitation, this work proposes a multifaceted processing scheme that extracts valuable insights from raw vibration data using a combination of advanced signal processing techniques, anomaly detection algorithms, alarm generation mechanisms, and SCADA data integration. The dedicated in-house developed hardware enables simultaneous measurements of high-frequency vibration data of multiple sensors located over the drivetrain. Acquired signals are further processed with state-of-the-art pre-processing techniques, and effective condition monitoring indicators are extracted. SCADA data integration provides contextual information about the operating environment and facilitates root cause analysis. Machine learning algorithms are then employed to detect anomalous patterns and classify fault types, while anomaly detection algorithms aim to identify deviations from normal operating conditions [10, 11]. Alarm generation mechanisms trigger notifications in real-time, alerting operators to potential malfunctions. This holistic approach ensures comprehensive fault detection and diagnostics, enabling proactive maintenance and minimizing downtime.

## 2. Measurement Hardware & Edge Device

In order to enable the advanced hybrid signal processing-machine learning pipeline proposed in this paper, adequate measurements are necessary. Firstly, modern day wind turbines have many slow rotating components implying the need for very long measurements to achieve a sufficient frequency-domain resolution. Secondly, many mechanical fault events are amplified by high-frequency resonances or have a very short time duration. To properly capture such events, a high sampling rate is required. Thirdly, to track the health state of the turbine, a sufficiently small time period between measurements is recommended during normal operating conditions. The data acquisition system used to facilitate the workflow is capable of meeting all three of these requirements with standard settings that allow for 3 minute long data sets measured every 15 minutes at 30 kHz on 8 different channels. Given that the entire hardware chain is developed in-house, measurement parameter changes can be made if necessary, e.g. even longer measurements, different sample rates, or more frequent measurements. The full hardware measurement setup can be seen in Fig. 1. It includes a casing for the power and networking, four casings for the sensor modules, and one for the edge computing device. The in-house data acquisition system was primarily developed due to the observation that many industrial monitoring systems out there do not offer two key points: large memory buffers to capture a sufficient number of rotations in (very) slowly rotating systems, and accurate low-frequency acquisition to not distort low-frequency vibrations through overlapping analog low-frequency high-pass filters. To ensure the proper monitoring of such slow rotating systems, we thus developed our own in-house multi-channel data acquisition module that manages to continuously sample data at a high sampling rate without suppressing any low-frequency content.

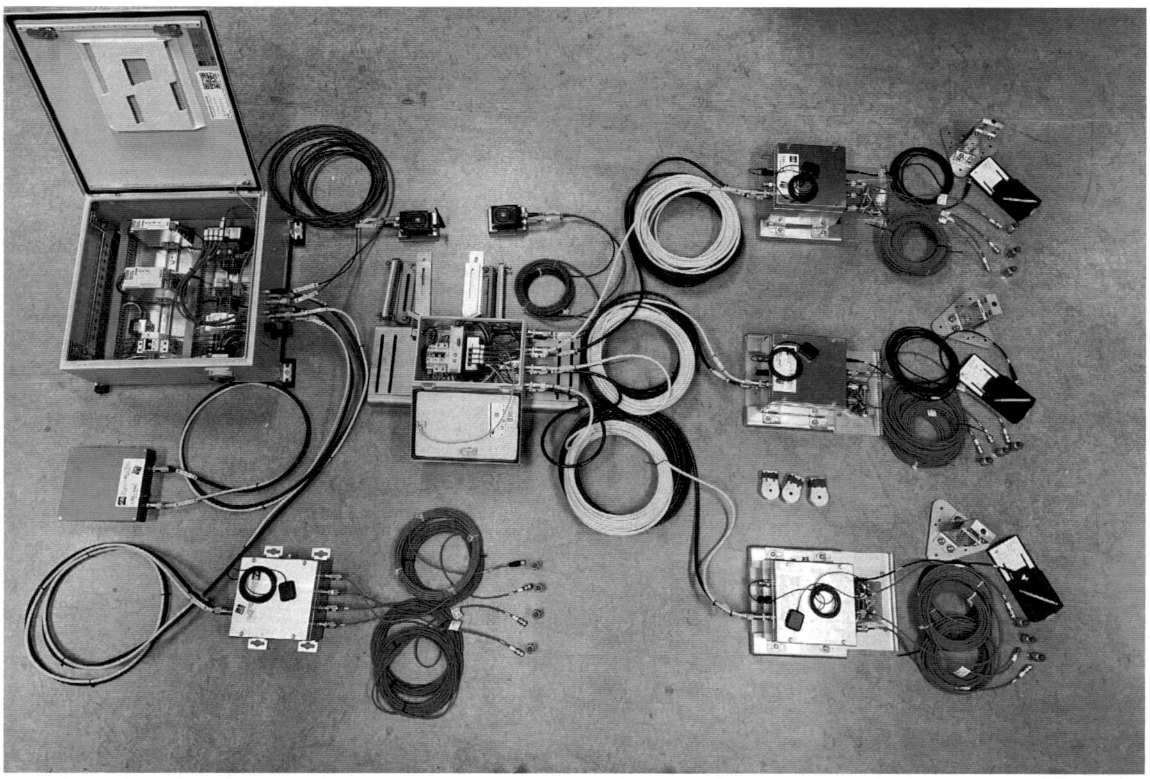

Figure 1: Full measurement setup for offshore wind turbine drivetrain monitoring using 8 high-end accelerometers.

Besides the data acquisition systems, the presence of an edge device also enables the local processing of data to do a first data quality check and low computational cost calculations. This reduces drastically the required bandwidth to transfer data to the centralized servers as well as the computational load on these servers as part of the analysis has been done already.

## 3. Signal Processing

Conventional SCADA-based health monitoring often relies on precise damage criteria for the failure mode of interest, limiting its effectiveness. Vibration-based monitoring complements SCADA-based monitoring, providing insights for fault detection and diagnosis. This paper presents a comprehensive vibration analysis framework that utilizes SCADA data and incorporates statistical, spectral, and cyclostationary analysis techniques. Preprocessing techniques enhance data quality, while SCADA data integration provides contextual information. This hybrid approach enables early fault detection, accurate diagnosis, and informed maintenance decisions.

### *3.1. Data quality assessment*
A critical step in any automated vibration-based industrial monitoring procedure is to evaluate whether a new measurement merits analysis. Ideally, measurements should exhibit quasi-steady-state behavior relative to operational parameters such as instantaneous rotational speed and load. Measurements captured during transient events like run-downs or control actions such as curtailment should be discarded to avoid complicating both the analysis process and the interpretation of its outcomes [12]. To ensure consistently high data quality, an automated quality assessment is performed on the vibration data. This analysis examines both SCADA and vibration data to determine the data's integrity. The SCADA data provides insights into transient events and control actions, while the vibration data verifies the presence of expected mechanical component signatures in both the time and frequency domains. Apart from using the SCADA data as an input for the data quality assessment module, the SCADA data is also used as a monitoring tool itself as well as an input for the machine learning models on the vibration-based condition indicators. An example of the former is the development of a temperature tracker which allows for farm-wide anomaly detection using purely SCADA data [13]. The latter increases the accuracy of the machine learning models by taking into account the operating regime of the wind turbine when predicting the health state of the turbine based on new vibration measurements. For more in-depth information on how SCADA is employed in our pipeline, the reader is referred to Refs. [11, 14, 15].

### *3.2. Speed estimation*
Once the data quality has been established, the next crucial step is to obtain an accurate instantaneous angular speed estimate. If an angle encoder or tachometer is readily available, this speed estimate can be directly derived from the measured pulse signal. In the absence of a high-resolution pulse signal, the SCADA data can potentially serve as an alternative if a coarse speed parameter is available for a gear stage or the generator in the drivetrain. This coarse speed estimate can then be refined using a vibration-based speed estimation algorithm, such as the Multi-Order Probabilistic Approach [16, 17] or the Multi-Harmonic Demodulation [18, 19]. The Multi-Order Probabilistic Approach employs a time-frequency distribution of the vibration signal to create a probability density map of the instantaneous angular speed. The Multi-Harmonic Demodulation, on the other hand, utilizes the phase of multiple harmonics to determine the speed. If no coarse speed can be extracted from the SCADA data, a preliminary rough speed estimation must be performed on the vibration data using a set of parameters that allow for wide-range speed estimation. Once the rotational speed is estimated, the vibration data is then order tracked, resulting in stationary signals in the angle domain. This transformation facilitates further analysis, eliminating the influence of varying time scales due to non-stationary rotating speeds.

### *3.3. Vibration pre-processing*
The angular domain vibrations undergo further pre-processing to ensure effective drivetrain condition monitoring. Typically, monitoring schemes aim to track both periodic (e.g. gear vibrations) and random events (e.g. bearing impulses) [5, 20]. However, these two types of vibrations necessitate distinct analysis approaches. Periodic vibrations can generally be monitored using conventional spectral analysis techniques [21, 22], while stochastic signals

often require a more thorough data cleaning step to achieve an adequate detection [5, 23]. In modern wind turbine drivetrains, this distinction between the two signal types is quite challenging since most gearboxes incorporate at least one planetary gearset. This configuration leads to fluctuating periodic vibrations and a complex transfer path due to the rotary motion of the planet gears and planet carrier. This rotary motion needs to be taken into account in order to optimally monitor such drivetrains. To achieve the necessary data cleaning for stochastic event monitoring, advanced deterministic-stochastic signal seperation techniques can be employed such as cepstrum editing and Discrete/Random Separation (DRS) [23, 24]. These methods effectively correspond to a filter that separates the signal into two distinct signals, one containing the periodic signal content and the other the stochastic signal content. By leveraging these separation techniques, vibration signals can be more effectively analyzed enhancing the pipeline's fault detection capability. By leveraging these separation techniques, vibration signals can be more effectively analyzed enhancing the pipeline's fault detection capability. An example of a wind turbine gearbox vibration spectrum before and after this separation is shown in Fig. 2.

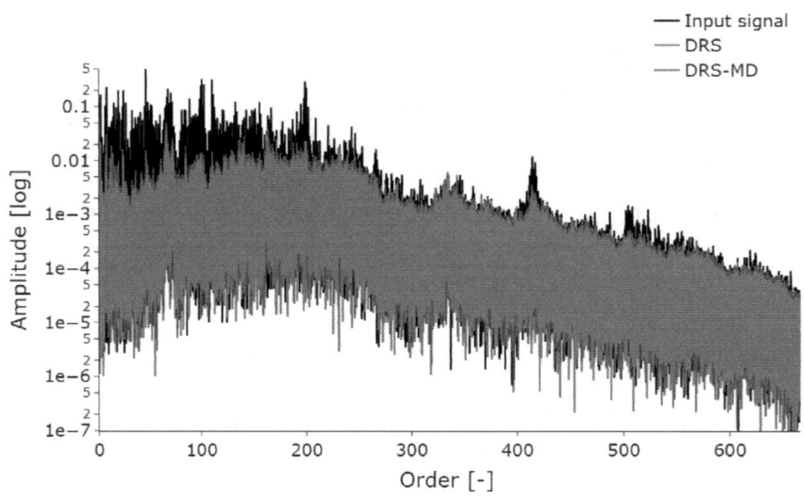


Figure 2: Example of the residual amplitude spectra of a wind turbine gearbox vibration signal after deterministic-random separation with two different techniques, DRS and DRS-MD [25]. The reference order used for the x-axis is the high-speed shaft speed.

Additional to the signal separation, several other pre-processing approaches exist and can be used in conjunction with it. Since many mechanical faults manifest as narrow-band phenomena, their Signal-to-Noise Ratio (SNR) can be boosted through adaptive frequency-based filtering techniques [26, 27, 28]. Consequently, such filtering is often recommended before the calculation of any component-specific condition indicators.

### *3.4. Condition indicators*

Signal processing based condition indicators are then calculated from the cleaned, quasi-stationary vibration data. These indicators typically involve the calculation of numerous time-domain statistics, spectral amplitudes and modulation features. As the measured vibration signals are pre-processed and transformed into multiple signals, the total number of potential condition indicators can increase significantly. This problem is compounded when tracking different frequency bands, as an infinite number of frequency bands can be defined. Consequently, the combination of a rigorous pre-processing phase and a complex state indicator computation step results in a potentially large number of indicators [29, 30, 12]. Manual analysis of such a large set of indicators is impractical. In addition, these indicators are dependent on operating conditions, as they have not incorporated operational information beyond speed compensation.

## 4. Machine Learning

The signal processing techniques highlighted in the previous sections enable the identification of fault trends in wind turbine components. However, they do require the continuous monitoring of indicator trends by domain experts to determine faults. This quickly becomes infeasible when scaled to large wind farms. This work proposes a hybrid fault detection method, which incorporates domain knowledge with the computational power of machine learning to label these indicators as healthy or faulty state simplifies the task for wind farm operators when monitoring a large fleet of wind turbines. This combination of advanced signal processing techniques with machine learning is not common industry practice, which typically only focuses on basic statistics and a standalone spectral and envelope analysis. The machine learning also goes beyond the typical statistical indicator thresholding which is still common practice in large-scale industrial monitoring schemes and leverages the physical insights provided by the signal processing.

The proposed normal behaviour machine learning model is trained on historical healthy data, to learn the normal behaviour of wind turbine components for different sensors. This model can then determine deviations from the learned normal behaviour to label potential faults. The wind turbines operate under different operating conditions, primarily due to variable wind speeds. The changing operating conditions complicate fault detection because a model trained on data from a lower wind speed regime may incorrectly identify a fault when the wind turbine is operating in a higher wind speed regime. To mitigate this issue, an essential operating condition independence step is used to avoid the impact of different operating regimes. K-means clustering is applied to the wind turbine operating data, grouping it into distinct operating regimes. The operating parameters, such as active power and rotation speed, are used to define these operating regime clusters, and each data point is assigned to a specific cluster before determining the health state of its associated indicator. A separate normal behaviour machine learning model is trained for each operating regime, employing the active power and rotation speed as input features to predict the expected value of a particular signal processing indicator as a target variable [31]. For in-depth details about the used normal behavior model, the reader is referred to Ref. [15]. The health state of an indicator at a specific data point is then determined after it has been assigned to its respective operating regime cluster, using the two-sigma rule. If the deviation of the actual indicator value exceeds two standard deviations from the predicted value, it is labelled as a faulty state. Deviations falling between one and two standard deviations are labelled as warning states, while deviations less than one standard deviation are classified as healthy states.

The labelled condition monitoring indicators, categorized as healthy, warning, or faulty, can be effectively monitored by an expert engineer or wind farm operator, without the need for any in-depth vibration analysis knowledge. However, observing these indicators over an entire wind farm, where each wind turbine is equipped with multiple sensors, quickly becomes an unfeasible task. Therefore, a high-level health indicator is generated, by aggregating condition indicators derived from a single sensor. This high-level indicator provides a more comprehensive overview of the health status of the drivetrain as it is easier to interpret and is based on all lower level indicator trends. This method effectively presents a high-level health status overview to wind farm operators, while expert engineers can delve into individually low-level condition indicators to identify the specific type of fault.

## 5. Experimental Results

Vibration signals were acquired from the in-house developed acquisition systems installed at three wind turbines, designated as A, B, and C. However, turbine C showed signs of a potentially faulty gearbox component. This was later confirmed by a borescope inspection to be a bearing failure and led to the eventual decommissioning of the gearbox. Turbines A and B, on the other hand, showed no signs of deterioration and are therefore considered to be healthy. Due to confidentiality restrictions, no details are given regarding turbine identifiers and the date axes are anonymized. Unfortunately, we were only able to install a data acquisition system when the gearbox in turbine C was already showing signs of degradation, so there is no healthy data available for turbine C and the trends for turbine C miss the first 30 days of data compared to turbines A and B. This also means that the normal behaviour models used to alarm on the condition indicators of turbine C were trained on the healthy data from the sensors on turbines A and B instead.

To showcase the difference in condition indicators (CIs) and the performance of the pipeline, both the low-level condition indicator values and the high-level fused alarms are shown. Since hundreds of condition indicators are generated per measurement and per sensor, we can only show a couple of indicator trends to illustrate the pipeline. Figure 3 illustrates the low-level trend of a condition indicator after data cleaning and filtering of the measured vibration signals on turbines A, B, and C for the planetary stage sensor. A low-level condition indicator signifies a feature that is a direct result of the signal processing stage prior to any machine learning or aggregation. The gaps in the shown trends correspond to measurements that were excluded due to the quality assessment algorithm, e.g. during standstill. The color-coding of the indicators denotes the anomaly severity for each indicator value as assigned by the normal behavior model. Intuitively, green and yellow indicate normal and warning operating levels, whereas red signifies strong anomalous behavior. The pipeline successfully identified the anomalous behaviour of turbine C from the outset and continuously triggered alarms for over a month to notify the end user. While condition indicator #1 showcases a strong deviation from healthy behavior for nearly all operating conditions, other indicators exhibit less anomalies. Condition indicator #2, which is the result of a different set of pre-processing techniques, is shown in Fig. 4. While some anomalies are definitely present in this indicator trend, it is not as consistent as for the previous trend shown in Fig. 3.

The observation, that for most cases, not all indicator trends exhibit anomalous behavior, is the reason why the higher-level trends are more practical in an industrial monitoring setting as an end-user does not want to bother having to check all of the hundreds of trends each day, looking for a potential anomaly. In reality, this often translates to some sort of traffic light system where the end-user can see at a glance whether the asset is healthy or not. One level below such a boolean traffic light system are the high-level graphs of the fused alarms shown in Figs. 5, 6, and 7. Only one such trend is generated per sensor per turbine significantly reducing the number of graphs an end-user has to inspect, while still providing some insight and allowing for straightforward interpretation as to which sensor showcases the strongest anomalies and which indicator is most sensitive. If more detailed insights are then required, the end-user can delve deeper into the relevant lower level graphs of individual indicator trends with specific pre-processing combinations.

Figure 5 displays the high-level fused alarm trends of CI#1 for the planetary stage sensor on turbines A, B, and C. This graph essentially aggregates all the low-level CI#1 indicator trends, i.e. with all the different pre-processing techniques, into a single graph per sensor for easy interpretation. This is done by modelling for each low-level condition indicator a deviation from normal behavior and aggregating this into a normalized alarm value, thus representing an overall anomaly score. As can be seen from Fig. 5, the fused alarm trend for CI# is very clear in confirming that turbine C does indeed portray strong anomalous behavior and needs inspection.

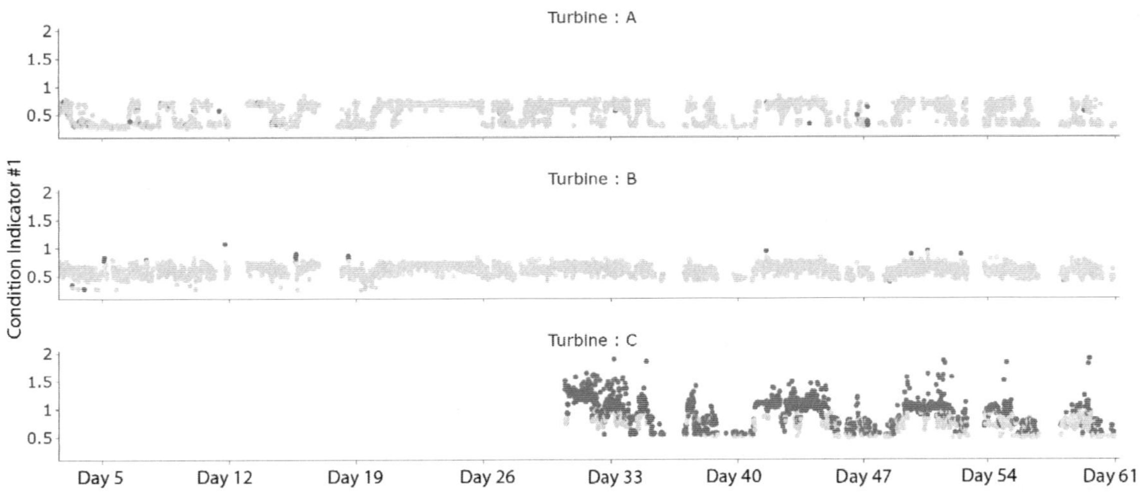


Figure 3: Trends of CI#1 for the planetary stage sensor data on turbines A, B, and C. The color code depicts the anomaly score for that specific indicator value, i.e. the deviation from normal behavior given the operating conditions.

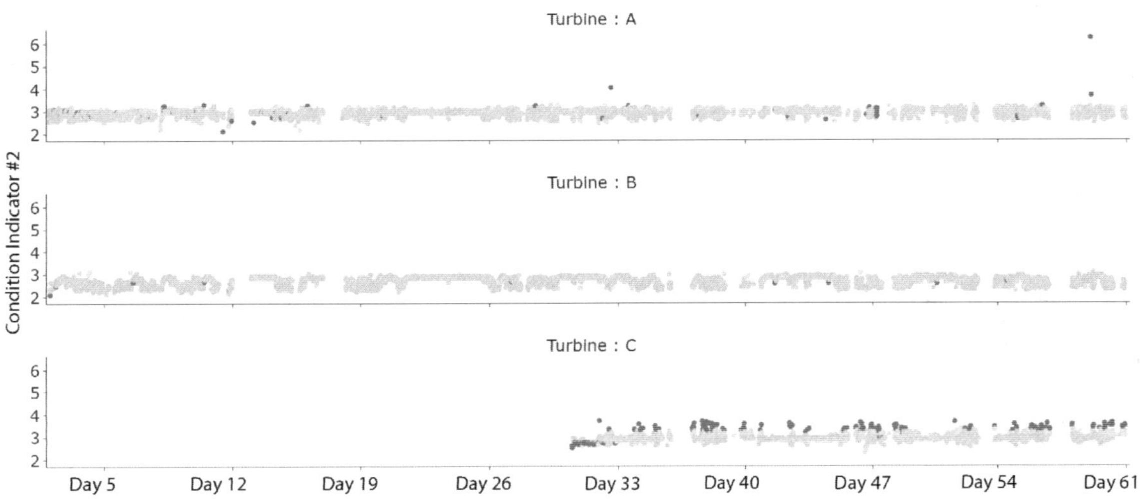


Figure 4: Trends of CI#2 for the planetary stage sensor data on turbines A, B, and C.

However, Fig. 6 which displays CI#2, is much less clear, with the alarm values only being slightly higher than those for turbines A and B. Depending on the type of indicator, an experienced end-user could for example deduce that such behavior can be representative of a fault that is more distributed rather than localized, e.g. a bearing fault that has passed the initial damage initiation phase or gear teeth with distributed wear patterns. Another example graph is shown in Fig. 7 for another condition indicator CI#3 which was calculated after a different set of pre-processing steps compared to CI#1 and CI#2. While this condition indicator alarm trend is less alarming than that of CI#1, it is still significantly more indicative of a fault than that of CI#2 for most operating conditions.

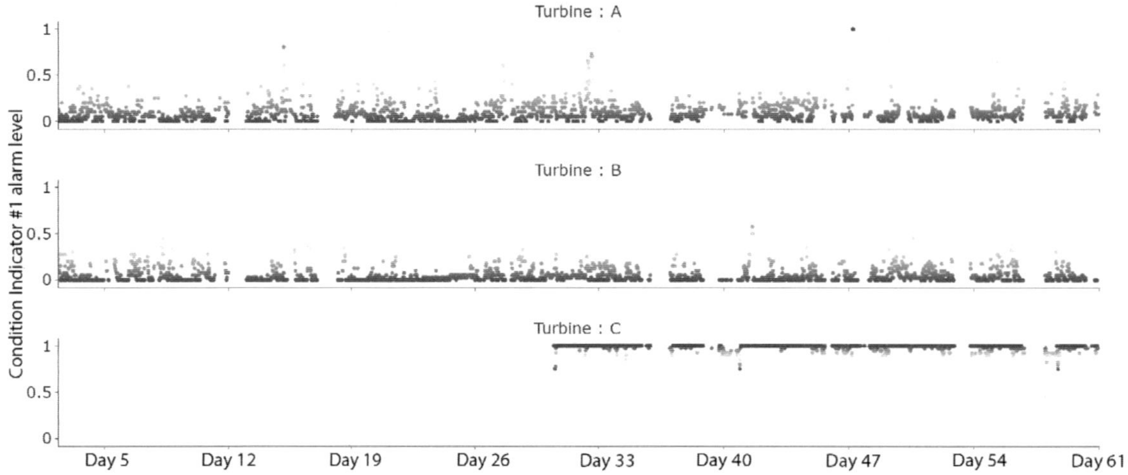


Figure 5: High-level alarm trends of CI#1 for the planetary stage sensor on turbines A, B, and C.

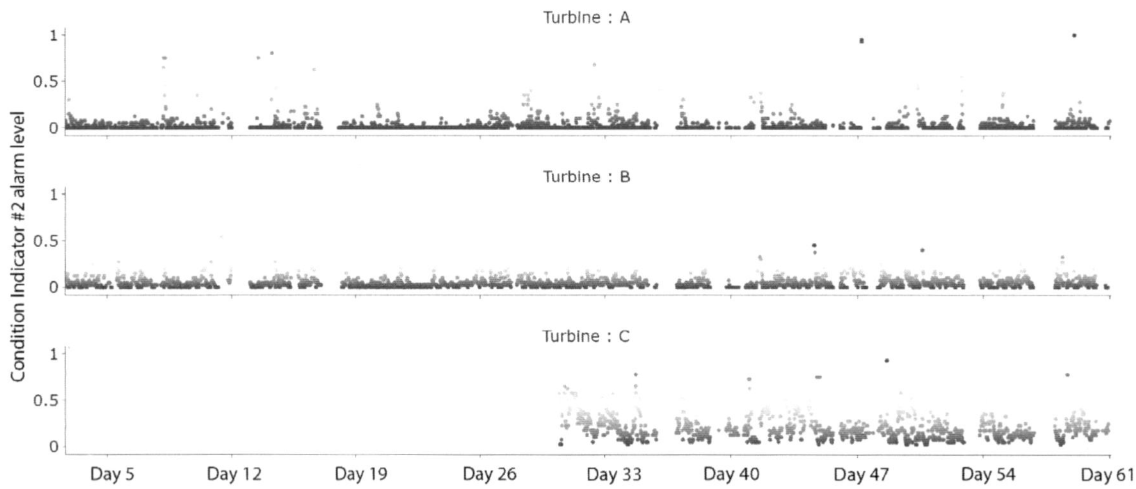


Figure 6: High-level alarm trends of CI#2 for the planetary stage sensor on turbines A, B, and C.

Lastly, a typical data inspection methodology involves pinpointing the physical fault location through a frequency-based analysis. Each mechanical component produces its own set of characteristic frequencies, thus identifying anomalous characteristic frequency peaks gives a good idea of which component is actually faulty. While the proposed pipeline already employs the spectra directly in its normal behavior models, it can still be useful to visually inspect them to corroborate the findings. In Fig. 8 a zoom of the autopower spectra is shown around the fault-related characteristic frequency of turbine C. All spectra are from measurements acquired

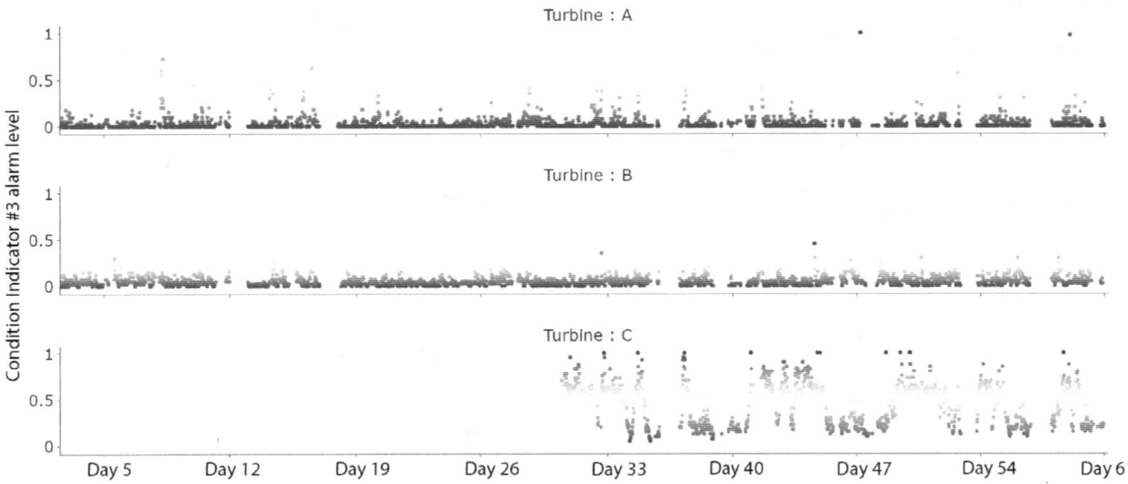


Figure 7: High-level alarm trends of CI#3 for the planetary stage sensor on turbines A, B, and C.

in similar operating conditions and the figures have all been standardized to have the same y-axes and color scales. Since turbines A and B had been measuring for a longer period of time, they have more measurements to compare to as evidenced by the x-axes. Similar to the previous figures, Fig. 8c showcases a clear deviation from turbines A and B due to the presence of a clear discrete peak at the characteristic fault frequency of the bearing. Turbine A does also exhibit some outlier measurements and harmonics spaced at 3P, but these do not trigger any alarms and are considered normal behavior.

## 6. Conclusions

This paper demonstrates the vibration-based condition monitoring framework developed at VUB. This framework goes all the way from sensorization and data acquisition hardware to automated data analysis and alarming. It is currently being used on several offshore wind turbines for drivetrain monitoring. Thanks to the advanced measurement hardware capabilities, very long duration signals can be acquired at high sample rates, high bit rates, and frequent intervals. These qualities are needed for an effective monitoring system of such large slow-rotating machines as you need a sufficiently high frequency resolution for most analyses whilst also being able to detect high-frequency peaks. In this work an overview is given of all the different data processing steps, starting with the integration of SCADA data for an automated data quality assessment step. After this data quality check, the signal processing part of the framework estimates the instantaneous angular speed, employs several data cleaning methodologies, and generates condition indicators from both time and spectral domain signal representations. Thanks to the employed edge device's computing capabilities, a majority of condition indicators can already be calculated offshore, drastically reducing data transfers and centralized server loads. The computed condition indicators are then fed to a machine learning part that focuses on learning the healthy behavior of the turbine in all operating regimes. Automated alarms are then sent to an end-user once the turbine's drivetrain starts deviating from its healthy behavior. This approach enables a straightforward tracking of the condition of an entire wind turbine farm.

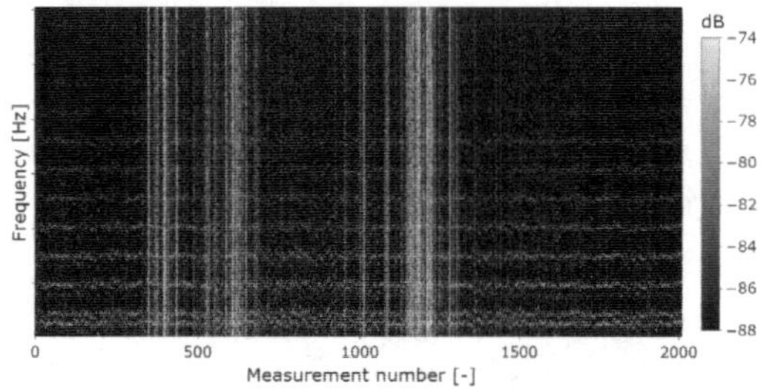


(a) Turbine A.

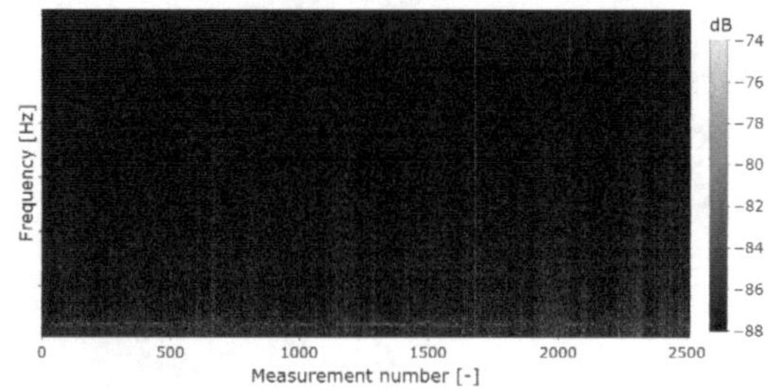


(b) Turbine B.

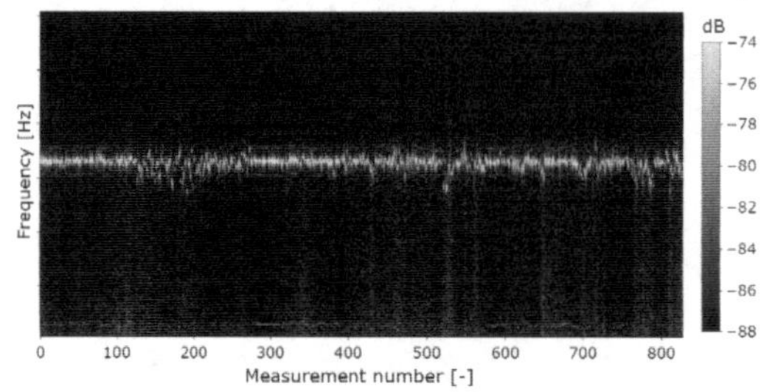


(c) Turbine C.

Figure 8: Autopower spectra over time for turbines A, B, and C. The shown spectra are from measurements in similar operating conditions. The axes and color scales of the figures are also identical to enable a fair comparison.

## 7. Acknowledgments

The authors would like to acknowledge FWO (Fonds Wetenschappelijk Onderzoek) for their support through SBO project Robustify (S006119N) and VLAIO in the context of the Blue Cluster ICON project Supersized 4.0. This research was also supported by funding from the Flemish Government under the "Onderzoeksprogramma Artificiële Intelligentie (AI) Vlaanderen" programme.

## 8. References

[1] Ramirez L, Fraile D and Brindley G 2020
[2] Short W, Packey D J and Holt T 1995 A manual for the economic evaluation of energy efficiency and renewable energy technologies Tech. rep. National Renewable Energy Lab.(NREL), Golden, CO (United States)
[3] Helsen J, Vanhollebeke F, Vandepitte D and Desmet W 2012 Some trends and challenges in wind turbine upscaling *Proceedings of ISMA International Conference On Noise And Vibration 2012* vol 6 (KATHOLIEKE UNIV LEUVEN, DEPT WERKTUIGKUNDE) pp 4345–4359
[4] Office W E T 2015 Statistics show bearing problems cause the majority of wind turbine gearbox failures
[5] Peeters C, Guillaume P and Helsen J 2018 *Renewable Energy* **116** 74–87
[6] Carroll J, McDonald A and McMillan D 2016 *Wind Energy* **19** 1107–1119
[7] de Azevedo H D M, Araújo A M and Bouchonneau N 2016 *Renewable and Sustainable Energy Reviews* **56** 368–379
[8] Lu B, Li Y, Wu X and Yang Z 2009 *2009 IEEE power electronics and machines in wind applications* 1–7
[9] Tchakoua P, Wamkeue R, Ouhrouche M, Slaoui-Hasnaoui F, Tameghe T A and Ekemb G 2014 *Energies* **7** 2595–2630
[10] Perez-Sanjines F, Peeters C, Verstraeten T, Antoni J, Nowé A and Helsen J 2023 *Mechanical Systems and Signal Processing* **185** 109760 ISSN 0888-3270
[11] Jamil F, Peeters C, Verstraeten T and Helsen J 2023 Wind turbine drivetrain fault detection using physics-informed multivariate deep learning *Surveillance, Vibrations, Shock and Noise* (Toulouse, France: Institut Supérieur de l'Aéronautique et de l'Espace [ISAE-SUPAERO]) URL `https://hal.science/hal-04166103`
[12] Helsen J, Gioia N, Peeters C and Jordaens P J 2017 Integrated condition monitoring of a fleet of offshore

wind turbines with focus on acceleration streaming processing *Journal of Physics: Conference Series* vol 842 (IOP Publishing) p 012052

[13] Chesterman X, Verstraeten T, Daems P J, Nowé A and Helsen J 2023 *Wind Energy Science Discussions* **2023** 1–41

[14] Jamil F, Jara Avila F, Vratsinis K, Peeters C and Helsen J 2023 Wind turbine drivetrain fault detection using multi-variate deep learning combined with signal processing *Turbo Expo: Power for Land, Sea, and Air* vol 87127 (American Society of Mechanical Engineers) p V014T37A003

[15] Jamil F, Verstraeten T, Nowé A, Peeters C and Helsen J 2022 *Renewable Energy* **197** 331–341

[16] Leclère Q, André H and Antoni J 2016 *Mechanical Systems and Signal Processing* **81** 375–386

[17] Peeters C, Leclere Q, Antoni J, Guillaume P and Helsen J 2017 Vibration-based angular speed estimation for multi-stage wind turbine gearboxes *Journal of Physics: Conference Series* vol 842 (IOP Publishing) p 012053

[18] Peeters C, Antoni J, Leclère Q, Verstraeten T and Helsen J 2022 *Mechanical Systems and Signal Processing* **167** 108533

[19] Peeters C, Antoni J, Gioia N, Guillaume P and Helsen J 2018 A novel multiharmonic demodulation technique for instantaneous speed estimation *Conference on noise and vibration engineering*

[20] Nejad A R, Keller J, Guo Y, Sheng S, Polinder H, Watson S, Dong J, Qin Z, Ebrahimi A, Schelenz R *et al.* 2021 *Wind Energy Science Discussions* **2021** 1–35

[21] Koukoura S, Peeters C, Helsen J and Carroll J 2020 Investigating parallel multi-step vibration processing pipelines for planetary stage fault detection in wind turbine drivetrains *Journal of Physics: Conference Series* vol 1618 (IOP Publishing) p 022054

[22] Soares M N, Gyselinck J, Mollet Y, Peeters C, Gioia N and Helsen J 2018 Vibration-based rotor-side-converter open-switch-fault detection in dfigs for wind turbines *2018 IEEE International Conference on Prognostics and Health Management (ICPHM)* (IEEE) pp 1–6

[23] Peeters C, Guillaume P and Helsen J 2017 *Mechanical Systems and Signal Processing* **91** 354–381

[24] Daems P J, Peeters C, Guillaume P and Helsen J 2022 *Mechanical Systems and Signal Processing* **165** 108329

[25] Peeters C, Antoni J, Daems P J and Helsen J 2020 Separation of vibration signal content using an improved discrete-random separation method *Separation of vibration signal content using an improved discrete-random separation method* (ISMA 2020) pp 1–9

[26] Peeters C, Antoni J and Helsen J 2020 *Mechanical Systems and Signal Processing* **138** 106556

[27] Kestel K, Peeters C, Antoni J, Leclère Q, Girardin F and Helsen J 2023 *Mechanical Systems and Signal Processing* **199** 110438

[28] Kestel K, Peeters C, Antoni J, Sheng S and Helsen J 2022 Bearing fault detection on wind turbine gearbox vibrations using generalized likelihood ratio-based indicators *Turbo Expo: Power for Land, Sea, and Air* vol 86137 (American Society of Mechanical Engineers) p V011T38A005

[29] Helsen J, Peeters C, Doro P, Ververs E and Jordaens P J 2017 Wind farm operation and maintenance optimization using big data *2017 IEEE Third International Conference on big data computing service and applications (BigDataService)* (IEEE) pp 179–184

[30] Helsen J, Peeters C, Verstraeten T, Verbeke J, Gioia N and Nowé A 2018 Fleet-wide condition monitoring combining vibration signal processing and machine learning rolled out in a cloud-computing environment *International Conference on Noise and Vibration Engineering (ISMA)* pp 17–19

[31] Helsen J, Perez F, Jamil F, Antoni J and Peeters C 2023 Signal processing informed deep learning for failure detection in a fleet of multi-stage planetary gearboxes with limited knowledge about characteristic frequencies *AIAC 2023: 20th Australian International Aerospace Congress: 20th Australian International Aerospace Congress* (Engineers Australia Melbourne) pp 663–668

Journal of Physics: Conference Series **2745** (2024) 012010 doi:10.1088/1742-6596/2745/1/012010

# Development and verification of a discrete event simulation tool for high-fidelity modelling of offshore wind and solar farm decommissioning campaigns

**Simone Mancini, Jesse Bloothoofd, Vinit Dighe, Harald van der Mijle Meijer**

TNO. Energy and Materials Transition Unit. Wind Energy group. Westerduinweg 3, 1755 LE Petten, the Netherlands

E-mail: simone.mancini@tno.nl

**Abstract.** The number of offshore wind turbines to be decommissioned is expected to grow in the next decade but some of the oldest wind farms in the North Sea region are already approaching the end of their service lives. This means that several detailed decommissioning plans will need to be prepared in the coming 2-5 years. So far, only small-scale wind farms consisting of relatively small turbines located in shallow, near-shore, or confined waters have been decommissioned. The limited experience and data available yield high uncertainty in the estimation of costs, duration, and environmental footprint of upcoming offshore wind farm decommissioning campaigns. High-fidelity logistic simulation tools as well as high-level cost models are both needed to support end-of-life decision-making. While several holistic cost models have been proposed, the literature lacks high-fidelity models to support the detailed planning and execution of offshore wind farm decommissioning projects. Given that offshore operations are the greatest source of costs, emissions, and safety risks in decommissioning campaigns, these require special attention when seeking optimization opportunities. In this work, a high-fidelity logistic simulation module for the decommissioning of offshore wind and solar farms has been developed within the discrete event simulation platform UWiSE. The new UWiSE Decommission module enables detailed assessment of alternative decommissioning scenarios taking into account offshore operation complexities like weather and resource dependencies, which are not included or heavily approximated in holistic cost models. The new tool has been verified against publicly available analytical cost model results from recent literature. A short assessment of the impact of weather delays on cost and duration estimates for the considered offshore wind farm decommissioning campaign has also been performed as an example of how UWiSE Decommission may bring value to the planning of offshore decommissioning projects.

## 1. Introduction

As some early commercial-scale offshore wind farms (OWFs) in the North Sea region are approaching their twentieth anniversary, offshore wind decommissioning and end-of-life (EoL) research is gaining more attention. Although the sharp increase in volumes is only expected in the next decade [1], a few detailed decommissioning plans will have to be prepared in the next 2-5 years. In the Netherlands for example, the 108MW Offshore Wind Egmond aan Zee (OWEZ [2]) and the 120MW Prinses Amalia [3] farms commissioned in 2007 and 2008, respectively, are nearing the expiry of their 20-year permits.

According to the 4C Offshore database [4], if single-turbine demo projects are excluded less than ten OWFs were decommissioned worldwide as of March 2023. The complete list is reported in Table 1. These past projects presented moderate technical challenges since most of the decommissioned wind farms consisted of relatively small turbines located in shallow waters near shore, with the only exception of two far offshore demo projects (i.e. Beatrice and Fukushima Forward), which anyhow consisted of two turbines only. Although there are many similarities with installation campaigns, the lessons learned from the decommissioning of the first hundred-megawatt scale OWFs will be key in reducing the cost and environmental footprint of future large-scale campaigns.

Table 1: List of OWF decommissioned worldwide as of March 2023 (excluding single-turbine pilot projects). Source: [4].

| OWF name | Country | WTGs | Farm capacity [MW] | Shore distance [km] | Water depth [m] | Comments |
|---|---|---|---|---|---|---|
| **Beatrice Demo** | UK | 2 | 10 | 25 | 45 | Far offshore wind demo project |
| **Blyth** | UK | 2 | 4 | 1.5 | 7.5 | - |
| **Fukushima Forward** II | JP | 2 | 12 | 20 | 120 | Floating wind demo project |
| **Irene Vorrink** | NL | 28 | 16.8 | - | 2.5 | In confined waters next to a dam |
| **Kemin Ajoksen** I&II | FI | 10 | 30 | 2.6 | 7 | Installed on artificial islands |
| **Lely** | NL | 4 | 2 | - | 3.5 | In confined waters |
| **Utgrunden** I | SE | 7 | 10.5 | 5 | 10 | - |
| **Vindeby** | DK | 11 | 5 | 1.8 | 3 | The world's first OWF |
| **Yttre Stengrund** | SE | 5 | 10 | 2.5 | 7 | - |

The limited experience and validation data available inevitably lead to high uncertainty in the estimation of OWF decommissioning cost [5, 6, 7]. Expected figures ranging from 250 k€/MW to 500 k€/MW have been proposed (e.g. [8, 9]), but the uniqueness of each project's specifications and requirements combined with the high sensitivity to vessel day rates make it hard to come up with generally applicable numbers. For example, foundations and balance of plant components may be removed fully, partially, or even left in situ depending on asset ownership, local regulations, and site-specific decisions. Therefore, estimates from different projects may be hard to compare even for farms of similar sizes and this further contributes to uncertainty. In this context, high-level holistic cost models and high-fidelity logistic tools are both needed to support OWF EoL decision-making and decommissioning campaign planning.

Several holistic cost models for OWF decommissioning have been developed and used in previous works. Two early examples are the EoL management support tool developed within the ODIN-Wind project [10], and the comprehensive cost model from DNV-GL [8], which was used to study the decommissioning requirements for potential offshore wind developments in the Great Lakes for the Ontario Ministry of the Environment and Climate Change [11]. A high-level Excel-based cost model was also developed to study the main operation parameters affecting offshore wind decommissioning activities [5]. Significant cost differences were found when comparing the model results to the estimates reported in the decommissioning programme of the 315MW Sheringham Shoal OWF [12] in the United Kingdom (UK). More recently, an economic assessment framework for OWF decommissioning based on a detailed cost breakdown structure was developed to help wind farm owners and public authorities arrange appropriate financial provisions during the development phase [9]. Finally, extensive research was carried out within the 5-year DecomTools project aimed at reducing the cost and environmental footprint of offshore wind decommissioning campaigns in the North Sea region [13]. A holistic bottom-up cost model was developed in this project [7], and it was then applied to estimate costs and emissions of future decommissioning activities for three medium-size OWFs that are now operational in the North Sea [14, 15]. The study found the emission footprint to be significant for all cases and the cost estimates to suffer from high uncertainty due to the lack of experience and data available.

Similar cost models and decision-support frameworks are particularly useful in early planning phases to either prepare high-level decommissioning plans, or to estimate the amount of guarantees that the wind farm owner should arrange to cover future decommissioning activities. However, as OWFs grow older and approach the end of their useful life, more detailed decommissioning plans have to be prepared by the wind farm owner to start the procurement process or comply with regulation requirements [16]. Since offshore operations are the greatest source of costs, emissions, and safety risks in a decommissioning campaign [9, 17], special attention should be devoted to the design and selection of offshore methods, resources, and activities to identify optimization possibilities when drawing up detailed decommissioning plans. A similar task requires high-fidelity logistic models able to evaluate the effects of weather dependencies (e.g. weather windows, operational limits) and resource constraints (e.g. shift patterns) on offshore activities. The latter are typically neglected by lower-fidelity models, whereas the impact of weather delays is typically included as a fixed percentage penalty applied to the campaign cost or duration (e.g. [7, 11]) relying on empirical assumptions that carry high uncertainty.

High-fidelity logistic modelling tools are frequently applied in the detailed design of OWF installation campaigns. This is not yet the case for decommissioning due to the limited market size and experience. To the best of the authors' knowledge, the only offshore logistic simulations concerning OWF decommissioning were carried out in the DecomTools project, where alternative decommissioning scenarios for the Horn Rev I OWF (in the UK) were evaluated, and the sensitivity of costs and emissions to supply chain logistic parameters was investigated [18]. However, since the focus was on overall supply chain logistics (including onshore activities), offshore operations were aggregated into macro-activities without modelling the detailed method statements and using relatively coarse resolution (24h) for the weather data.

In this work, a new high-fidelity discrete event simulation tool that enables detailed modelling of offshore decommissioning campaigns has been developed aiming to fill the present gap in the literature. The model has been implemented as a new module in the Unified Wind farm Simulation Engine (UWiSE [19]) software platform developed by the Netherlands organization for applied scientific research (TNO) to simulate and optimize offshore wind farms logistics.

The paper is organized as follows: Section 2 describes the new software module outlining its key inputs, outputs, and functionalities; Section 3 presents the results of a first model verification case comparing against estimates from the DecomTools project; Section 4 addresses the main conclusions and plans for future research.

## 2. Model description

The new UWiSE Decommission module has been developed within TNO's discrete-event simulation software platform UWiSE [19], which is introduced in Section 2.1. Then the UWiSE

Decommission module is described in Section 2.2 following its standard modelling workflow.

*2.1. The UWiSE software platform*

The Energy research Center of the Netherlands (ECN), which is now TNO, has been one of the pioneers of wind energy research and it has more than 20 years of experience in developing cost and decision-support models for offshore wind farm installation, and operation & maintenance (O&M). ECN's O&M modelling activities began in the early 2000s and led to the development of commercial tools like the ECN O&M Tool [20], the ECN O&M Cost Estimator (OMCE [21]), and the ECN O&M Access [22]. These models have been very popular within the industry and contributed to the rapid growth of the wind energy sector in the Netherlands and abroad.

In 2017, TNO developed the UWiSE software [19] to bring the main ECN logistic models into a single platform, unifying the software logic, improving the user experience, and streamlining future development activities. UWiSE was created to support the industry in optimizing OWF construction, operations, and maintenance activities to bring down the cost of offshore wind projects. The software combines a discrete-event simulation engine backend with an interactive graphical frontend (fig. 1). The main frontend and backend features are listed below.

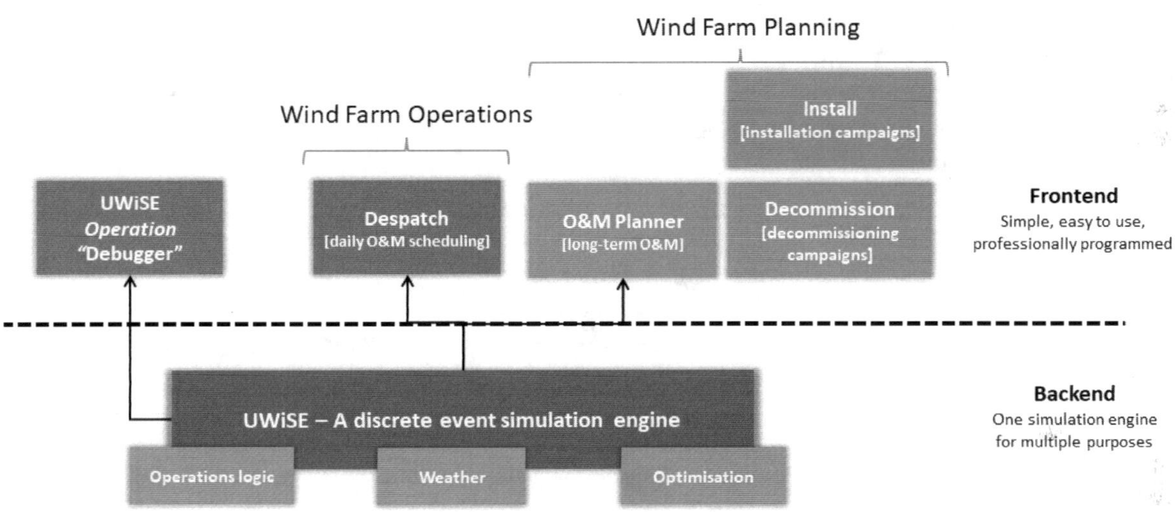


Figure 1: UWiSE software platform overview.

- **UWiSE backend**:
  - discrete-event simulation logic based on next-event time progression;
  - advanced work order scheduling logic accounting for activity dependencies, weather, and resource constraints (e.g. weather windows, operational limits, resource availability, shift patterns, permit periods, etc.);
  - power production logic including models for wind turbine generators and solar photovoltaic (PV) plants, which also take inter-array cable string connections into account;
  - stochastic event generation logic based on the Monte Carlo method to account for component failure modes and non-deterministic task durations;
  - metocean data import routines;
  - Excel output summary and simulation Gantt chart generation (see Section 2.2.3).
- **UWiSE frontend**:

– map-based graphical user interface (GUI) allowing visual review of the detailed input information (see Section 2.2.2);
– process editor whiteboard to create convenient block diagram descriptions of method statements, including task and resource dependencies, non-deterministic durations, operational limits, weather windows, iterations, learning curves, etc. (see Section 2.2.2);
– simulation setup tab to specify the run settings like simulation duration, random seeds, weather years, multi-core parallel execution, etc. (fig. 5b);
– Detailed, zoomable, and color-coded Gantt chart visualization of every simulation scenario.

UWiSE is constantly developed based on outcomes of joint research projects and feedback from industrial clients or partners. Four different application modules have been developed within the platform:

- **UWiSE Install** for high-fidelity modelling of offshore power plant installation campaigns. It can be used to evaluate the cost-benefit profile of alternative installation campaign scenarios (e.g. [23, 24, 25]).
- **UWiSE Decommission** for high-fidelity modelling of offshore power plant decommissioning campaigns. This new module is described in Section 2.2.
- **UWiSE O&M Planner** for high-fidelity modelling of the long-term operation and maintenance activities of offshore power plants. It allows, among other things, to evaluate and compare the operational expenditures (OPEX) associated with different O&M strategies (e.g. [26, 22, 27]).
- **UWiSE Despatch** for the optimal daily scheduling of maintenance activities on offshore wind farms. It can be used, for example, to automate the generation of daily transfer plans for an operating wind farm [28].

Each module in the UWiSE platform may be accessed via specific license agreements or joint development projects. Academic licenses may also be arranged for teaching or fundamental research purposes.

### *2.2. The UWiSE Decommission module*

The newly developed UWiSE Decommission module has been designed to enable detailed modelling of offshore power plant decommissioning campaigns. Being part of the UWiSE platform, the new module benefits from the frontend and backend functionalities described in Section 2.1 allowing activities, resources, and weather dependencies to be taken into account in the discrete event simulations. The current version supports the modelling of worldwide decommissioning campaigns of offshore wind turbines, both bottom-fixed and floating, solar PV systems, and balance of plant (BOP) components including but not limited to foundations, scour protection, substations, export and inter-array cables, and met masts.

UWiSE Decommission users are free to define the scope of their offshore decommissioning campaign. The tool provides them with the flexibility to simulate full or partial removal scenarios of any assets or BOP components depending on their needs. Users are also free to decide what level of detail to include in the method statement description to be provided when preparing a simulation scenario. This flexibility makes UWiSE Decommission useful for a different range of applications, from the high-level modelling of the decommissioning campaign for estimating financial guarantees during wind farm development, to the detailed simulation of decommissioning projects in their planning and execution phases.

As a result, the target user group of UWiSE Decommission is also quite broad and it may include different functional teams within asset owner organizations, who are accountable for the decommissioning of their assets and need to prepare budgets, service tenders, and plans for these campaigns, as well as offshore contractors who have to prepare service bids and detailed technical plans for offshore decommissioning projects. Research institutes (like TNO) may exploit the tool too, for example, to evaluate the impact of alternative processes or innovative equipment (e.g. pendulum versus feeder barge strategies, alternative cranes, low-emission floating heavy lift vessels, etc.) looking for ways to reduce the cost, duration, and environmental footprint of future OWF decommissioning campaigns.

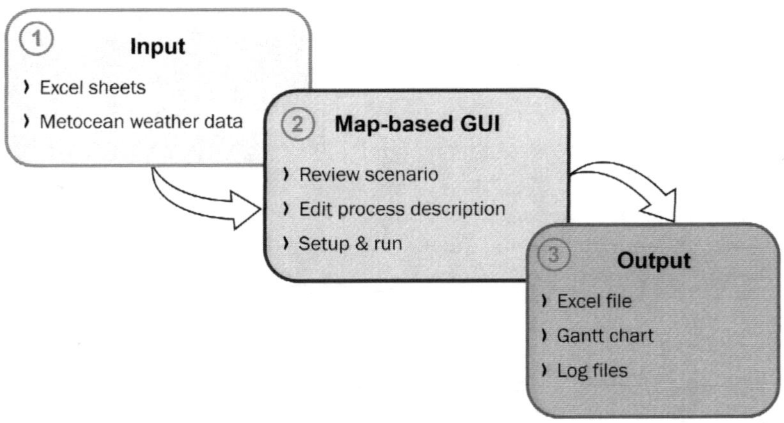


Figure 2: UWiSE Decommission modelling workflow.

Figure 2 depicts the standard simulation workflow for a generic UWiSE Decommission scenario. Such a workflow consists of three phases, and the following sections briefly describe each of them: Section 2.2.1 covers the initial input definition; Section 2.2.2 describes the use of the GUI to review the inputs and define the method statement; Section 2.2.3 covers the different outputs of the tool.

*2.2.1. Inputs* In the first phase of UWiSE Decommission's modelling workflow, two project input files have to be prepared:

(i) **the Excel input file**. For a specific decommissioning project, most input parameters, including the types of assets to be modelled and the resources available, are supplied through an Excel input file. This file comprises 14 sheets, each featuring predefined cells to be filled in by the user. Conceptually, these input sheets can be grouped into three main categories:
- *project information*, including basic project-specific inputs like the farm layout, available ports, metocean locations, project fixed costs, permit periods, and special vessel routes;
- *assets*, covering all input information on the energy-producing assets (i.e. wind turbines and solar PV systems) and BOP, including the component breakdown whose level of detail is entirely up to the user. It is also possible to specify whether any of the assets are still producing electricity at the start of the simulation and what is their power rating (i.e. de-rated assets can be modelled).
- *resources*, covering costs, performance, shift patterns, and availability information for the resources included in the scenario (i.e. vessels, technician teams, equipment, and storage locations).

The complete list of input sheets is given in fig. 3. Here it is noted that not all input fields are mandatory and the user can decide whether to include them or not depending on the scope of the analysis. All cells in the input Excel file are color-coded to distinguish necessary inputs from elective ones.

(ii) **The metocean parameters table**. This table (to be provided in a .csv format) contains the time series of all the weather parameters specified in the "Metocean" sheet of the Excel input file. These parameters can be used for the definition of operational limits (e.g. weather windows and safety constraints) as well as for the calculation of the energy generated by the active assets in the farm (if any). There is no length requirement for the weather input time series. The longer is the list compared to the simulation duration (specified through the GUI), the more starting weather years will be available for selection when tuning the run settings (step iii in Section 2.2.2). Also, as long as the metocean table format is respected, the tool accepts data from any (combination of) weather parameter sources regardless if it is forecast or hindcast data from models or measurements.

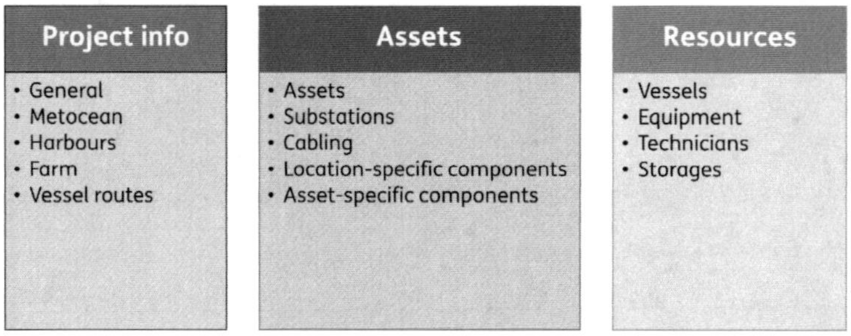


Figure 3: List of all the sheets in the UWiSE Decommission Excel input file.

*2.2.2. User interface: input review and process editor* Once the input Excel file and the metocean parameters table have been prepared, the user uploads them to the GUI and starts the second phase of the simulation workflow, which is carried out entirely through the GUI (alternatively, a command line interface is available for scripting but it shall be considered non-standard workflow and therefore it is not described here).

The UWiSE GUI is organized into 4 tabs (i.e. user view panels): the Projects tab listing all the open projects (top panel in fig. 4); the Inputs tab summarizing all input information provided by the user (left panel in fig. 4); the Process tab that opens a whiteboard where the block-diagram description of the method statement can be entered (see fig. 5a); the Simulations tab where the latest simulation runs are listed and the corresponding outputs can be opened.

This phase of the standard modelling workflow involves three main steps:

(i) **inputs review**. When the two input files of Section 2.2.1 are uploaded to the GUI, an interactive map-based view of the scenario is generated to facilitate the review of the main input parameters (fig. 4). This step is key when generating detailed scenarios as the large amount of data to be entered in the Excel input file makes the task prone to errors.

(ii) **Process description**. The Process tab in the GUI opens an interactive process editor, i.e. an editable whiteboard where block-diagram descriptions of offshore decommissioning campaigns can be conveniently created by adding, customizing, and interconnecting different process blocks (fig. 5a).

There are more than 15 types of blocks available, each one representing a specific activity (e.g. transits, work, transferring of personnel, etc.). To structure the process description,

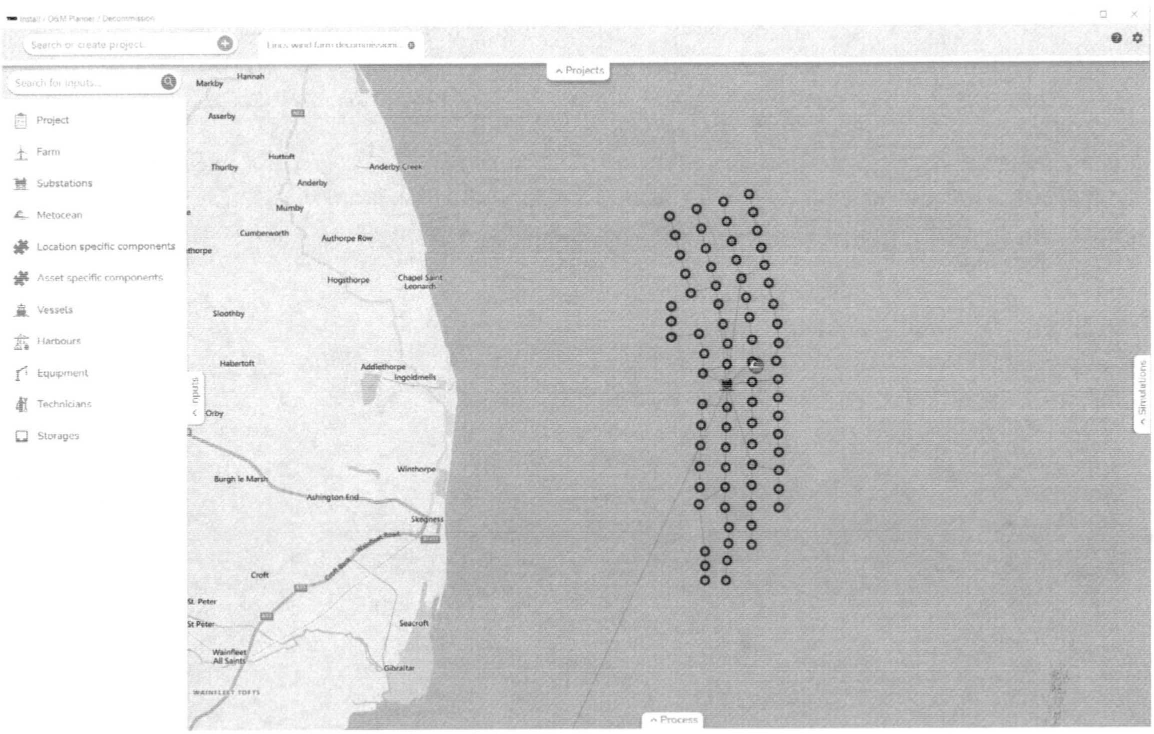


Figure 4: Snapshot of the Inputs tab in the UWiSE GUI for the Lincs OWF decommissioning scenario.

the user can define multiple levels in the process diagram by grouping chains of activities into parent blocks. In fig. 5a, for example, three parent blocks are defined in the first level of the diagram by grouping the wind turbine, foundation, and substation removal activities and providing a detailed description of each campaign using child blocks in lower levels of the diagram.

Blocks are editable and depending on their type (i.e. which activity they represent) a number of block-specific input parameters can be specified, only some of which are mandatory. Typical block inputs are:

- *process step parameters* like activity durations (which can be stochastic or deterministic) or process iterations (repetitive activities for which learning curves can be defined);
- *resource dependencies*, i.e. what entities (vessels, technicians, equipment) among those specified in the Excel input file are involved in the activity;
- *block-specific restrictions* due to weather (e.g. operational limits, weather windows) or permit constraints.

Process interdependencies (e.g. serial or parallel activities) are specified by connecting different blocks. No constraint is set to the number of blocks to be defined in the process diagram. Therefore, any method statement, however detailed, can be described by assembling individual blocks in the process editor whiteboard.

Note that the block diagram defines the scope of the decommissioning campaign to be simulated. Hence, the user has to specify which assets or BOP components are removed by adding specific blocks in the process editor.

(iii) **Tuning the simulation settings**. Once the process description is finalized, the user may tune the final settings before simulating the scenario (fig. 5b). These include:

- *random seed settings*, which are only relevant when non-deterministic activity durations are specified in the process blocks, to ensure results reproducibility;
- *simulation settings*, like the simulation length, the target date for identifying the campaign's latest possible start week, the number of threads for parallel execution, and the weather starting years to be considered in the simulations. The weather years available for selection depend on the length of the metocean parameters table uploaded (Section 2.2.1).

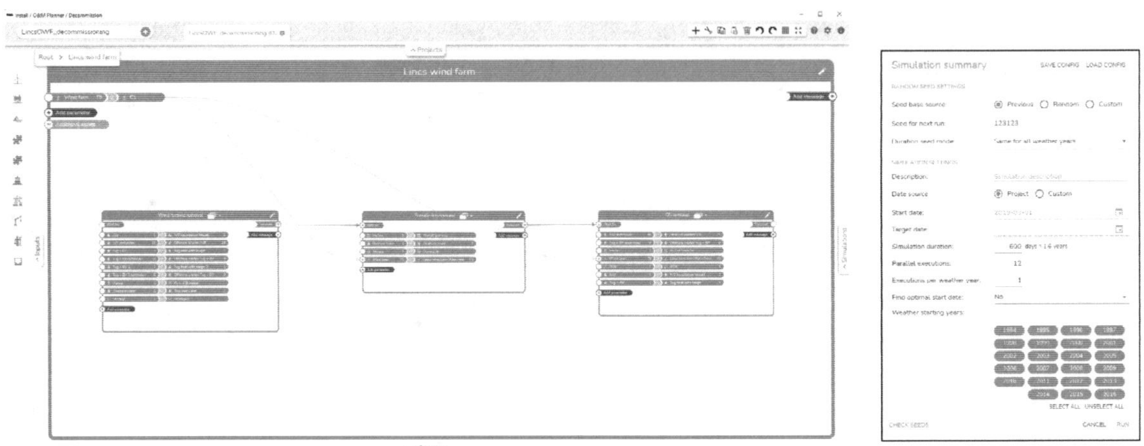


(a) (b)

Figure 5: (a) Snapshot of the first level of blocks defined in the process description of the Lincs OWF decommissioning verification scenario. (b) Snapshot of the simulation settings tab of UWiSE Decommission.

*2.2.3. Outputs* The third and final phase of the standard modelling workflow involves the processing of the model outputs. For every simulation, UWiSE Decommission generates three different types of output, which may or may not be relevant to the user depending on the analysis to be performed. These are:

- **Simulation log files**. For each individual run (i.e. for each random seed and weather starting year considered), a log file recording all discrete events that occurred in the simulation is generated to ensure full transparency and traceability of results. Although log files are human readable and might be useful to check specific simulation details (e.g. for debugging purposes), the large amount of events recorded requires post-processing scripts to extract relevant information from the data stored in these files. For example, the two model outputs described below are both automatically derived from the log files at the end of each simulated scenario. Whenever custom information that is not included in the Excel summary output file or the Gantt chart is needed, this can be obtained by processing the data contained in the log files.
- **An Excel summary output file**. For each simulated scenario, the tool generates a detailed summary output file in Excel format containing an extensive list of indicators including detailed breakdowns of the project costs, duration, resource utilization, and delays. The file consists of three overview sheets providing aggregate results from the full spectrum of simulations performed, plus additional summary sheets characterizing each

weather starting year considered separately. The aggregate quantities in the overview sheets are calculated from all the random seeds and weather years so that average output values are accompanied by statistical parameters (including minimum, maximum, P50, and P90), which define the uncertainty associated with each output quantity.

- **Interactive Gantt charts**. For each individual run (i.e. for each random seed and weather starting year considered), a detailed simulation Gantt chart is generated and can be opened through the GUI. Figure 6 shows a Gantt chart snapshot from one of the Lincs scenario simulations discussed in Section 3. In the GUI, the chart is color-coded to aid the visual interpretation of the events occurring along the timeline, and it can be zoomed from the full campaign duration down to the hourly scale. The tool also allows the timeline to be broken down into individual activities (i.e. specific blocks defined in the process description) for detailed reviews.

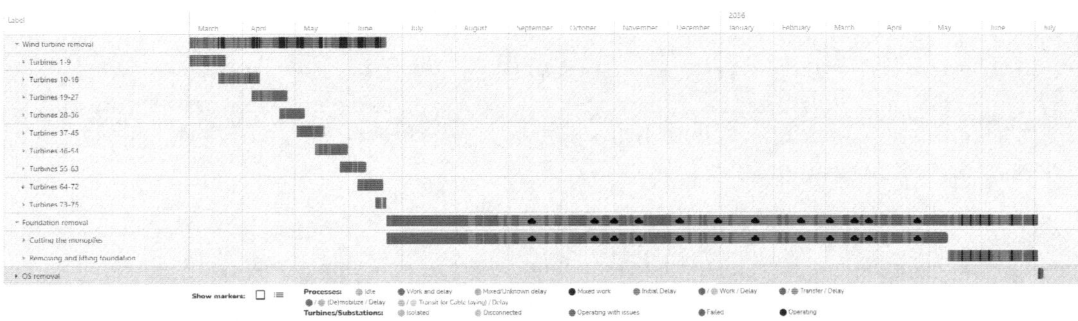


Figure 6: Snapshot of the interactive Gantt chart from one of the weather starting years simulated in the Lincs OWF decommissioning verification scenario.

## 3. Results

The first version of UWiSE Decommission has been released in October 2023. Being a new module, validation and verification of its results is paramount to enable its exploitation. Due to the lack of validation data publicly available, a first model verification case has been considered in this work. By verification it is meant checking that the discrete event algorithm implementation works as intended [29]. In practice, this entails comparing against other numerical model results to verify whether the new UWiSE Decommission module yields the expected output.

A well-documented decommissioning scenario for the Lincs OWF has been considered as the benchmark for this first verification case. The Lincs OWF was installed in 2013 about 8km off the coast of Skegness in the east coast of England and it consists of 75 x 3.6MW turbines with a 120m rotor diameter. The farm is currently operational but a high-level decommissioning scenario for the wind turbine generators (WTGs), foundations, and the offshore substation (OS) was devised and modelled in the DecomTools project [13], starting from the publicly available decommissioning plan prepared during wind farm development [30]. Cost and emission footprint for this scenario were estimated using a holistic analytical cost model that was developed in the same project [7].

Although the official decommissioning plan was used as a guideline, the DecomTools scenario applied several cost and process parameter variations that resulted in substantial differences between the two sets of estimates (>50% variations in vessel utilization and up to ∼30% in costs [15]). The official decommissioning plan did not provide a detailed enough description of the inputs and outputs of the cost model to replicate its results, therefore only the DecomTools project scenario described in [15] has been used as a benchmark for the verification of UWiSE Decommission results.

It is important to point out that the benchmark decommissioning process description from the DecomTools project is rather high-level, in line with the holistic perspective of the analytical model that was used. For example, generic vessels and equipment were considered without accounting for specific technical limitations that might impose changes to the proposed method statements. Moreover, the report considers a vessel transit distance of just 8km as if the decommissioning port was located in the nearest town of Skegness, whereas the closest industrial harbours are located more than 40km away from the farm (e.g. in King's Lynn, Grimsby, or Hull). These simplifications combined with the uncertainty associated with the assumed vessel day rates and activity durations make the results presented in this section solely intended for numerical model verification. The decommissioning campaign costs and duration figures hereinafter reported should be treated with care and not taken as reference values for the Lincs OWF decommissioning. More detailed simulations would be needed to get more accurate and realistic estimates.

The results of the Lincs OWF verification case are discussed below in Section 3.1. Subsequently, Section 3.2 presents a brief sensitivity analysis as an example of how the new UWiSE Decommission module's capabilities may be leveraged to support detailed planning of OWF decommissioning campaigns.

### *3.1. Results verification*

As a first verification case for the UWiSE Decommission module the removal of WTGs, foundations, and the OS of the Lincs OWF have been considered. To maximize consistency between the DecomTools model parameters and the UWiSE Decommission inputs, the same method statements, activity durations, and vessel day rates as reported in [15] have been prescribed in the UWiSE Decommission simulations.

Due to the different fidelity levels of the benchmark model and UWiSE not all differences could be eliminated. An attempt has been made, however, to minimize these differences by neglecting all elective input details that are typically included in UWiSE simulations (e.g. activity durations have been modelled as deterministic, technicians have been assumed always available for work, no costs other than vessel day rates have been considered, permit restrictions have been neglected). The modelling of the Lincs OWF layout (see fig. 4) and of the weather delays were the main outstanding sources of discrepancy between UWiSE Decommission and the DecomTools estimates. While in DecomTools weather delays were modelled as a fixed percentage increase (i.e. 20%) of the estimated campaign duration, UWiSE calculates them based on user-specified metocean parameter time histories (Section 2.2.1), weather windows, and operational limits that can be applied to the activities defined in the block-based process description (Section 2.2.2).

Luckily, for the sake of model verification, such a difference could be avoided by comparing results for the "perfect weather" (PW) case, i.e. a reference case where all weather dependencies are neglected and thus no weather delay occurs.

Figure 7 compares the total duration of each sub-campaign obtained with UWiSE Decommission (plotted in orange) with the DecomTools estimates (plotted in blue). For each sub-campaign, the lower pair of bars represents the PW case.

The largest duration difference of the PW results has been found for the WTGs removal where UWiSE estimates a 1.7% longer campaign duration than the benchmark. This discrepancy can be explained by the fact that the DecomTools calculation neglected the additional duration due to the offloading of the last WTGs at the port, which are critical path activities and thus have been included in the UWiSE Decommission estimate. WTG offloading activities added about 70.5h to the UWiSE estimate, 36h due to the unloading of the last three WTGs plus 34.5h of

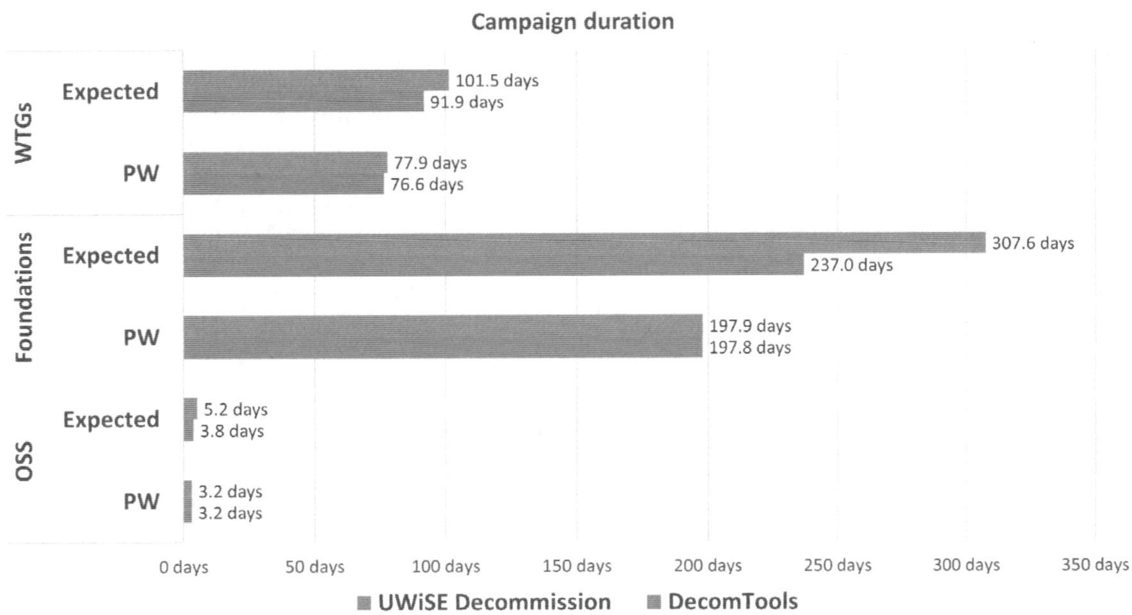


Figure 7: Comparison of the Lincs OWF decommissioning sub-campign duration estimates from UWiSE Decommission and the DecomTools results [15] for the expected (i.e. including weather delays) and perfect weather cases (PW, i.e. excl. weather delays).

lag caused by the unloading of the previous batch of 9 WTGs not being finished before the other barge vessel (BV) with the last 3 WTGs reached the quayside. The remaining difference (~5h) is a result of the transits to/from the port and between individual turbines in the farm, which have been considered in UWiSE and neglected in DecomTools. Similarly, the PW OS removal duration estimated with UWiSE Decommission has been found ~1h longer than the benchmark due to the two transits of the jack-up vessel (JUV) considered in UWiSE only. Finally, the PW foundations removal results have also been affected by similar differences, namely the unloading of the last batch of monopiles at the port and the transits of the JUV being accounted for in UWiSE Decommission only. However, these have been largely compensated by the lower overall transit duration of the offshore support vessel (OSV). The benchmark assumed OSV transits to add 0.25h per WTG (i.e. 18.75h in total) whereas in UWiSE, where transit durations are calculated based on the farm layout coordinates and vessel speed specified in the Excel input file, these accounted for just ~5h in total (as in the WTGs removal campaign).

Besides the PW case, fig. 7 also reports the expected duration of each sub-campaign accounting for weather delays. While the DecomTools estimates are obtained by simply increasing the PW duration by 20%, the UWiSE Decommission bars depict the expected value of twenty-three simulations[1] using different hourly hindcast wind and wave time series from year 1994 to 2016. Local wind speed data was sourced from the ERA5 database [31], and the time histories of significant wave height at the Lincs OWF location were obtained from the North Sea Wave Database [32].

To estimate the impact of weather delays in UWiSE, generic operational limits have been specified for the three vessel types used in the reference decommissioning scenario, i.e. a JUV, an OSV, and two BVs towed by tug boats (TB). The assumed weather limits per vessel type are

[1] The total run time for the whole simulation set (i.e. PW + 23 weather year) was less than 1min on a 2022 laptop featuring a 12-threads processor.

reported in Table 2. Moreover, an arbitrary but customary 50% relative weather window has been applied to all offshore activities (meaning that a 1.5h window of favorable weather would be needed for a 1h activity to start) with the only exception of the monopile cutting, where a fixed weather window equal to the task duration has been applied due to the multi-day span of the activity.

Table 2: Generic vessel weather limits used in UWiSE Decommission simulations.

| vessel type | max wind (@10m) [m/s] | max $H_s$ [m] |
|---|---|---|
| JUV | 16 | 3 |
| OSV | 16 | 2.5 |
| BV + TB | 16 | 1.5 |

Here it is noted that, in UWiSE Decommission, the expected duration of the individual sub-campaigns reported in fig. 7 have been estimated by simulating each campaign independently from the others and always assuming March $1^{st}$ as the start date. This start date was selected as reasonably representative of the whole weather year based on the analysis of Section 3.2.

The expected sub-campaign durations comparison reveals how, for the given site and decommissioning process description, the assumption of a 20% duration increase due to weather delays is likely too optimistic. The UWiSE Decommission estimates based on hindcast weather data exceed the DecomTools report values by ~10.5% for the removal of WTGs, ~38% for the OS, and ~30% for the foundations. Of course, these estimates are strictly linked to the somewhat arbitrary weather limits and windows that have been assumed in the simulations, but it is deemed unlikely that expected duration estimates will reduce substantially once method statements are refined and specific resource limitations are taken into account.

Figure 8: Comparison between the UWiSE Decommission and DecomTools report [15] estimates for the Lincs OWF decommissioning scenario, in terms of: (a) expected sub-campaign costs; (b) use of vessels in the foundations' removal sub-campaign.

The longer durations directly translate into greater expected costs for each sub-campaign as shown in fig. 8a since the resources (only vessels in this simple scenario) have to be hired for longer periods. In particular, the largest cost increase (of 34.6%) compared to the DecomTools estimates is found for the foundations' removal as a result of a significantly higher number of working days expected for the JUV (+43.8%) and OSV (+27.3%) used in that campaign (fig. 8b).

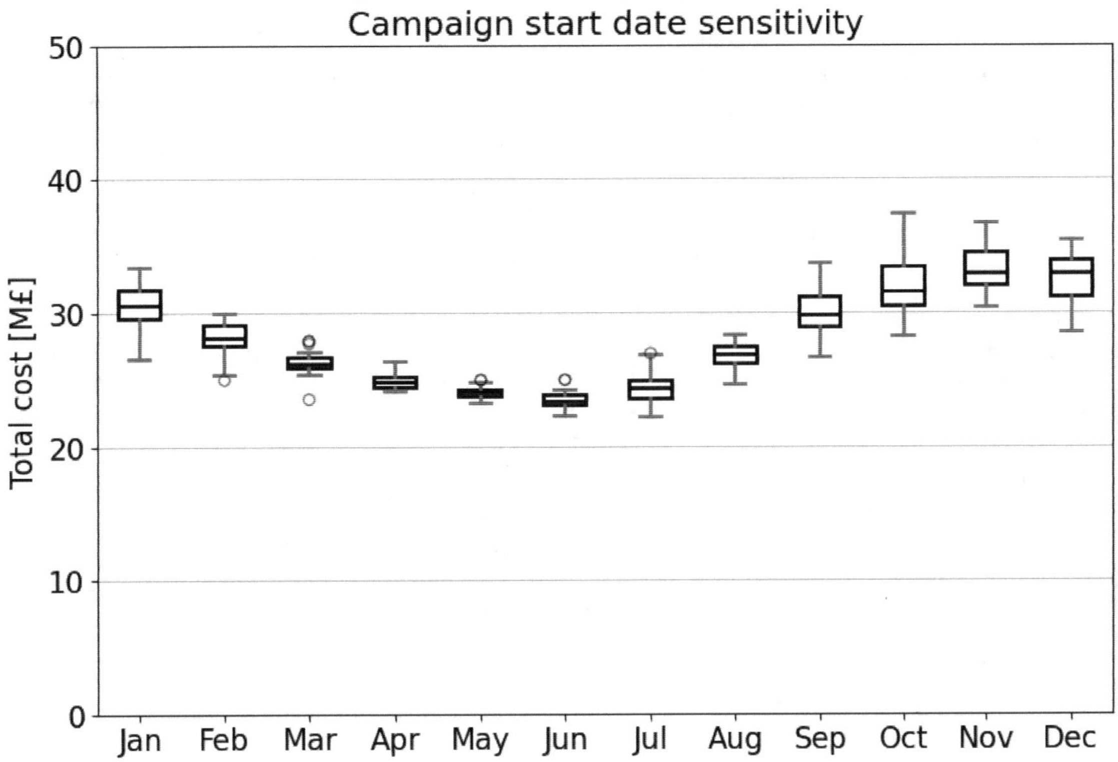


Figure 9: Variation of the total expected cost of the Lincs OWF decommissioning scenario as a function of the campaign start date.

### *3.2. UWiSE Decommission application example: cost sensitivity to the campaign start date*

After having verified the model results for the Lincs OWF decommissioning scenario in the PW case, the sensitivity of total expected decommissioning cost estimates to the campaign's start date has been investigated. This analysis has been included here as a brief example of a possible UWiSE Decommission use case, which is representative of a real-life application.

For this analysis, the same decommissioning scenario as presented in the previous section has been simulated, again considering the same 23 years of local wind and wave data. The only difference compared to the previous results is that the full Lincs OWF decommissioning campaign has been modelled in each simulation, assuming that the three sub-campaigns are performed in series. Note that this has no effect on the PW results but it affects the estimates that include weather delays, as each sub-campaign starts as soon as the previous one finishes and thus it is subject to different weather conditions than the isolated case (when the start date of each simulation was set to March 1st).

The full campaign simulation, consisting of 23 individual runs (one per weather year), has been repeated 12 times to investigate the total expected decommissioning cost sensitivity to different start dates. Each time the simulation start has been set to the first day of a different month.

The statistical results are shown in the box plot in fig. 9, where for each starting month:

- the box encompasses the cost range between the first and third quartile (of the 23 weather years output dataset);
- the line inside the box represents the median of the dataset;
- the circles represent outliers, which are defined as data points more than one inter-quartile range lower than the first quartile or higher than the third quartile;
- the whiskers indicate the dataset range excluding the outliers.

The box plot shows that decommissioning cost estimates for this specific scenario are highly sensitive to the period in which the activities are carried out. In this case, the best starting period is between April and June where not only the expected costs are the lowest, but the variance of the results between the different weather years is also minimal. Low cost but slightly higher variability is found starting in March and July. The last three months of the year are considerably worse as a large portion of weather-critical operations (part of both WTGs and foundations' removal campaigns) occur in winter time and thus cause long and expensive delays. In all three months, the median of the total expected decommissioning cost is about 30% higher that in the most favourable months.

## 4. Conclusions

A new discrete event simulation tool has been developed to support high-fidelity modelling of offshore decommissioning campaigns. The tool has been devised to support asset owners and contractors in the preparation of detailed decommissioning plans for offshore wind and solar farms, as well as research institutes to evaluate innovative concepts to reduce the cost, duration, environmental impact, and risk associated with these campaigns. The new module, named UWiSE Decommission, has been implemented within TNO's offshore logistic software platform UWiSE [19], which uses a next-event time progression logic to minimize computational costs while retaining accuracy.

Unlike the holistic cost models available in the literature, UWiSE Decommission simulates offshore campaigns in time with high fidelity, allowing uncertainties due to weather, activities, and resource availability to be taken into account when estimating costs, duration, and environmental impact of offshore decommissioning projects. In its current version, the module supports the modelling of worldwide decommissioning campaigns of offshore wind turbines, both bottom-fixed and floating, solar PV systems, and balance of plant (BOP) components including but not limited to foundations, scour protection, substations, export and inter-array cables, and met masts. The tool also allows accounting for energy production revenues from any potential asset still operational.

UWiSE Decommission has been designed to be flexible so the user has complete freedom to define the scope of the decommissioning scenario under consideration so that full or partial removal of any asset or BOP components can be simulated. The level of detail can also be tailored to the user's needs, as the offshore method statement to be simulated can be aggregated into large work packages or broken down into individual activities. This makes the tool useful for a wide range of applications, from the high-level modelling of decommissioning campaigns for budgeting to the detailed simulation of offshore decommissioning projects in their planning and execution phases.

As a first verification case, UWiSE Decommission results for a high-level decommissioning scenario of the Lincs OWF in the UK have been compared against publicly available analytical cost model results from the DecomTools project [15]. The comparison of the WTGs, foundations, and OS campaign duration in perfect weather (i.e. neglecting weather delays) showed a good agreement, with maximum differences below 2%. These minor discrepancies are attributed to the modelling of the last offloading activities at port and of vessel transits, both of which differed between the models. A further simplification of the UWiSE scenario (i.e. removing offloading activities from the method statement and condensing the farm layout to a single point location) would eliminate these differences.

Expected decommissioning campaign cost and duration estimates including weather delays have also been compared for the Lincs OWF. While a flat 20% duration increase was used in DecomTools to account for weather uncertainty, the UWiSE Decommission estimates have been obtained by defining generic but customary operational limits and weather windows to the offshore activities, and by considering 23 years of hourly hindcast wind and wave data at the Lincs OWF site. The comparison results suggest that the 20% duration increase assumption is likely too optimistic for this site, with the expected cost estimates obtained with UWiSE Decommission being from 10.5% to 38% higher than the reference, under the assumptions made.

Finally, the sensitivity of the expected decommissioning cost estimates to the campaign start date has been investigated as an example of a possible UWiSE Decommission use case. The results have shown that, under the assumptions made, the total decommissioning campaign costs vary more than 30% between starting on the second or the last quarter of the year. As expected, the uncertainty bounds around the cost estimates also increase significantly in the least favourable starting months.

Although UWiSE is a mature software platform with a long validation and verification track record, the UWiSE Decommission module has just been developed and the validation of its results against field data from actual OWF decommissioning projects is key to enable its exploitation and it is therefore set as a priority for future work. While waiting for the right opportunity to get access to validation data, a more detailed decommissioning baseline scenario will be defined for the Lincs OWF to then estimate the potential impact of innovative process and equipment variations on the overall cost, duration, and emission footprint of this reference campaign.

## Acknowledgments

The work described in this paper has been carried out within the EoLO-HUBs project (https://www.eolo-hubs.eu/) co-funded by the European Union (Agreement No. 101096425 - EoLO-HUBs - HORIZON-CL5-2022-D3-01). Views and opinions expressed are however those of the author(s) only and do not necessarily reflect those of the European Union or CINEA. Neither the European Union nor the granting authority can be held responsible for them.

## References

[1] M. Kruse *et al.*, "Market analysis," report, DecomTools project, 2019.
[2] "Offshore wind farm egmond aan zee (owez)."
[3] "Prinses amaliawindpark." ENECO.
[4] "Global offshore wind farm database and intelligence." 4C Offshore.
[5] E. Topham and D. McMillan, "Sustainable decommissioning of an offshore wind farm," *Renewable Energy*, vol. 102, pp. 470–480, 2017.
[6] G. Smith and G. Lamont, "Decommissioning of offshore wind installations - what we can learn," in *presentation at Offshore Wind Energy 2017*, (London, UK), 6–8 June 2017 2017.
[7] S. Jalili, A. Maheri, and A. Ivanovic, "Cost modelling for offshore wind farm decommissioning," report, DecomTools project, 2021.
[8] G. Smith, C. Garret, and G. Gibberd, "Logistics and cost reduction of decommissioning offshore wind farms," in *presentation at EWEA Offshore 2015*, (Copenhagen, Denmark), 10–12 March 2015 2015.
[9] T. Adedipe and M. Shafiee, "An economic assessment framework for decommissioning of offshore wind farms using a cost breakdown structure," *The International Journal of Life Cycle Assessment*, vol. 26, pp. 344–370, 2021.
[10] J. F. Gjødvad and M. D. Ibsen, *MARE-WINT: New Materials and Reliability in Offshore Wind Turbine Technology. Chapter 22: ODIN-WIND: An Overview of the Decommissioning Process for Offshore Wind Turbines.* Springer Cham, 2016.

[11] DNV-GL, "Assessment of offshore wind farm decommissioning requirements," report, The Ontario Ministry of the Environment and Climate Change, 2016.
[12] S. O. Energy, "Decommissioning programme sheringham shoal," report, Statoil, 2010.
[13] "Decomtools project." Interreg North Sea Region.
[14] S. Jalili, A. Maheri, and A. Ivanovic, "Cost and emission analyses of decommissioning of offshore wind farms using reverse installation method: Cases of lincs limited, gunfleet sands, and horns rev i wind farms," report, DecomTools project, 2022.
[15] S. Jalili, A. Maheri, A. Ivanovic, R. Neilson, M. Bentin, S. Kotzur, R. May, and I. Sünner, "Economic and environmental assessments to support the decision-making process in the offshore wind farm decommissioning projects," *Renewable and Sustainable Energy Reviews*, vol. 190, p. 114080, 2024.
[16] H. Korporaal *et al.*, "Gap analysis: Rules and regulations," report, DecomTools project, 2022.
[17] M. Shafiee and T. Adedipe, "Offshore wind decommissioning: an assessment of the risk of operations," *International Journal of Sustainable Energy*, vol. 41, no. 8, pp. 1057–1083, 2022.
[18] M. Hillers *et al.*, "Simulation analysis: Comparison of various logistical concepts for offshore wind farm decommissioning," report, DecomTools project, 2023.
[19] "Uwise by tno." TNO.
[20] P. Eecen, H. Braam, L. Rademakers, and T. Obdam, "Estimating costs of operations and maintenance of offshore wind farms," in *presented at the EWEC 2007 conference*, (Milan, Italy), 7–10 May 2007 2007.
[21] R. Pieterman, H. Braam, T. Obdam, and L. Rademakers, "Operation & maintenance cost estimator (omce): Estimate future o&m cost for offshore wind farms," in *presented at the DEWEC 2010 conference*, (Bremen, Germany), 17–18 November 2010 2010.
[22] A. Dewan and M. Asgarpour, "Reference o&m concepts for near and far offshore wind farms," Tech. Rep. ECN-E–16-055, ECN, Petten, Netherlands, 12 2016.
[23] A. Dewan, M. Asgarpour, and F. Savenije, "Commercial proof of innovative offshore wind installation concepts using ecn install tool," Tech. Rep. ECN-E–15-007, ECN, Petten, Netherlands, 03 2015.
[24] G. Katsouris and A. Marina, "Cost modelling of floating wind farms," Tech. Rep. ECN-E–15-078, ECN, Petten, Netherlands, 03 2016.
[25] N. Saraswati, "Integrated project logistics and costs calculations for gravity based structure," in *presented at the EERA Deepwind 2019 conference*, (Trondheim, Norway), 16–18 January 2019 2019.
[26] M. Asgarpour and R. Pieterman, "O&m cost reduction of offshore wind farms - a novel case study," Tech. Rep. ECN-E–14-028, ECN, Petten, Netherlands, 07 2014.
[27] N. Saraswati, T. Stehly, A. Dewan, and A. Delmarre, "Operation and maintenance map of u.s. offshore wind farms," Tech. Rep. ECN-E–17-028, ECN, Petten, Netherlands, 11 2017.
[28] C. Stock-Williams and S. K. Swamy, "Automated daily maintenance planning for offshore wind farms," *Renewable Energy*, vol. 133, pp. 1393–1403, 2019.
[29] P. Veers, C. L. Bottasso, L. Manuel, J. Naughton, L. Pao, J. Paquette, A. Robertson, M. Robinson, S. Ananthan, T. Barlas, A. Bianchini, H. Bredmose, S. G. Horcas, J. Keller, H. A. Madsen, J. Manwell, P. Moriarty, S. Nolet, and J. Rinker, "Grand challenges in the design, manufacture, and operation of future wind turbine systems," *Wind Energy Science*, vol. 8, no. 7, pp. 1071–1131, 2023.
[30] Centrica, "Lincs offshore wind farm decommissioning plan," Report LN-D-CE-013-0117-000000-025, Lincs wind Farm Limited, 2010.
[31] "Era5." ECMWF.
[32] G. Lavidas and H. Polinder, "North sea wave database (nswd) and the need for reliable resource data: A 38 year database for metocean and wave energy assessments," *Atmosphere*, vol. 10, no. 9, 2019.

Journal of Physics: Conference Series **2745** (2024) 012011 doi:10.1088/1742-6596/2745/1/012011

# Experimental Development and Field Validation of Rock Anchors for Sustainable Onshore Foundations

**Jan Suchorzewski[1], Mathias Flansbjer[2], Ojas Chaudhari[1] and Natalie Williams Portal[2]**

[1] Department of Infrastructure and Concrete Technology, RISE Research Institutes of Sweden, Gothenburg, Sweden
[2] Department of Chemistry and Applied Mechanics, RISE Research Institutes of Sweden, Gothenburg, Sweden

*E-mail: jan.suchorzewski@ri.se

**Abstract.** The development of an innovative rock anchor prototype manufactured using high strength steel sheets produced locally in Sweden is the core of the PROWIND concept. Steel sheets provide a design freedom to easily manufacture complex geometries, which can be advantageous to enhance the shear force transmission in the bond-length segment of the anchor. The underlying challenge of this concept has been to design a solution which meets the design requirements of today and future technological advancements, all while keeping conventional installation practices in mind. The project followed a 4-step development process: (1) concept analysis and modelling, (2) small-scale prototypes testing and (3) large scale lab-validation and lastly (4) field validation. The performance of the developed rock anchor prototype and grouting material was experimentally quantified on both small and large-scale test specimens and also validated in full scale in the field concerning installation process, proof-loading and maintaining the prestress over time. The PROWIND anchors with the end feature with ribbed design have 4-5 times higher load bearing capacity. The experience from the anchor installation proved that the developed grout and anchors are faster and easier to install. The field test in two different geological conditions has proven that the news design is reducing the required anchorage length to just 1 meter. The restressing of anchors is fully possible with the proposed lock-off solution with a nut. All of those contribute to lower costs of installations and possibly longer service-life.

## 1. PROWIND concept

Onshore wind is facing future technological developments and challenges, as the industry is trending towards larger wind power units. To meet this demand, both stronger and more efficiently designed foundations will also need to be implemented. Another key aspect to consider is the environmental impact of the foundation itself and how this could be further minimized to, in turn, render the onshore wind infrastructure more sustainable as a whole.

The base of an onshore wind turbine is typically anchored using the following foundation solutions: 1) gravity force foundation, 2) cabled foundation and 3) piled foundation. A gravity foundation, the most common solution for onshore wind, can be designed in several different ways to primarily resist the forces from the tower and support reaction forces. One of the most promising gravity foundation designs in terms of sustainability is the rock anchored type. This solution is reported to reduce CO2 emissions by 58 % compared to traditional gravity foundations [1], as a result of less excavation and concrete. The only apparent shortcoming of this solution is that it can only be implemented in locations with suitable geological conditions.

In Sweden, rock anchored foundations make up 70% of SR Energy's onshore wind parks, and there is a continued demand and interest for this solution to be applied in upcoming projects [1]. PROWIND [2], a research project funded by the Swedish Energy Agency (SWEA), has developed and tested in the field an innovative and sustainable rock anchor prototype along with grouting material to support the future technological and environmental challenges being faced by the onshore wind sector.

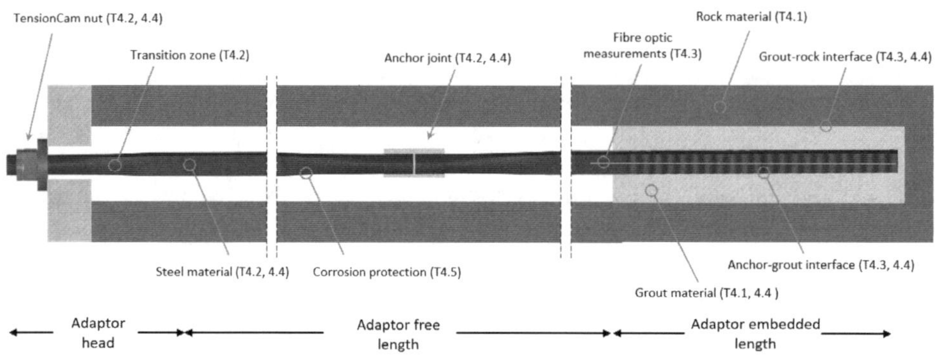


**Figure 1.** The PROWIND rock-anchor concept, components developed and tested in the project.

## 2. Components development

### *2.1 Grout*

All grout products available on the market are based on Ordinary Portland Cement, which has high CO2-emission associated with its production (700-900 kg CO2eq/ton). Ground Granulated Blast Furnace Slag (GGBFS) is a byproduct of the iron industry that can be used as a cement substitute which radically lower carbon footprint (20-40 kg CO2eq/ton).

The application in question is securing of rock-bolt anchors in bore holes for wind turbine installations putting requirements for the grout formulation concerning filling the voids and cracks, and providing a strong bond between the grout and the anchor bolt, as well as between the grout and rock mass. The basic requirement for the grout is also strength development, low shrinkage or small expansion and durability. After a thorough analysis of the benchmark requirements and comparison with commercially available grouts, cement replacement with GGBFS by 40 % weight was chosen for PROWIND formulation.

The results of laboratory testing indicated that the PROWIND grout formulation met the required performance standards for the application in question. The grout had a low marsh cone time (43 sec), high compressive strength (63 and 58 MPa within 28 days for 40 and 60%

Journal of Physics: Conference Series **2745** (2024) 012011 doi:10.1088/1742-6596/2745/1/012011

replacement respectively) and low shrinkage (-0.2 % and 0.7 % without and without expansive agent), making it suitable for use in rock-bolt anchor applications (Figure 2).

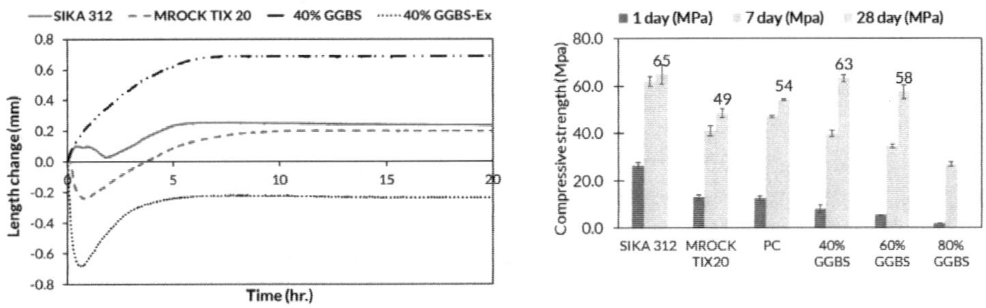


**Figure 2.** Testing of PROWIND expansive grout with GGBFS against commercial products and pure Portland Cement: a) Shrinkage deformations measured by shrinkage cone and b) strength development.

### *2.2 Anchor*

The PROWIND rock anchor is intended to be primarily used to stabilise onshore wind turbine rock foundations designed to transfer the gravity loads to the bedrock. The anchor consists of three main parts: embedded length, the free length, and the anchor head, as presented in Figure 1. The anchors are manufactured by Swedish company Proplate, from high strength steel plates (SSAB Strenx960) by plasma cutting. This process ensures fast cutting, low heat affect, minimum material loss and high dimension precision. Today, the anchors can be manufactured with a length of up to about 15 m, but can be extended with an anchor joint. The design of the anchor head and anchor joint will not be discussed further in this paper. The free length is characterised by a straight rectangular cross-section, while the embedded length is accentuated by a special geometric configuration to optimize the bond behaviour of the anchor with surrounding grout and rock mass. Three different geometric configurations concepts were preliminary evaluated experimentally and numerically in the early stages of the project: a) straight (reference), b) snake and c) diamond.

### *2.3 Prestress force monitoring*

One of the objectives of PROWIND project was to use TensionCam (TC) measurements to enable prestress load measurement in an efficient way at wind power sites. The TC lock nut concept was developed and incorporated into the anchor head design. The nuts were provided with six notches and measurement cavities at three of the six faces. The notches are used to magnify the deformations at the cavities and thereby increase the measurement accuracy. The prestress load is determined based on the diameter reduction of the cavities, obtained from image analysis, as the nuts are loaded. Each cavity has a unique identification pattern that is automatically detected by the system and the results are transferred to a cloud-database. Errors like eccentricity and misalignment can be compensated for in the analysis. The relation between prestress load and cavity diameter is established by a laboratory calibration procedure. Two dimensions of TC nuts were produced and verified within the project, namely M48 and M72.

## 3. Laboratory testing

The laboratory test included the following: Grout tests of fresh and hardened properties at mortar level, upscaling of grout production and adjustments, small scale pull-out tests, large scale pull-out test.

### *3.1 Prototyping in small scale*

The preliminary experimental phase included small-scale pull-out tests to be able to establish a robust testing method, investigate a larger set of parameters and screen through the most promising design solutions for the bond-length segment and grout selection. The test specimens consisted of anchors (diameter 12x12 mm) downscaled bond-length segments cast in an artificial borehole (steel tube with diameter of 150 mm and 300 mm long) with grout. Various embedment lengths and anchor end features (flat, snake shaped, and diamond shaped) were studied to be able to characterize the bond-slip behaviour (different failure modes – anchor failure, bond-slip between anchor and grout and grout splitting) and verify the anchorage requirements. The free length of the anchor, i.e., protruding from the steel tube, was loaded with a tensile force until clear anchor slip was achieved, or failure attained. Moreover, multiple types of grouts (commercial products as a reference, Portland cement-based grout and grout with different slag content up to 80 %) were tested in the lab. Applied load, together with the anchor slip at active and passive side were measured. Further, the strains at the confining steel tube were measured by strain gauges and Digital Image Correlation. Figure 3 presents

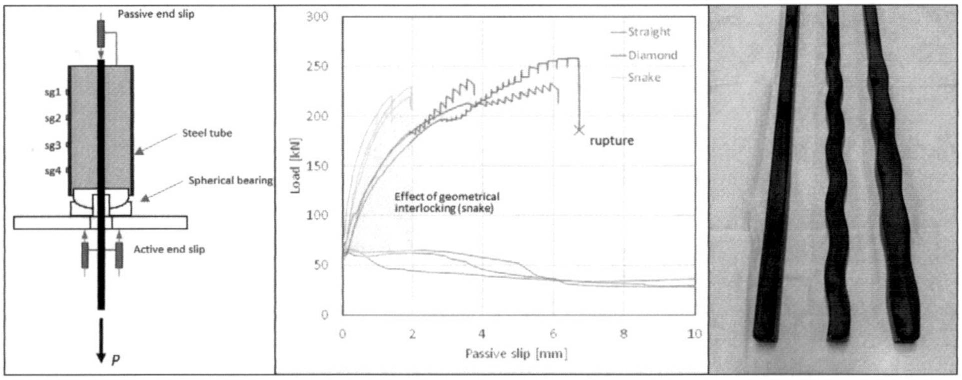


**Figure 3.** The pull-out tests a) scheme, b) Load vs. anchor slip and c) tested anchor designs.

load-slip relationships from pull-out tests on specimens with flat, snake-shaped and diamond shaped anchor ends.

The tests indicated that the specimens with deformed end feature (snake and diamond shaped) had 4 to 5 times higher pull-out capacity than a standard flat end feature due to mechanical interlocking and thereby stress redistribution in the grout. The flat specimens were debonded from the grout, while the failure of the deformed specimens took place in the grout as local crushing and shear cracks along the ribs. Moreover, the strain gauges installed on the circumference of the pipe indicated much higher strain for the deformed anchor specimens, i.e. the steel tube, representing the bore hole, entails positive confining effect on the anchor.

*3.2 Full-scale lab validation*

Large-scale pull-out tests were subsequently performed on the most promising design solution determined from the small-scale tests. The large scale test was performed in steel tubes with a length of 500 and 1000 mm and diameter of 150 mm and with anchor specimen cross-section of 48x48 mm. Again, two types of end features (flat and ribbed) were tested. Applied pull-out load and global deformation were measured, but to gain more detailed measurements, fibre-optics were affixed on the bond-length segment of the anchor specimen to measure the strain distribution during testing. The fibre-optics applied in the laboratory were used as a basis for monitoring strains of a similar test setup in the field validation. Moreover, the lateral pressure built up by e.g., the bond-length segment features during the deformation was monitored via strain measurements on the outside of the artificial borehole.

The result in these large-scale experiments confirmed that the specimen with the ribbed end features exhibits 4 times higher load bearing capacity, see Figure 4 . The fibre-optics measurements were able to monitor curing of grout (temperature changes) and indicated a linear distribution of strains along the anchor.

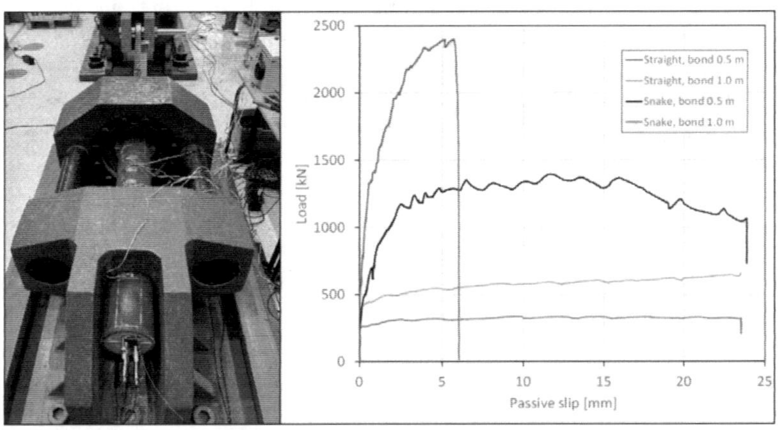


**Figure 4.** Laboratory tests of PROWIND rock anchors by large scale pull-out tests.

## 4. Field validation

The final experimental phase included full-scale field validation of the developed rock anchor solution and grout on the Swedish west-coast. A total of 16 boreholes with a diameter of 160 mm and depth of 10 meters were drilled in two sites with different rock mass types (granite and paragnejs). The standard water loss measurements were performed before placing the anchors. 9 meter long anchors with snake shaped anchorage length of 1, 2, 3 and 4 meters were placed in the boreholes and filled with the developed expansive grout with GGBFS. The anchors were tested after 4 weeks of curing according to the standard EN1537 [3] and then left for 3 months for monitoring of prestress loss in the anchors by fibre-optics and a TensionCam lock nut sensors [4].

*4.1 Site description and installation*

One drill core was extracted from each site for rock mass geological analysis. After that, the surface for the rock adaptors testing was prepared by casting evening concrete to obtain support for prestressing. Standard concrete material was used for this purpose used in wind towers foundation works. Thereafter, the boreholes were drilled through the concrete slabs to the required depth of 12 meters. The boreholes quality was evaluated with BIPS scanning (using digital camera) for detecting internal rock fractures. Moreover, the standard water loss measurement was used to check the possible filtration through cracks in the rock mass. All boreholes exhibited cracks of different size, but the water loss measurements were acceptable for grouting without a risk of grout leakage.

The installation process of the rock adaptors followed current practices given in e.g. SS-EN 1537:2013 [3] where the quality will be monitored via measurements during the grouting. The rock adaptor was prepared before installation with an optical fibre for measuring strain distribution in the anchored fixed section.

*4.2 Testing plan*

A summary of the testing plan is presented in Table 1. At site 1, the anchors were placed in two rows, numbered 1-4 and 5-8, with a free distance of 2 m in both directions, while all anchors were place in one row at site 2, numbered 1-8. The embedment length ($l_b$) of the different anchors was as follows, anchor number: 1 & 5 – 4 m, 2 & 6 – 3 m, 3 & 7 – 1 m and 4 & 8 – 1 m. The anchors were prestressed in the order from 1 to 8 at both sites. The different borehole configurations and loading scheme were chosen to study neighbourhood effects as individual anchors were prestressed.

The anchor load was applied by a hydraulic jack and the load was measured by a pressure sensor. The anchor displacement was measured between the locking nut and the steel plate on top of the concrete slab. In addition, the anchor strains were measured along the embedded length by fibre optics in the same way as in the laboratory tests.

The anchor prestressing followed a standard procedure: 1) loading in steps up to a specified proof load ($P_{proof} = 0.8P_{nom}$) and note any creep deformations, 2) unloading, 3) loading up to lock-off load ($P_{lock\text{-}off} = 0.6\ P_{nom}$ or $0.7P_{nom}$),
4) tighten lock nut, 5) unload the hydraulic jack. The initial prestress load after locking was also evaluated from the load-displacement relationship. The anchor prestress was also measured by the TensionCam system (Figure 6) at regular occasion between the time of prestressing and the load capacity test to determine the prestress loss. In the load capacity test the anchors were loaded to maximum load ($P_{max}$) of about 90 % of the nominal anchor capacity ($P_{nom}$ = 2100 kN) or to a clear slip deformation was observed.

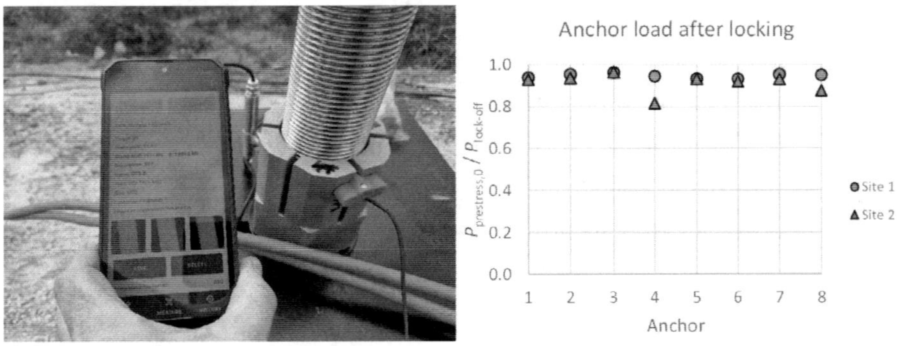


**Figure 5.** load measured at the locking nut by TensionCam mobile device (left) and prestress load in the rock anchors after installation, in relation to lock-off load (right).

**Table 1.** Summary of testing plan.

| | | | Pre-tensioning | | | | Maximum loading | | | |
|---|---|---|---|---|---|---|---|---|---|---|
| | $l_b$ | $l_{free}$ | $P_{proof}$ | $P_{lock\text{-}off}$ | $P_{prestress,0}$ | $P_{prestress,0}/P_{lock\text{-}off}$ | $P_{prestress,1}$ | $P_{prestress,1}/P_{lock\text{-}off}$ | $P_{max}$ | $P_{max}/P_{nom}$ |
| | [m] | [m] | [kN] | [kN] | [kN] | [-] | [kN] | [-] | [kN] | [-] |
| *Site 1* | | | | | | | | | | |
| S1A1 | 4 | 6 | 1716 | 1470 | 1379 | 0.94 | 1325 | 0.90 | 1847 | 0.87 |
| S1A2 | 3 | 7 | 1729 | 1460 | 1392 | 0.95 | 1350 | 0.92 | 1883 | 0.89 |
| S1A3 | 2 | 8 | 1743 | 1450 | 1393 | 0.96 | 1389 | 0.96 | 1892 | 0.89 |
| S1A4 | 1 | 9 | 1714 | 1450 | 1367 | 0.94 | 1351 | 0.93 | 1881 | 0.89 |
| S1A5 | 4 | 6 | 1724 | 1260 | 1173 | 0.93 | 1152 | 0.91 | 1896 | 0.89 |
| S1A6 | 3 | 7 | 1734 | 1250 | 1162 | 0.93 | 1092 | 0.87 | 1900 | 0.89 |
| S1A7 | 2 | 8 | 1722 | 1255 | 1196 | 0.95 | 1183 | 0.94 | 1880 | 0.88 |
| S1A8 | 1 | 9 | 1737 | 1238 | 1193 | 0.96 | 1197 | 0.97 | 1904 | 0.90 |
| *Site 2* | | | | | | | | | | |
| S2A1 | 4 | 6 | 1763 | 1480 | 1374 | 0.93 | 1311 | 0.89 | 1896 | 0.89 |
| S2A2 | 3 | 7 | 1775 | 1460 | 1366 | 0.94 | 1308 | 0.90 | 1912 | 0.90 |
| S2A3 | 2 | 8 | 1765 | 1470 | 1415 | 0.96 | 1330 | 0.90 | 1900 | 0.89 |
| S2A4 | 1 | 9 | 1721 | 1420 | 1157 | 0.82 | 1090 | 0.77 | 1810 | 0.85 |
| S2A5 | 4 | 6 | 1751 | 1280 | 1191 | 0.93 | 1183 | 0.92 | 1914 | 0.90 |
| S2A6 | 3 | 7 | 1772 | 1260 | 1158 | 0.92 | 1137 | 0.90 | 1885 | 0.89 |
| S2A7 | 2 | 8 | 1751 | 1260 | 1171 | 0.93 | 1151 | 0.91 | 1887 | 0.89 |
| S2A8 | 1 | 9 | 1601 | 1265 | 1106 | 0.87 | 950 | 0.75 | 1883 | 0.89 |

*4.3 Results*

From the results in Figure 5, one can observe that the initial anchor prestress load after locking were as high as about 95% of the lock-off load in all cases, except for the anchors with the shortest embedment length of 1 m, number 4 and 8 at Site 2. The latter two also exhibited some slip during the prestressing. Furthermore, the lock nut force sensor and fibre-optics indicated (Figure 6) that the prestress losses, from the occasion of prestressing to the load capacity tests, were below 5% for all anchors.

At the load capacity test, only small irreversible anchor slip was registered for all anchors at the maximum prescribed load, except for the anchors 4 and 8 at Site 2, were major slip occurred already at lower load levels due to the insufficient bond. Figure 6 shows the strain distribution from the fibre-optic measurement for anchor with different embedment length (anchor 5-8 at Site 1). In general, the strain gradient is rather linear over the embedment length in the prestressed stage and in the same range for the anchors with 2-4 m embedment lengths, while the gradient is larger in the for the shortest embedment length of 1 m. During the load capacity tests, the strain gradients become more non-linear and shifts towards the anchor end.

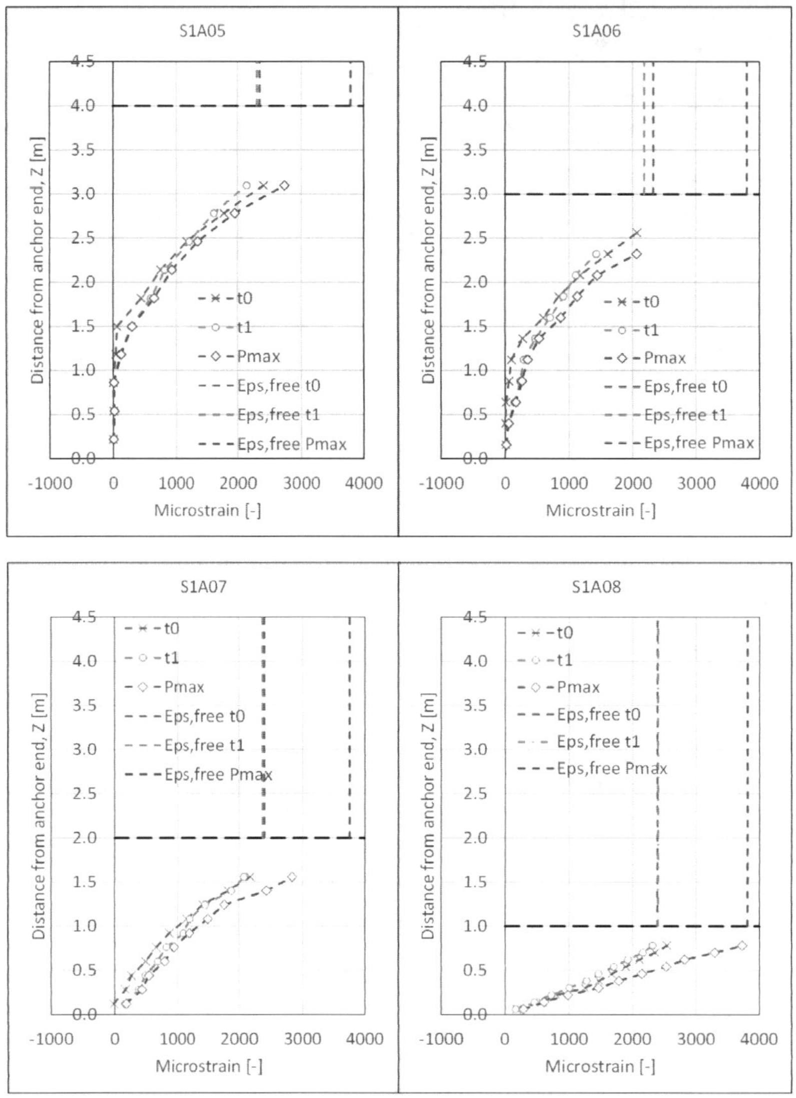


**Figure 6.** Example of strain distribution at rock anchor with different embedded length.

## 5. Conclusions

The PROWIND rock anchors development and validation proved that:

- The anchors with the end feature with ribbed design have 4-5 times higher load bearing capacity which leads to possible shorter anchorage, reducing the costs and time of foundation construction.
- The experience from the anchor installation proved that the developed grout and anchors are faster and easier to install. Moreover, the prestressing solution used enables adjustment of the prestressing of anchors after installation.
- The field test in two different geological conditions has proven that the new design is reducing the required anchorage length to 1 meter. The restressing of anchors is fully possible with the proposed lock-off solution with a nut. The lock nut force sensor and fibre-optics indicated that the prestress losses were minimal (below 5%). All of those contribute to lower costs of installations and possibly longer service-life.
- The tested design is limited to 10 meter length of the anchor which is a drawback, however the consortium continues the development of a technical solution to provide much longer anchors in the near future.

## 6. Acknowledgments

The research presented in this paper has been financed by The Swedish Energy Agency (Energimyndigheten) within the project PROWIND "PROWIND rock-adaptors for wind power" (51382-1) in years 2022-2024.

Contribution of the PROWIND project partners: BESAB – Rock drilling in filed test, MBCC Group – provision of additives for grout development, PROPLATE – production and design of anchors, Rabbalshede Kraft – provision of test site, RISE – coordination, testing and design support, SREnergy – expertise in rock-anchor solutions, SSAB – production of steel sheets, SWECEM – provision of GGBFS and premixing of grout, TensionCAM – Testing of prestressing force, VEIDEKKE – installation and loading anchors in the field.

## References

[1] SR Energy AB, private communication, August, 2023.
[2] PROWIND. [Online]. Available: www.prowind-project.com
[3] Standard SS-EN 1537:2013 Execution of special geotechnical works – Ground anchors. CEN 2013
[4] TensionCam sensor. [Online]. Available: https://tensioncam.com/

# Full scale testing of fatigue resistant composite joints for offshore wind Jacket and Floating structures

**M Pavlovic[1,2], M Koetsier[1], J Yang[1], V Mylonopoulos[2], M Wolters[2], M Segeren[2], J Van Wittenberghe[3], P Thibaux[3] and M Vanderschueren[3]**

[1] Department of Engineering Structures, Delft University of Technology, Stevinweg 1 2628CN Delft, The Netherlands
[2] Tree Composites B.V., Paardenmarkt 1A 2611PA Delft, The Netherlands
[3] OCAS N.V., Technologiepark Techlane A2, building 48, 9052 Zwijnaarde, Belgium

E-mail: m.pavlovic@tudelft.nl

**Abstract.** Complex welds in the joint region reduce the fatigue resistance of structural joints of circular hollow sections with factors that results in increased wall thicknesses of the tubes and overspending of steel in the multi-membered part of the jacket and floating support structures for offshore wind turbines. The wrapped composite joint is a breakthrough technology utilising bonding and fatigue resistant composite material to connect steel tubular members instead of welding. The main technical advantage is that the superior fatigue life of wrapped composite joints. Full-scale wrapped composite joint is tested by cyclic out-of-plane bending loading in resonance-based Cronos testing rig at OCAS N.V. in Belgium. Results reveal exceptional fatigue life of wrapped joints to combined loading when compared to welded joints, showcasing a potential that can radically reduce steel use for offshore structures.

## 1. Introduction

Foundations for offshore Wind Energy Convertors are dominantly made with monopiles. Jackets is an alternative foundation utilizing structural efficiency of Circular Hollow Sections (CHS, a.k.a tubes) in multi-membered structure, but is facing challenges in terms of costs and manufacturing speed. The dominant design criterion for foundations is fatigue. In the case of jackets, large stress concentrations factors at the tubular joints are requiring larger member thickness. If the stress concentrations at the welded tubular joints could be reduced, a significant thickness reduction could be applied, up to 50%.

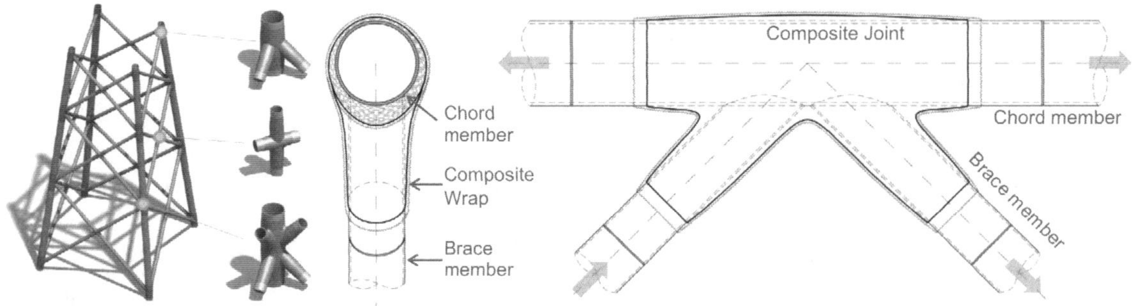


**Figure 1.** a) Visualization of offshore wind turbine jacket supporting structure with wrapped composite joints; b) Schematics of a wrapped composite joint type K-joint.

If the fatigue resistance for these joints can be increased, the wall thicknesses of the legs (chords) and brace members can be reduced and can sections in joint regions can be omitted, resulting in significant lighter structures. If the complex joints can be prefabricated the production rate of the jackets structures can increase by 50-100%.

TU Delft and Tree Composites, together with their partners, are investigating the behavior of wrapped composite joints to solve this challenge for lattice type welded structures. The breakthrough of the concept is that the load is transferred through a dedicated composite wrap and not trough the small section of the weld. The brace members (diagonals) are bonded to the chord members (legs) through a coupling formed by a sufficiently large composite wrapping, see figure 1.

Within the WrapNode-I project (2021-2024) financed by Dutch Ministry of Enterprise (RVO) the full-scale wrapped composite joints are tested and understood under influence of multi-axial loads and offshore environment and the technological and economic feasibility in offshore jackets is validated on critical points. Project is led by TU Delft conducting experimental and numerical research with Tree Composites B.V. as largest executing partner developing engineering and production of composite joints, and other contributing partners: HSM offshore, Enersea, AOC, Siemens Gamesa Renewable Energy, Parkwind, Shell and Vattenfall. Project aim is to develop wrapped composite joint that in full-scale and offshore environment will have static and fatigue resistance exceeding the resistance of the steel jacket tubes that it connects.

The paper presents the first full-scale tests of wrapped joints that was recently performed. First, the specimen design will be presented, followed by the experimental set-ups. Then the results of the tests will be presented and discussed. This paper also presents estimation of fatigue life of an equivalent welded joint for comparison reasons, as well as overview of validation of wrapped composite joints for static loads and multi-axial behavior and outlook for future research.

## 2. Specimen design and manufacturing

One full-scale X-joint with a 90° angle, see figure 2, was prepared to investigate the fatigue performance of a wrapped joint. The specimen length was 12 m. The braces were chosen to be sufficiently long to obtain an adequate resonance frequency and limit the shear forces. Glass Fibre Reinforced Polymer (GFRP) composite wrap was used to make the composite joint. There was no welding involved in connecting the steel tubes. The composite wrapping length was close to 3m.

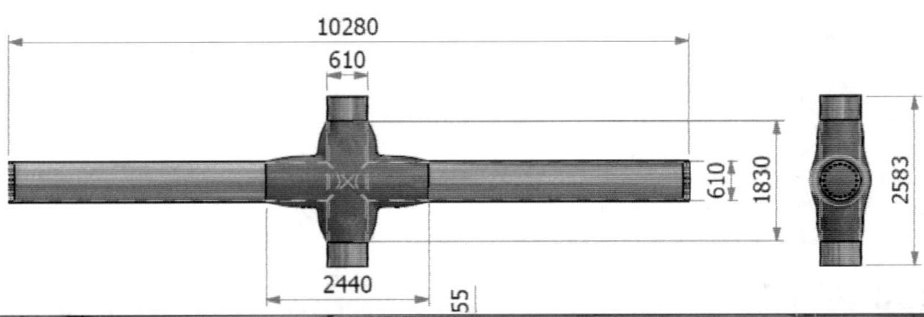


**Figure 2.** Full-scale wrapped composite joint geometry.

**Figure 3.** Full-scale specimen of wrapped composite joint after production presenting the X-joint of braces in offshore wind turbine jacket supporting structure.

The full-scale 90° wrapped X-joint was designed based on a case study design for a representative jacket foundation supporting a 14MW turbine and located in the North Sea, off the coast of the Netherlands in a water depth of 40 meters. The structure uses tubes of 610 diameter and 16mm thickness for the braces and chord. To enhance the fatigue life of the specimen steel tubes and prevent fatigue failure prior to damage of the composite wrap, the design was modified, using ø610x25mm tubes. Figure 2 illustrates the geometry of the full-scale sample. The maximum composite thickness at the center of the joint was approximately 150mm. Figure 3 shows the full-scale specimen after production.

## 3. Experimental set-up

The fatigue test was performed in the Cronos test rig at OCAS, Belgium. The principle of loading is to excite the specimen by a loading at a frequency close to the specimen eigen frequency. Shakers are positioned at the specimen's ends generating the excitation force. More details about the development and operation of the of the set-up and utilization in testing of welded tubular joints can be found in [1], [2] and [3]. The orientation of the excitation was chosen to be out-of-plane, which is controlled by selecting the direction of the shaker's excitation, see figure 4a and figure 4b. The amplitude of the loading was controlled by measuring the strain amplitude measured on the braces and adjusting accordingly the excitation frequency. The specimen is supported at the positions where the eigen mode is not causing any displacement. In this specific case, the test was performed with an out-of-plane excitation.

The specimen denominated X90-Fs-FLS was subjected in total to 2.8 million cycles. The first 1 million cycles of damage equivalent load (DEL) corresponds to 25 years operation of the turbine and the support structure subject to wind and waves. The latter 1.8 million cycles were tested with tripled load (corresponding to storm load level), see figure 4c for loading regime that was applied. The trippled load was applied with the motive to push the join to the limits and impose damage that can later be used to estimate the total life and validate numerical models for damage prediction in a later stage of the research program.

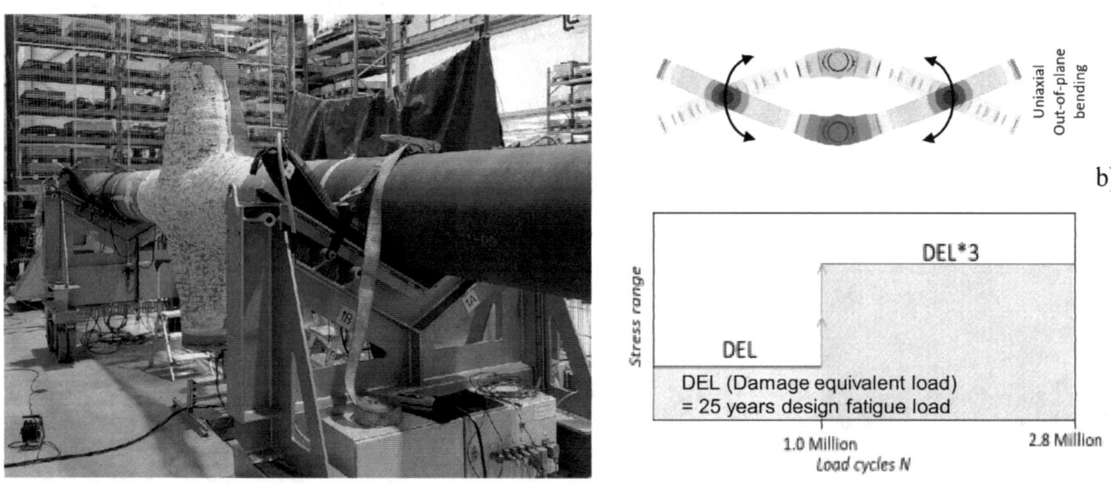


a) c)

**Figure 4.** a) Test setup of full-scale joint specimen X90-Fs-FLS subjected to out-of-plane bending fatigue test; b) Loading priniciple; c) Load history of the fatigue tests given stress range of DEL and tripled DEL.

### *3.1. Damage Equivalent Load*

The constant amplitude damage equivalent load (DEL) on the joint was calculated based on load histories generated for the full-scale jacket design in 40m deep water supporting 14MW turbine. The load histories result in combination of variable joint loads in form and axial forces and in-plane and out-of-plane bending moments. However, only bending load can be applied in the Cronos set-up. To this end, the relevant combination of the design FLS axial forces and bending moments on the join were converted to a constant amplitude out-of-plane bending load that has equivalent damage as the variable amplitude multiaxial load combination. The damage equivalence was based on strain energy release rate (SERR) at the debonding crack front inside the joint which was calculated using Finite Element Models and VCCT approach in Abaqus software package [4]. For the analyzed joint the resulting constant amplitude damage equivalent load (DEL) applied through out-of-plane bending moment was found to be 250 kNm for 1M cycles.

### *3.2. Measurements set-up*

Fatigue experiments on full-scale were supported by real-time 3D Digital Image Correlation (DIC) measurements. DIC was utilized for global deformation and surface strain measurements. Two GOM Aramis adjustable systems and one GOM Aramis SRX system equipped with two 12- megapixel cameras were used. The 3D DIC measurement technique indirectly identifies debonding between the steel and FRP by measuring surface strains, as suggested by He and Pavlovic [5].

The adjustable systems were positioned to capture each brace, while the SRX system was placed to measure the chord. Measurements were taken at the maximum and minimum bending moments during the experiment. The arrangement of the DIC systems is shown in figure 5. The SRX (high-speed system) positioned in the middle, captured 6 pictures at the top of the load cycle and 6 pictures at the bottom of the cycle. Based on the time delay between the first picture and the picture at the top of the cycle, the adjustable systems (normal-speed systems) were triggered to take only one picture at the top and bottom of the load cycle. This sequence was repeated every 100 seconds (equivalent to approximately 1500 cycles). The sequence is illustrated in figure 6.

**Figure 5.** Three 3D DIC systems measuring displacements and strains simultaneously during cyclic test at 15 Hz.

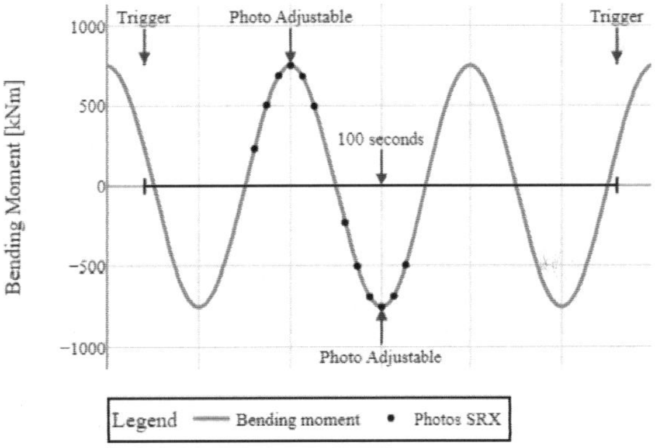


**Figure 6.** Measurement sequence imposed to DIC systems during full-scale experiment to capture the peak load points without stoping the test.

### *3.3. Full-scale FLS testing protocol*

In total the FLS test lasted for 3 weeks in period of February and March 2023. The test was performed under supervision of an operator. Hence the test was paused overnight and during the weekend. The FLS test was divided in 3 separate stages defined below:

- In Stage 1 the constant amplitude load level corresponded to the design Damage Equivalent Load (DEL) of an offshore wind jacket structure with a 25-year design life. For the FLS sample geometry this DEL corresponds to 1 million cycles at 250 kNm bending amplitude. Considered on the level of a equivalent welded steel joint the maximum bending stress (amplitude) in the centre of the joint would be 39 MPa.

- In Stage 2 the load amplitude was multiplied by a factor of 3, roughly corresponding to ULS design load i.e. the storm load level with 50 year return period. The sample is cycled for 1.8 million additional cycles and is stopped when a crack started propagating through the surface of the composite wrap. The 3x250=750 kNm bending amplitude corresponds to a maximum bending stress amplitude of 116 MPa.
- Stage 3 was performed at the same loading amplitude as Stage 1 for 170 000 cycles to investigate the behaviour of the sample at the FLS design load level in presence of the crack that appeared in Stage 2.

In stage 1 the test was interrupted several times for inspections, adjustments and overnight stops. The testing frequency at the start of the Stage 1 test was 14.9 Hz which is about 1 Hz below the theoretically calculated value for the eigenfrequency of the out-of-plane bending mode. At the end of Stage 1 the excitation frequency was slightly reduced to 14.8 Hz which might have been caused by a small stiffness reduction due to limited self-heating.

The stage 2 loading with three times higher testing amplitude required higher starting testing frequency of 15.2 Hz because stiffness degradation of the specimen was expected. During inspection of the paused test, it was noticed that the FLS sample outside temperature had increased. The increased temperature will have caused a certain stiffness reduction which resulted in a frequency drop and a deflection increase. The testing procedure was altered, and the outside temperature of the specimen was monitored using an infrared thermometer and a thermographic camera. The test was interrupted more frequently, and pauses are introduced allowing the sample to cool down. During the test intervals of Stage 2, the testing frequency gradually decreased. Over the total duration of the test a decreasing trend of the test frequency and an increasing trend of the displacement amplitude was visible indicating a stiffness reduction likely related to internal damage at the steel-composite bonded interface and in the composite wrap.

## 4. Full-scale FLS test results

### *4.1. Stiffness degradation*

After 1 million cycles of the DEL load (stage 1), no stiffness degradation was observed as shown in figure 7a. The conclusion is that the load level corresponding to damage equivalent load results in no damage. As expected, the tripled load level of stage 2 resulted in stiffness degradation and limited damage. Stiffness reduction results at tripled DEL are shown in figure 7b. After the 1.8 million cycles at the increased load level, the measured stiffness decreased by by 11% with no signs of loss of structural integrity of the joint.

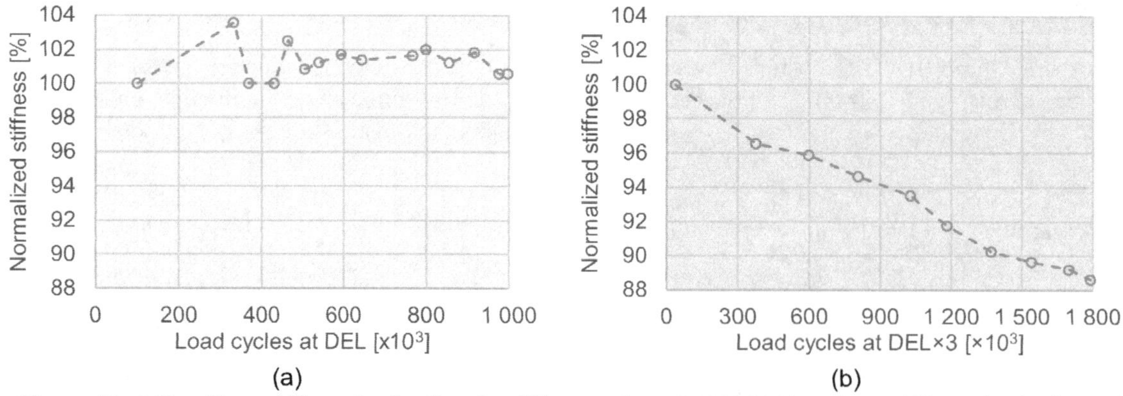


**Figure 7.** a) Bending stiffness in the first 1 million cycles of DEL;b) Bending stiffness in the lateral 1.8 million cycles of tripled DEL.

### *4.2. Interpretation of full-scale wrapped composite joint FLS test result*

Stage 1 loading, corresponding to design load of 25 years operation of the structure supporting 14MW turbine, shows no signs of damage. The estimation is that the fatigue life of the joint at this load level is below threshold of damage initiation and therefore able to practically reach the cut-off limit, i.e. to have infinite fatigue life. This is attributed to to favourable fatigue crack retardation that is found in wrapped composite joints on smaller scale as presented in [6]. Further estimation of the fatigue life at the design FLS load level, with help of literature published on wrapped composite joints so far, is given in chapter 4.2.

The full-scale joint survived at least 1.8 million cycles in test with loads ranging from +750kNm to -750kNm. During the first 1 million cycles of the increased load the stiffness degradation is primarily driven by steady debonding crack growth between the steel CHS brace members and the composite wrap. Subsequently, in the next 0.8 million cycles, a through-thickness crack initiated in the composite material. The total life of the joint in this case was considered to be 1.8 million cycles, even though complete failure of the joint was not reached.

To prove that the structural integrity of the joint is still at very high level the flow-up loading in stage 3 was corresponding to FLS design load level, i.e. +/-250 kNm. During the stage 3 loading there were no signs of further propagation of the identified surface crack, nor stiffness degradation over additional 170 thousand cycles.

## 5. Comparison of wrapped composite and equivalent welded joint performance

An analysis was performed for a welded joints, considering the same loading scenario as the one applied for the tested wrapped joint. The goal is to analytically reproduce what would be the fatigue life of the equivalent welded joints if they were tested under identical loading conditions and setup. In the first instance I load range considered for the comparison to the welded joint corresponds to loading stage 2 of full-scale FLS test of wrapped composite joint. Only at this very high load level it was possible to impose damage and test the fatigue life of the wrapped composite joint. Ithe appearance of the surface crack after 1.8 million cycles of stage 2 loading in the test of the wrapped composite joint is considered as conventional end of life to be compared to fatigue life estimate of the equivalent welded joint. Therefore, the bending moment amplitude of 750 kNm, with fully reversed load (R=-1), was used to estimate fatigue life of the equivalent welded joint. Here the equivalent is considered to be the welded joint that has the same diameter and thickness of the tube as it was the case for the tested wrapped composite joint. The out-of-plane bending moment range is therefore 1500 kN that leads to the nominal stress range at the joint root of 232 MPa for the tube diameter 610 mm and thickness 25mm.

### *5.1. Fatigue Life Estimation of equivalent full-scale welded joint*

Fatigue life estimation of the equivalent full-scale welded X-joint is done according to the Hot Spot stress approach in accordance with DNV-RP-C203 [7], using the T-curve (tubular joint in air)]. HowI, the S-N curves provided in DNV-RP-C203 [7] are the characteristic S-N curves, namely reduced towards 2.3% percentile to account for uncertainty. For a fair comparison between the fatigue life of wrapped composite joint obtained by test and the fatigue life of the equivalent welded joint obtained by calculation , the average curves, not the characteristic ones need to be used for calculation of fatigue life of welded joints at the load level of +/-750 kNm. The S-N curves provided in DNV-RP-C203 [7] originate from fatigue tests on offshore tubular joints summarized in OTH-92-390 [7]. Therefore, the average S-N curves from OTH-92-390 [8] are used. The average S-N provides approximately 3 to 4 times longer fatigue life than the characteristic curve.

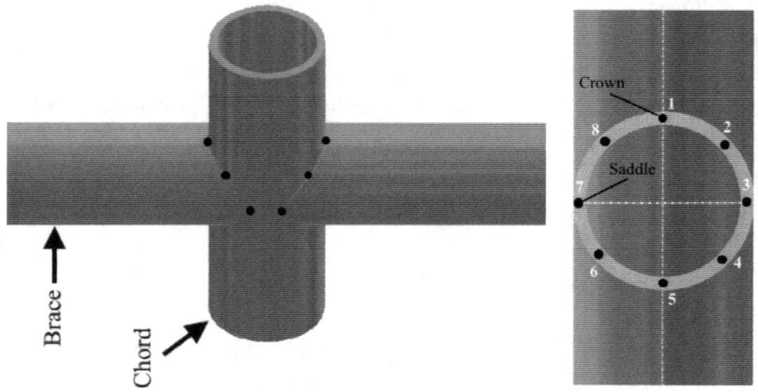


**Figure 8.** Hot spot stresses extraction points for equivalent welded joint.

For the equivalent welded full-scale joints, the hot spot stresses can be determined as the nominal stresses in the connection, multiplied by a corresponding stress concentration factor (SCF), calculated using empirical equations as per DNV-RP-C203 [7]. In general, the hot spot stresses are evaluated in 8 different points around the joints circumference, including the crown and saddle positions as depicted in figure 8. The corresponding SCF for the joint loaded by out-of-plane bending SCFs are calculated for the saddle location at chord and the brace, $SCF_{\mathrm{MOP,br}} = 1.42$ and $SCF_{\mathrm{MOP,ch}} = 2.68$, respectively. The critical location is the weld toe at the saddle location on the chord, resulting in hot spot stress of 624 MPa due to the range of the out-of-plane bending moment of 1500 kN. The c tilizinactise in offshore welded jackets is that the welds connecting braces to chords and braces to braces in X-joints are made as double-sided to prevent weld root crack which is difficult to predict and has very poor performance. The welds are made from both outside and inside of the full-scale joints composed of short cans. Decision to use single or double-sided welding is project specific and can depended on tube diameter, SCFs, etc. In this comparision the the assumption is taken that the welded joint is made with double-sided welds to have as long as possible fatigue life. tilizinI The hot spot stresses are highest at the weld toe on the outside of the weld. Using the average T curve from the S-N curves provided in OTH-92-390 [8],the fatigue life for the stress range of 232*2.68 = 624 MPa is only 26.055 cycles. This is the estimated life of an equivalent joint of same diameter and thickness of the tubes with double sided welds if it would be tested under the same +/- 750 kNm cyclic load as the tested wrapped composite joint.

### 5.2. *Discussion of the results*

The fatigue life of the tested wrapped composite joint at +/- 750 kNm out-of-plane bending moment range is at least 1.8 million cycles. The estimated fatigue life of an equivalent welded joint with same dimeter and thickness of the tube (ø610x25mm) is 26 thousand cycles. The conclusion is that wrapped composite joint has 70 times longer fatigue life.

Such substantial improvement of the fatigue life demonstrated by the wrapped composite joints is based on comparisons at a very high load level, the tripled damage equivalent load that approximately corresponds to the ULS loads, or in other words to a storm with 50 years return period. Such high load level was chosen in order to be able to impose damage in full-scale wrapped composite joint fatigue test. Much lower load levels, factor 3 to 5 lower, are representative for actual cyclic loading in the real joints of tubular members in offshore structures. For example, the damage equivalent load in this study is 3 times lower. Feng and Pavlovic [9] showed that the expected slope of the S-N curve of wrapped composite joints is in range of m=8 to m=10. Comparably the slope of the S-N curve of welded joints according to state-of-the-art research and design codes such as DNV-RP-C203 [7] is in range m=3 to m=5. This means that the difference in the fatigue life of wrapped composite and welded join would be even bigger at the lower, realistic fatigue load levels. Rough estimate of such difference can be made based on the slopes of the S-N curves and the fatigue life difference found in this study. The estimate of the fatigue life difference is extrapolated to the damage equivalent load which is 3 times lower than the tested load to depict this. For the sake of comparison, the slope m=8 and m=4 is assumed for wrapped composite and welded joints, respectively. The estimated fatigue life of wrapped composite joint at the lower load level would be $11{,}8 \cdot 10^9$ cycles ($=1{,}8 \cdot 10^6 \cdot 3^8$). The estimated fatigue life of welded joint at the lower load level would be $2{,}11 \cdot 10^6$ cycles ($=26{,}055 \cdot 10^3 \cdot 3^4$). The wrapped composite joint would have approximately 5.000 times longer fatigue life that the equivalent welded joint at the load level corresponding to DEL in this study, or in terms of stress range, at a load level corresponding to 68 MPa nominal stress in the brace member. This shows that transferring loads trough composite material is far superior and that steel consumption in the jacket can be optimized. Such result opens new ways of thinking about design of jacket structures and opens ideas on re-use and re-purpose of full structures. Fatigue life of joints in jacket made with composite wrapping have potential to reach cut-off limit, i.e. infinite fatigue life. Threfore, jacket structures with lifetime of 100 years or more are thus possible without any or small weight increase.

## 6. Validation of wrapped composite joints static resistance at full-scale

Two full-scale X90-joints joints were loaded in uniaxial tension to investigate the static performance in the ultimate limit state. Figure 9 shows the test setup of the full-scale joint specimen. To investigate the interfacial bond capacity between steel and wrapped composite, the design thickness of steel braces in tension was increased from 16 mm to 25 mm to avoid steel yielding. The ULS tests are carried out in the OCAS Tubular Testing System (TTS). This test rig has a capacity of 1580 kN (3.5 Mlbs) in both tension and compression. The system is installed in an explosion-resistant testpit which allows for safe operation given the high loads that are applied. In case of tubular test specimen the system allows for testing with high internal liquid or gas pressures, but this is not utilized in X90 full scale wrapped composite joint ULS test.

**Figure 9.** Test setup of full-scale specimen X90-Fs-ULS-01 in the test pit with blue light for 3D DIC cameras used for real-time full-filed strain and displacement measurement.

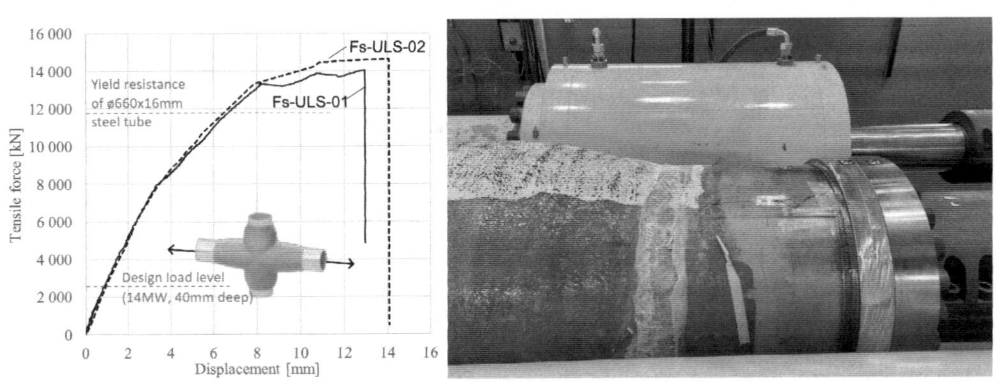


**Figure 10.** Static performance of two X90-Fs joints in uniaxial tensile tests with the imposed debonding failure.

Journal of Physics: Conference Series **2745** (2024) 012012 doi:10.1088/1742-6596/2745/1/012012

The static tensile tests show that the interfacial bond capacity between steel and composite could reach 14.4 MN on average which is 22% higher than the yield resistance of the reference steel brace member (ø660 × 16 mm) in the middle bay of full-scale jacket designed specifically in the project to support 14 MW turbine in 40 m deep water. Figure 10 shows the load-displacement curves of the two specimens tested and the failure mode dominated by debonding which results in pull-out of the steel tube form the composite wrap.

## 7. Outlook of the development of the wrapped composite joints

In summary, the result of the presented tests in this paper demonstrate that the wrapped composite joint can outperform fatigue performance of complex welds. Testing is followed by the development of finite element and analytical based prediction models that show very good predictions of joints performance across the scales and for different load directions.

Research under the WrapNode-I project focuses on de-risking the most unknown failure mode of the wrapped composi–e joints - the debonding of the steel-composite interface. To this end, all the test utilizied usage of over-designed thickness of steel tubes and thickness of the composite material, to allow the debonding failure to exhibit and to be studied in terms of size-effects and multi-axial behaviour. Optimizing of the composite design and realizing thinner composite material will be used in real deployment of wrapped composite joints which is part of the scope of investigation in the WrapNode-II project (2023-2026).

## 8. Conclusions

Wrapped composite joint is a breakthrough tutilizing bonding and fatigue resistant composite material to connect steel tubular members instead of welding. The main advantage is that the superior fatigue life of wrapped composite joints leads to a more economical design of the jacket as significant steel quantities can be reduced.. To verify this, full-scale fatigue test on one X-joint resembling joint in wind turbine jacket support structure was tested in Cronos testing rig at OCAS laboratory in Belgium. The tested wrapped composite joint connects steel tubes (ø601x25 mm) without any welding. Damage equivalent load on the joint in jacket structure corresponding to 25 year operation of 14MW turbine was applied within 1 million cycles. In addition, tripled damage equivalent load was applied with aim to impose damage and characterize failure life of the joint. Following conclusions and key findings are drawn:

- The composite joint results demonstrate that it outperforms a similar welded connection with a factor 5000. No damage was observed in the joint after 1 million cycles at damage equivalent load level (+/- 250 kNm our-of-plane bending) corresponding to 25 year operation of 14MW turbine. A welded joint with same diameter and thickness of steel tubes would have consumed its lifetime.
- When the cyclic load level is tripled, which roughly corresponds to ULS load levelthe wrapped composite joint can survive more than 25 years. Joint survived 1.8 million cycles at the tripled, high, load level of +/- 750 kNm resulting in 236 MPa nominal stress range in steel tube. The tests at this level showed that the fatigue life of the wrapped composite joint outperforms by approximately two orders of magnitude (70 times) the fatigue life of equivalent welded joint at full-scale which is expected to fail at 26 thousand cycles due to stress concentration around the welds.
- With these excellent fatigue resistances of the composite joint the use of steel in offshore renewable foundations can be drastically reduced. This initiates a more sustainable use of materials and more economical foundation design. The reduction of foundation weight also requires smaller cost effective installation vessels.
- With the demonstrated extensive lifetime the composite joint in there-use of complete structures or extension of lifetime of structures to 100 years is suddenly a realistic goal.

- Such outstanding fatigue performance of wrapped composite joints demonstrated in the full-scale test opens the path to unprecedented designs and optimization of jacket support structures for offshore wind and the floating support structures that are even more governed by fatigue.

With fatigue issues out of the way for connecting tubular members, the use of high strength steel can be realized and utilized in its strength, and hybrid structures, combining (tubular) members made of other materials with steel, can be realized when wrapped composite joints are used.

**Acknowledgments**

The authors would like to express their gratitude to RVO for the financial support with Topsector Energiesubsidie van het Ministerie van Economische Zaken through WrapNode-I project, and Tree Composites B.V. for the production of specimens. We want to thank our partners Enersea, Vattenfall, Shell, Siemens Gamesa Renewable Energy, AOC, BUEFA, HSM Offshore Energy, Smulders and the GROW foundation of their support.

**References**

[1] Van Wittenberghe, Coste A. "Fatigue Testing of Large-Scale Steel Structures in Resonance with directional Loading Control" Procedia Structural Integrity p41-48 (2019)

[2] Thibaux P. Van Wittenberghe J., Van Pottelberg E. Van Poucke M., De Baets P., De Waele W. "Efficient Fatigue Testing of Tubular Joints", Proceedings of OMAE 2015, OMAE2015-41740 (2015)

[3] Van Wittenberghe J., Thibaux P., Van Poucke M. "Large-Scale Resonant Fatigue Testing of Welded Tubular X-Joints for Offshore Jacket Foundations", Proceedings of OMAE 2019, OMAE2019-96371 (2019)

[4] Abaqus User Manual "Abaqus 2021." Corp., Dassault Systèmes Simulia. (2021). URL https://help.3ds.com/2021/English/SIMULIA/ ABAQUS/2021/HTML/Default.htm.

[5] He P and Pavlovic M. "Failure modes of bonded wrapped composite joints for steel circular hollow sections in ultimate load experiments." *Engineering Structures* Vol. 254 (2022): p. 113799. DOI 10.1016/J.ENGSTRUCT.2021.113799.

[6] Feng W. Pavlovic M. Koetsier M. Veljkovic M.. Interfacial fatigue debonding retardation in wrapped composite joints: Experimental and numerical study. Composite Structures, Vol. 319 (2023): p. 117146

[7] DNV-RP-C203 "Fatigue design of offshore steel structures." DNV GL AS (2021).

[8] OTH 92 390 "Offshore Technology Report - Background to New Fatigue Guidance for Steel Joints and Connections in Offshore Structures." Health & Safety Executive, UK (1999).

[9] Feng W and Pavlovic M "Fatigue behaviour of non-welded wrapped composite joints for steel hollow sections in axial load experiments." *Engineering Structures* Vol. 249 (2021): p. 113369. DOI 10.1016/J.ENGSTRUCT.2021.113369.

Journal of Physics: Conference Series **2745** (2024) 012013 doi:10.1088/1742-6596/2745/1/012013

# An Onshore Deployment of Advanced Dual-Doppler Radar for Wind Energy Applications

**B D Hirth[1], J L Schroeder[2], and J G Guynes[1].**

[1] National Wind Institute, Texas Tech University, Lubbock, TX, USA
[2] Atmospheric Science Group, Texas Tech University, Lubbock, TX, USA

**Abstract:** Two advanced X-band radars have been developed and deployed by Texas Tech University to provide continuous wind measurements serving project AWAKEN, including the construction of near real-time dual-Doppler synthesized wind fields covering a large 35 km x 35 km analysis domain. This deployment represents the first overland long duration utilization of the technology, which was configured to document a wide range of wind energy relevant flows spanning turbine inflow/wake flow to regional wind flows and wind plant interactions. The overland environment coupled with technology innovation has led to very high data availability yielding a growing archive of measurements documenting diverse atmospheric phenomena including numerous instances of wind plant interaction. Through the first few project months, data availability greater than 90%, 80% and 70% has been assessed at the ranges of 8 km, 16 km, and 25 km, respectively. Robust data availability coupled with fast radar scan speeds and high along beam measurement resolution position advanced Doppler radar technology to broadly contribute to diverse wind energy measurement applications covering domains sizes that far exceed existing industry standard measurement systems.

## 1. Introduction

The need for comprehensive measurements of wind energy relevant flows is greater than ever. On one hand, the community is striving to better understand the impacts of cluster wakes at the wind plant scale impacting performance for neighboring plants. On the other end of the spectrum, there remains a need to better characterize turbulence scales that impact individual turbine performance and loading. Measurements of these varied phenomena are necessary to inform and validate engineering models and ultimately contribute to next generation controls strategies to optimize turbine and plant performance.

Specialized Doppler radar is uniquely qualified to simultaneously address these industry needs due to fast scan speeds, excellent along-beam resolution, and extended maximum range [1]. Recent offshore commercial deployments of dual-Doppler (DD) radar have highlighted its capability to document turbine inflow, wake flow and interaction, as well as regional-scale flows such as wind plant wakes [2], coastal gradients, and roughness-induced variabilities [3]. These R&D focussed efforts have provided important insights that have improved wind plant wake modelling efforts as well as the understanding of the wind structure observed in certain offshore environments. DD radar derived wind fields have also been applied to develop new methodologies to generate short-term turbine inflow forecasts which can be related to forecasts of energy production [4], further demonstrating the diverse contributions radar-derived wind fields can provide to the industry.

Building from the knowledge gained from previous offshore deployments, two advanced X-band radars have been deployed by Texas Tech University (TTU) in Oklahoma as part of the United States Department of Energy (DOE) funded American WAKE experimeNt (AWAKEN) project [5]. A pre-existing DOE funded X-band system [1] has been upgraded and paired with a newly developed counterpart representing the most advanced radar technology currently available to measure wind energy relevant wind flows. The radar systems will collect measurements between May 2023 and October 2024 allowing for documentation of diurnal and seasonal influences and variabilities. This

Journal of Physics: Conference Series **2745** (2024) 012013 doi:10.1088/1742-6596/2745/1/012013

paper details several aspects the AWAKEN deployment including the DD measurement domain setup and data collection parameters, the data processing procedure, and technology advancements including an initial data availability assessment.

## 2. Deployment Approach

Two customized X-band radar systems have been deployed at sites east and southeast of Enid, OK; dubbed TTU-X1 (south site) and TTU-X2 (north site). For each system, a 20-foot sea container housing all command-and-control infrastructure is affixed to set of poured concrete footings. A steel exoskeleton structure straddles the container and supports the rotating pedestal, antenna and protective radome (Figure 1). The systems are powered by a land connection and contain a UPS battery system for backup. Table 1 outlines the operational specifications used for the AWAKEN project. The radars make use of a 10 μs compressed pulse which allows for a very high along-beam resolution of 9 m, while the 4.9 m diameter antenna (dish) results in an across-beam resolution of 0.5° in the horizontal (azimuth) and vertical (elevation) dimensions. Azimuthal and elevation oversampling are conducted at 0.35° and 0.1°, respectively, the highest resolution available as dictated by the signal processor and radar operations software. Since the primary objective for the radars during AWAKEN is to document wind plant interactions, the systems are deployed in a configuration allowing for the construction of DD wind fields over a large domain. To accommodate the necessary coverage, a pulse repetition frequency of 4820 Hz is employed yielding a maximum range of 35 km.

Figure 1. (Left) TTU-X1 and (Right) TTU-X2 deployed southeast and east of Enid, OK, respectively.

**Table 1.** TTU-X radar operational specifications used for AWAKEN.

| Parameter | Specification |
|---|---|
| Frequency | TTU-X1: 9330 MHz<br>TTU-X2: 9430 MHz |
| Power | TTU-X1: 10,495 W<br>TTU-X2: 10,447 W |
| Pulse Width | 10 μs |
| Blanking Range | 1500 m |
| Pulse Repetition Frequency | 4280 Hz |
| Maximum Range | 35 km |
| Scan Speed | 30° $s^{-1}$ |
| Range Resolution | 9 m |
| Azimuth/Elevation Resolution | 0.5° |
| Azimuthal Oversampling | 0.35° |
| Elevation Oversampling | 0.1° |

The two radar systems are deployed with a separation of 11.4 km, providing the appropriate geometry to construct DD wind fields across much of the Breckinridge, King Plains and Armadillo Flats wind plants (Figure 2), though influences from two nearby wind plants to the north are also observed. Each radar is configured to horizontally scan a 145° sector sweep across 18 different elevation tilts, covering the entire eastern DD lobe through a large vertical depth. The resulting volume revisit time is roughly 124 seconds and the DD analysis domain spans nearly 35 x 35 km which includes 250 individual wind turbines.

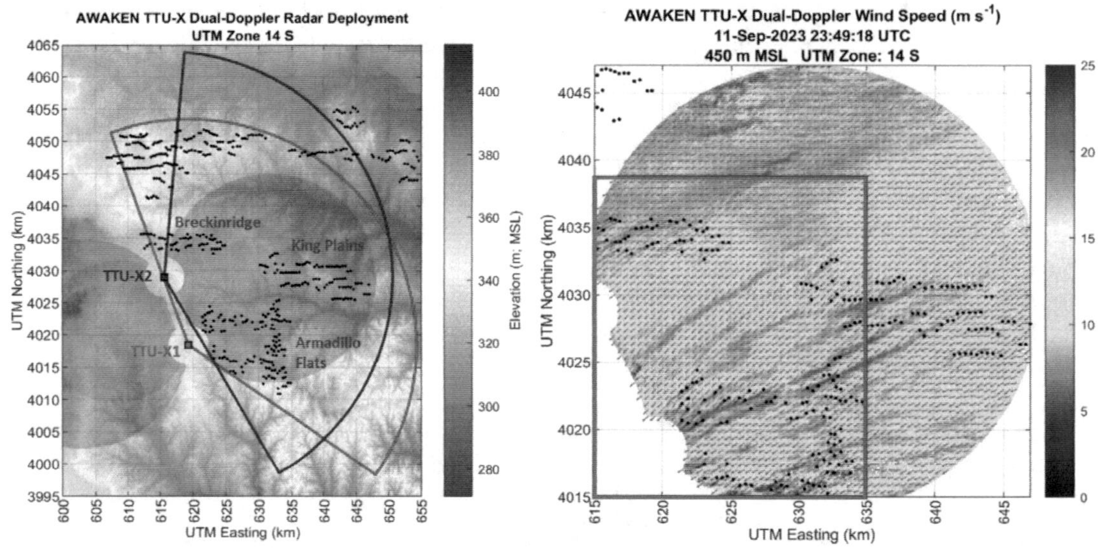


Figure 2. (Left) AWAKEN TTU-X radar deployment schematic showing the radar locations (red and blue squares), the DD analysis region (shaded) and the sectors scanned (rad and blue outlines). Turbine locations (black dots) and background elevation (MSL) are also shown. (Right) Example resulting DD wind speed map at 400 m MSL from 11-Sep-2023 23:49 UTC with the boundaries of the fine analysis grid shown with the red rectangle.

## 3. Data Processing and Examples

Line-of-site (radial) velocity measurements from each volume undergo real-time quality control and interpolation locally using a custom software suite developed by TTU. The software makes use of the information contained within the various radar moments output by the signal processor to perform thresholding and various statistical analyses tuned to search for and remove erroneous radial velocity returns. The software also performs automated horizontal alignment checks and corrections based on the total return power signatures relative to the locations of known point targets (i.e., turbines).

The edited radial velocity fields are then interpolated from their native polar coordinate system into a pre-defined Cartesian grid. For AWAKEN, a nested Cartesian grid is implemented where 25 m horizontal grid spacing is used within the fine grid (red rectangle in the right panel of Figure 2) while 50 m horizontal grid spacing is used within the course grid (full domain in the right panel of Figure 2). Ten vertical levels are used between 375 – 600 m AGL with a 25 m spacing for both the fine and course grids. This nested grid approach is intended to preserve higher detail at ranges closer to the radar where the beam spread is less pronounced, while allowing for computational efficiency to keep up with processing such a broad domain in real-time. A total of roughly 10 M unique grid points exists across both the fine and course grid. Raw and processed volumes are then transmitted in near real-time using the local cellular network to a central server at TTU where corresponding Cartesian volumes are paired, and DD wind field synthesis is conducted. The raw and DD data fields are then uploaded to the DOE public archive for immediate community access.

The generated dual-Doppler wind fields simultaneously serve a diverse range of wind flows scales. At the regional scale, details of the background flow can be documented as well as the impact and behavior of cluster wakes and plant to plant interactions. Within the wind plant, turbine interactions and inter-plant flows can be evaluated. Drilling down further, details of individual turbine inflow and wake flows can be assessed, including the characterization of shear, veer and turbulence scales across the rotor sweep. The over-land environment consists of highly variable wind flows due to factors like diurnal cycles characterized by substantial stability changes as well as transient, sometimes significant, meteorological events. Multiple examples of diverse wind fields documented within the AWAKEN domain are provided in Figure 3. Diurnal flow evolution has been frequently observed, as background stability drives the scales of gust and lull features found embedded within the background flow as seen when comparing the wind field during the early afternoon (Figure 3, top left) versus near sunset on the same day (Figure 3, top right). The diurnal evolution of turbulence scales plays an important role on how individual turbines and neighboring wind plants interact with each other. Multiple thunderstorm outflow events have been documented, including severe thunderstorm outflows with embedded gusts of 35-40 m $s^{-1}$ impacting the DD domain (Figure 3, bottom left). Localized transient wind features such as thunderstorm downbursts have also been observed directly impacting turbine groupings within the analysis domain as well (Figure 3, bottom right). Numerous examples of frontal passages, gravity waves, and other transient flow features also exist within the growing AWAKEN data archive. These are the type of multi-scale measurements needed to properly inform and validate engineering models and ultimately contribute to the next generation of controls strategies to optimize wind turbine and plant performance.

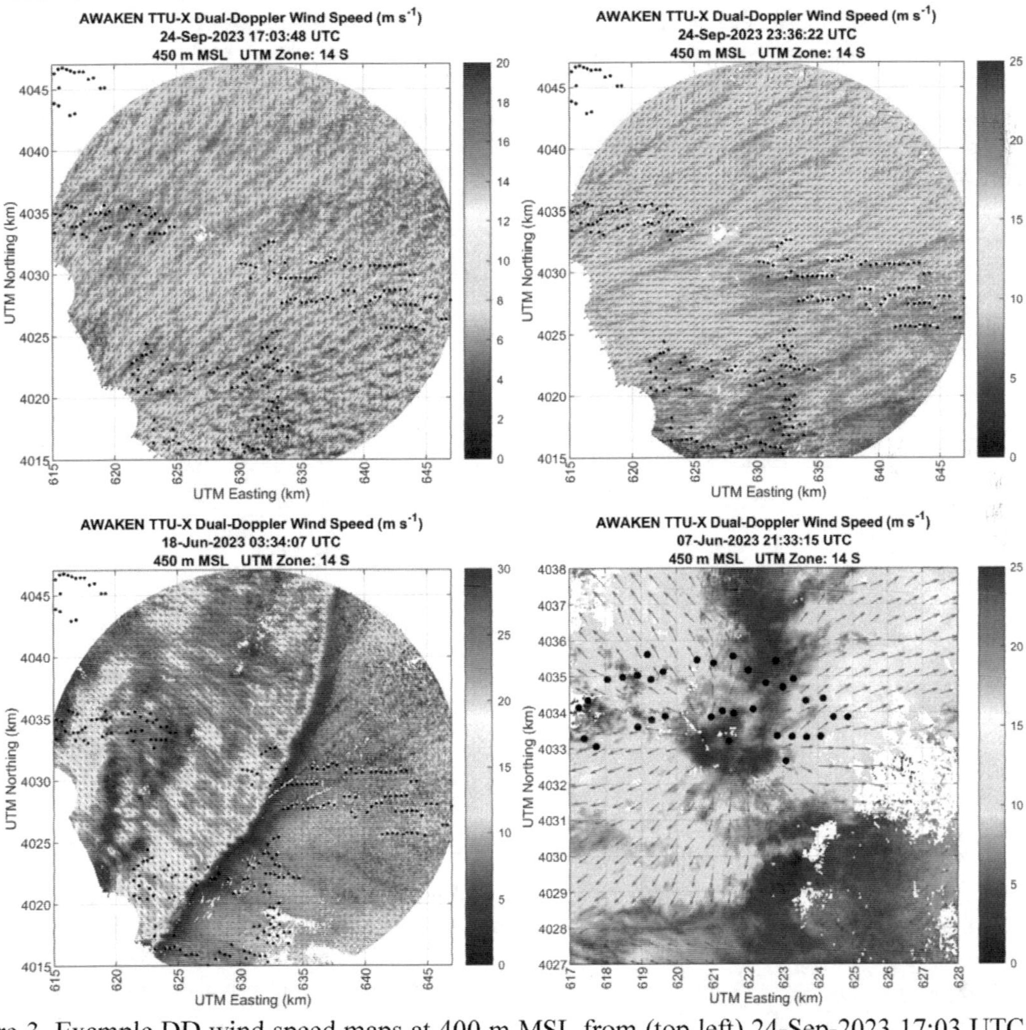


Figure 3. Example DD wind speed maps at 400 m MSL from (top left) 24-Sep-2023 17:03 UTC, (top right) 24-Sep-2023 23:36 UTC, (bottom left) 18-Jun-2023 03:34 UTC and (bottom right) 07-Jun-2023 21:33 UTC.

## 4. Technology Advancement and Data Availability Assessment

The TTU-X deployment for AWAKEN is also the first to leverage recent innovations developed at TTU focussed on enhancing data availability and artifact reduction. Specifically, advancements to the receiver chain, including increased dynamic range and sensitivity coupled with high-order analog spectral rejection of turbine artifacts, have led to a marked increase in data availability while simultaneously reducing the negative influence of turbine artifacts. Knowledge gained from initial AWAKEN measurements has also led to improved tuning of the real-time processing algorithms yielding increased data quality for the overland deployment.

Data availability was assessed over the initial project months spanning 01-June through 31-August 2023 using the operational scanning strategy previously discussed. This period was selected because TTU-X1 was operating with the prior version of the receiver chain while TTU-X2 operated with the advanced version allowing for a direct assessment of performance improvement. TTU-X1 was subsequently upgraded resulting in both systems maintaining the same level of performance. Through this test period, 56,828 and 65,240 individual data volumes were considered for analysis from TTU-X1 and TTU-X2, respectively. First order data thresholding/editing was conducted, consistent with the methods used by the operational processing software, to provide baseline quality control before assessment. Any data removed, regardless of reason (ground clutter, point target, turbine artifact, etc.), counted against the data availability calculation suggesting the final statistics represent a conservative estimate.

To construct bulk data availability statistics, line-of-site velocity data were first assessed for each sweep within a given volume. Each 145° horizontal sector oversampled at 0.35° contained roughly 414 data bins in the azimuthal dimension for each along beam range gate. The percentage of available radial velocity data at each range arc was calculated, and that percentage was aggregated over the full collection of volumes for all 18 elevation tilts. A 500 m along-beam moving filter was then used to smooth out spikes. The dependency of availability for TTU-X2 on range, elevation angle, and height can be seen in Figure 4.

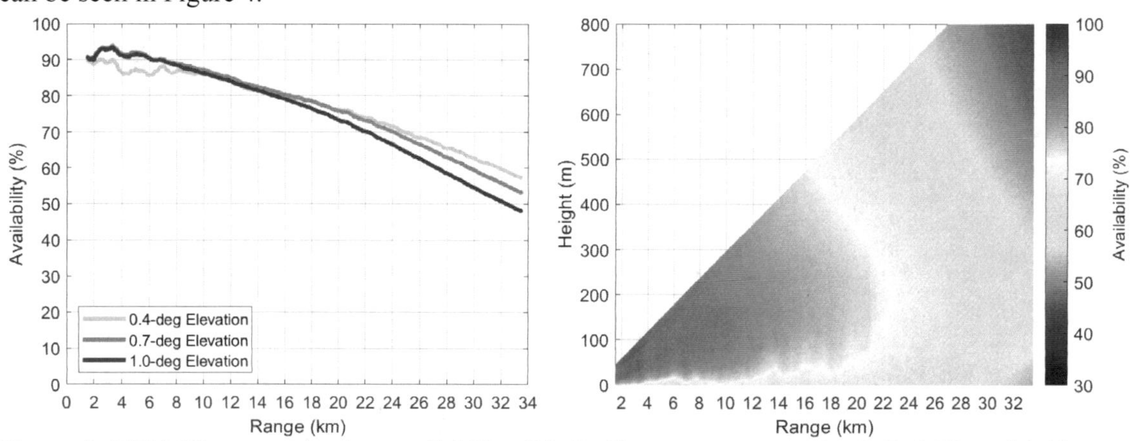


Figure 4. TTU-X2 composite data availability (%) (Left) versus range at the 0.4°, 0.7°, and 1.0° elevation tilts and (Right) versus range and height using all 18 elevation tilts.

Data availability exceeds 90% out to roughly 8 km range except for the lowest elevation tilts. Lower availability at these levels is a result of data editing owing to terrain complexities and associated ground clutter. These influences are generally negligible above 50 m and high availability exists at hub height and above. Peak availability remains above 80% out to roughly 16 km, and above 70% to 25 km, and 60% to 32 km range. Availability dependence on elevation angle is also evident, where slightly lower availability is observed for close ranges at lower tilts, while higher tilts result in lower availability at farther ranges. Lower tilts are subject to ground clutter interference at close ranges and therefore greater data removal while higher elevation tilts are likely to extend above the atmospheric boundary layer at farther range, where scatterers become sparse. Regardless, these robust statistics coupled with fast scan speeds and extensive data coverage (thousands of square kilometers) place advanced radar in excellent position to broadly contribute to diverse wind energy measurement applications. To assess the impact of the receiver chain advancements, data availability curves between TTU-X1 and TTU-X2 were compared at the 0.7° elevation tilt (Figure 5). Distinct availability improvements are seen at all ranges, especially as range increases. Data availability increases of 5% exist within 10 km, while increases of 10% at 20 km and 18% at 30 km are found. These results confirm a meaningful performance improvement due to the latest technological advancements of TTU-X2. It is worth noting that seasonal variations in data availability are expected, and availability values are anticipated to be lower during the cold season when atmospheric scatterers supportive of radar measurements will be less prevalent (i.e., insects, pollen, etc.). Further receiver chain advancements and future custom signal processing algorithm innovation are expected to contribute to further meaningful data availability gains in the coming years.

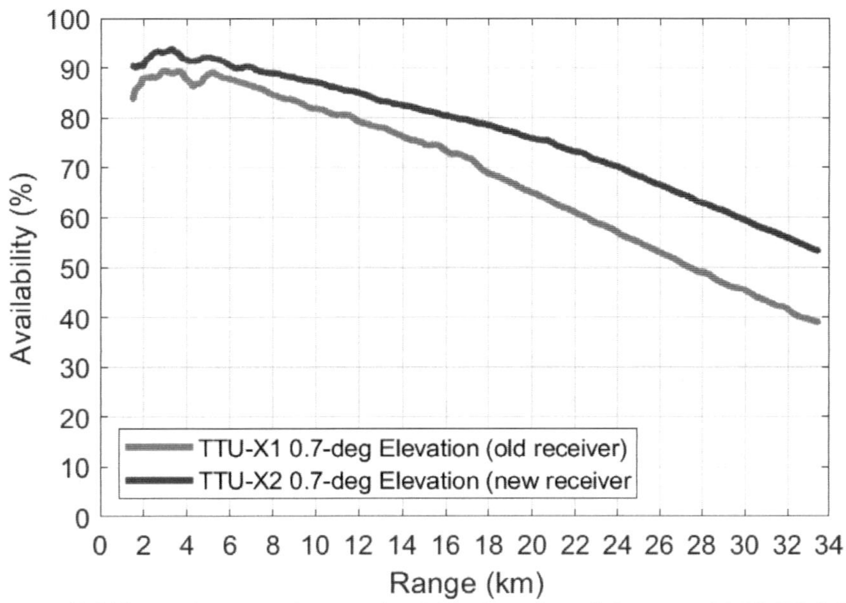


Figure 5. Data availability (%) comparison at the 0.7° elevation tilt between TTU-X1 (old receiver chain) and TTU-X2 (new receiver chain)

**5. Summary and Conclusions**

TTU has developed and deployed two advanced X-band radars to provide continuous wind measurements serving project AWAKEN. This deployment represents the first overland long duration utilization of the technology, which was configured to document a wide range of wind energy relevant flows. The AWAKEN DD coverage region spans nearly 35 x 35 km and contains 250 wind turbines belonging to three separate wind plants. The overland environment across the AWAKEN domain has yielded very high data availability during the first three project months due to the presence of favorable atmospheric scatterers supporting valid velocity measurements. Technology advancements have further enhanced availability resulting in radars that can consistently measure to ranges that far exceed existing industry standard measurement systems. A near real-time data quality control, grid interpolation, and DD processing system has also been developed and implemented to process the very large radar data volumes in an effective manner to serve the community.

Among the strengths of the DD radar system is its ability to simultaneously document various wind phenomena at diverse scales. At present, multiple datasets are already being used by the research community to explore a wide range of topics spanning regional wind flows, cluster wakes, and wind plant interactions down to inflow and wake flows at the individual turbine scale. Knowledge of the complex, multi-scale wind flows that impact wind plant arrays and their performance will continue to grow as the archive of diverse datasets builds.

More comprehensive data availability and measurement validation studies are ongoing and will be featured in future work. Multiple comparison studies between both the line-of-sight velocities and synthesized DD wind fields are being conducted relative to the measurements of multiple industry accepted lidar systems. These studies will provide greater context concerning the strengths and weaknesses of using DD radar for over land wind energy studies to further bolster the use cases for the technology. It is expected that the AWAKEN DD radar deployment will continue to run through October 2024 which will yield at least 18 months of measurements available to serve diverse R&D interests within the wind energy community.

**References**

[1] Hirth, B.D., J.L. Schroeder, J.G. Guynes. (2017) *Diurnal evolution of wind structure and data availability measured by the DOE prototype radar system.* Journal of Physics: Conference Series, 926, 012003.

[2] Nygaard, N.G., A.C. Newcombe. (2018) *Wake behind an offshore wind farm observed with dual-Doppler radars.* Journal of Physics: Conference Series, 1037, 072008.

[3] Hirth, B.D., N.G. Nygaard, J.L. Schroeder. (2022) *Dual-Doppler radar measurements of the coastal gradient impacts of an offshore wind farm.* NAWEA Wind Tech 2022 Conf. Proc., Newark, Delaware.

[4] Valldecabres, L., N. G. Nygaard, L. Vera-Tudela, L. Von Bremen, M. Kühn. (2018) *On the use of dual-Doppler radar measurements for very short-term wind power forecasts*. Remote Sensing, 10, 1701.

[5] Debnath, M., A.K. Scholbrock, D. Zalkind, P. Moriarty, E. Simley, N. Hamilton, C. Ivanov, R.S. Arthur, R. Barthelmie, N. Bodini, A. Brewer, L. Goldberger, T. Herges, B.D. Hirth, G.V. Iungo, D. Jager, C. Kahl, P. Klein, R. Krishnamurthy, S. Letizia, J.K. Lundquist, D. Maniaci, R. Newsom, M. Pekour, S.C. Pryor, M.T. Ritsche, J. Roadman, J.L. Schroeder, W.J. Shaw, J. Van Dam, S. Wharton. (2022) *Design of the American Wake Experiment (AWAKEN) Field Campaign.* Journal of Physics: Conference Series, 2265, 022058.

**Acknowledgments**

Funding for the TTU-X radar participation in AWAKEN is provided by the National Renewable Energy Laboratory and the US Department of Energy.

Journal of Physics: Conference Series **2745** (2024) 012014 doi:10.1088/1742-6596/2745/1/012014

# Scalable SCADA-Based Calibration for Analytical Wake Models Across an Offshore Cluster

**Diederik van Binsbergen [1,2], Pieter-Jan Daems[2], Timothy Verstraeten[2], Amir Nejad[1] and Jan Helsen[2]**

[1] Department of Marine Technology, Norwegian University of Science and Technology (NTNU), Otto Nielsens veg 10, 7052 Trondheim, Norway

[2] Department of Mechanical Engineering, Vrije Universiteit Brussel, Pleinlaan 2, Brussels, 1050, Belgium

E-mail: dirk.w.van.binsbergen@ntnu.no

**Abstract.** A multi-level hyperparameter optimization framework is performed to calibrate analytical wake models in the context of multiple wind farms within the Belgian-Dutch offshore cluster. The calibration, applied on the TurbOPark model with Gaussian wake profile, is performed on different scales. Initially, calibration focused solely on internal wake effects, followed by a calibration for individual wind farms, considering both internal and external wake effects, and finally performing the calibration using SCADA data from multiple wind farms within the concession zone, with and without accounting for blockage. It was observed that calibrating wakes for freeflow wind directions using internal wakes only results in tuning parameters similar to the calibration involving both intra-and-inter farm wake effects. Minor variations are noted between wind farms, with dependencies on wind speed and wind direction across all cases. When the tuning parameter is calibrated using SCADA data from multiple wind farms, a significant reduction in the tuning parameter was observed, compared to calibration that focuses on one wind farm per calibration. Analyzing the model residual error for wind coming from the north-west reveals that the calibrated wake model effectively accounts for cluster wake effects. Furthermore, a consistent and substantial presence of heterogeneous inflow perpendicular to the wind direction originating from the south-west is observed, which is not captured by the modeling framework, which assumes homogeneous inflow. The observed trend of inflow heterogeneity suggests that the discrepancy cannot be attributed solely to blockage effects.

## 1. Introduction

With the increasing demand for renewable energy, the offshore wind energy sector has been experiencing significant growth [1]. This growth is partly due to the increasing size of wind turbines and wind farms, leveraging the benefits of economies of scale. However, the increasing scale of wind turbines and wind farms presents new challenges, especially concerning the flow physics of wind farms [2, 3, 4, 5]. Wind turbine wakes, characterized by a velocity deficit and increased turbulence behind the wind turbine, can lead to considerable losses in downwind turbines.

Analytical wake models offer a balance between low computational time and reasonable accuracy. These models have significantly evolved over time, starting with simpler concepts like the Jensen model [6, 7], featuring linear wake expansion and a uniform radial profile, to more complex version that incorporate self-similar Gaussian wake profiles and non-linear expansion [8, 9, 10, 11, 12, 13, 14, 15]. Despite advancements, these models initially relied on calibration from either CFD models, wind tunnel experiments, or data from smaller turbines and wind farms.

Recent advancements in turbine and wind farm sizes have highlighted two primary discrepancies in traditional models: the underestimation of farm-wide blockage effects and the cluster wakes and far wakes [16]. These factors often lead to an overestimation of the predicted energy yields. To address these issues, newer models, such as the TurbOPark [11, 14], Cumulative Wake [13], and the Zong wake model [10] have been introduced, refining either the wake expansion downwind and the wake superposition. Additionally, various studies have developed blockage models [11, 17, 18, 19] to address the blockage effect.

The shift towards larger wind turbines necessitates increased turbine spacing, challenging the previously accepted assumption of uniform inflow conditions in offshore environments with a fully developed wind field [20]. The complexity of this issue becomes more apparent in larger wind farms, where farm-wide blockage leads to a more varied wind speed and direction profile [21, 22]. Additionally, as wind turbines grow larger and therefore reaching higher in the atmosphere, their interaction with the atmospheric boundary layer becomes significant, since this interaction can actively influence the development of the atmospheric boundary layer, which is largely dependent on atmospheric stability and diurnal and annual cycles. Additionally, factors like wind veer and shear also become more pronounced across the rotor planes of larger turbines, adding to the complexity of accurate wake modeling.

Historically, when wind turbines were smaller and wind farms more spaced apart, an operator only needed to consider the wake effects within their own wind farm for yield assessment. However, with the increasing number of closely spaced wind farms, the impact of neighbouring wind farms (inter-farm effects) on each other's yield has become increasingly apparent. As a result, it is now crucial for wind farm operators to accurately estimate not just the internal wake losses but also the losses caused by neighbouring wind farms, as these losses directly impact the expected yield and the levelized cost of energy [23]. The growing number of offshore wind farms means that calibrating a single wind farm without accounting for the influences from neighbouring wind farms can lead to inaccurate results. This is particularly evident when the wind originates from the direction of these neighbouring wind farms [24]. Therefore, including the entire offshore cluster in the calibration process is essential to ensure accuracy.

Calibrating analytical wake models on a cluster of wind farms involves unique challenges on its own. It requires detailed information from multiple wind farms, such as thrust and power curves. Equally important is the identification of any abnormality in the turbines being considered within the calibration framework, which is vital to prevent skewed results. Despite these challenges, successfully overcoming them can lead to an accurate estimate of the wake losses and can further increase the understanding of losses associated to large clusters of wind turbines.

## 2. Objectives

The primary aim of this work is to implement the calibration framework developed by Binsbergen et al. (2023) [24] on an upscaled model of the Belgian-Dutch offshore zone. This framework employs a hyperparameter tuning framework for robust calibration of tuning parameters in analytical wake models, utilizing SCADA data from an offshore wind farm. By applying this framework on an entire offshore cluster, the objectives are:

(i) To analyze the intra-farm and inter-farm wake effects in the Belgian-Dutch offshore zone, providing insights into the dynamics and interactions of wind turbines within and between wind farms.

(ii) To accurately determine tuning parameters for analytical wake models, with a specific focus on large cluster of wind farms where interactions are more complex and pronounced.

(iii) To quantify the performance accuracy for cluster wake effects, enabling a better understanding and prediction of these effects in large wind farm clusters.

(iv) To measure and quantify the magnitude of non-uniform inflow conditions across the wind farms, enhancing the understanding of variable wind patterns for the case study wind cluster.

## 3. Methodology

In this study, the calibration of the wake model is conducted using turbines located in the Belgian offshore zone, as illustrated in Figure 1, while computation of the wake model encompasses all wind turbines in both the Belgian and Dutch offshore zones, as shown in Figure 2. The Belgian offshore zone is characterized by its high power density, at 9.5MW/km$^2$, with a rated capacity of 2.2GW. In contrast, the Dutch offshore zone has a lower power density of 4.8MW/km$^2$ and a rated capacity of 1.5GW. Combined, these zones reach a total rated capacity of 3.7GW. This configuration offers a comprehensive view of wakes in a densely populated wind farm region, allowing for a detailed analysis of both intra-and inter-farm wake effects for one of the world's largest concession zones to date.

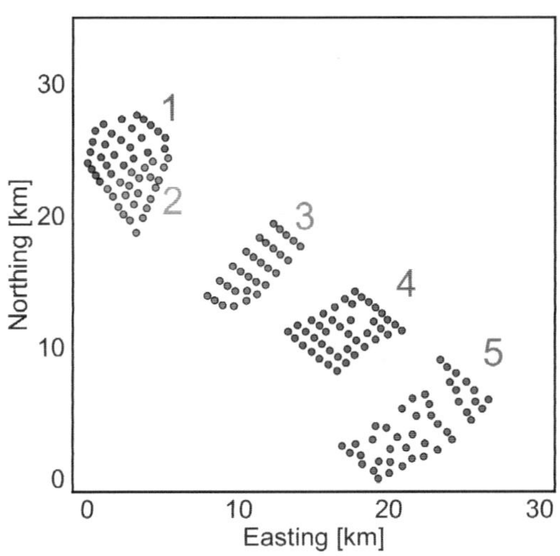


**Figure 1.** Wind turbines within the Belgian offshore cluster used to perform the calibration on.

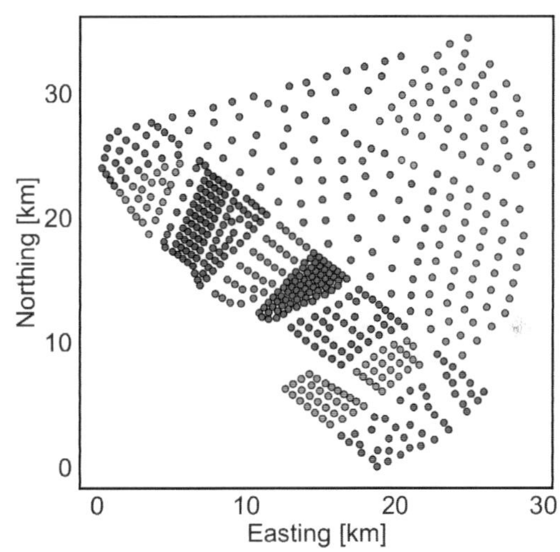


**Figure 2.** Wind turbines within the Belgian-Dutch offshore cluster used within the wake calculation.

The optimization is performed on different scales:

(i) Internal wake calibration: Initially, the calibration focuses solely on the internal wake effects of each wind farm and is performed for each wind farm depicted in Figure 1.

(ii) Internal and external wake calibration: At this stage, the calibration accounts for both the internal wake effects within each wind farm and the external wake effects from neighbouring wind farms, as depicted in Figure 2. This calibration is performed on the wind farms as shown in Figure 1.

(iii) Cluster wake calibration: In this phase, calibration takes into account both internal and external wake effects, similar to the previous step. Additionally, the cost function is modified to encompass all wind farms from Figure 1 within one cost function, instead of having separate cost functions per wind farm.

(iv) Cluster wake calibration including blockage: This calibration considers internal and external wake effects, along with a blockage model. The cost function is, similar to the previous step, adapted to include all wind farms in a single evaluation.

Systematically scaling up the calibration framework enables the identification of variations in the distribution of the tuning parameter, the inflow conditions, and the residual error of the cost-function.

The cost-function utilized in this study, as described by Binsbergen et al. (2023) [24], aims to minimize errors at two levels: turbine-specific and overall farm-or-cluster-level between SCADA data and wake model predictions and remains consistent across different optimization scales. For further details, Binsbergen et al. (2023) [24] can be consulted.

The wake model considered within this work is the TurbOPark model with a Gaussian wake profile, as described by Pedersen et al. (2022) [14]. Additionally, the blockage model considered is the model proposed by Troldborg and Meyer Forsting (2017) [18]. the TurbOPark model has been selected for its computational efficiency and effectiveness in modeling far-wake scenarios [25]. Furthermore, the TurbOPark model is characterized by a single tuning parameter, $A$, which reduces uncertainty in the optimization outcomes, as it eliminates correlations between multiple tuning parameters.

### 3.1. Filtering procedure

One year of SCADA (Supervisory Control and Data Acquisition) data is used for the analysis. The mean and variance of relevant variables are calculated for 10-minute intervals. The data is filtered based on active power, wind speed, wind direction, and nacelle position of each wind turbine. Power-and-thrust curves for each wind farm within the calibration framework are known. SCADA data involving grid curtailment or turbine derating are excluded, and identified by power curve mismatches. Misalignments up to 15 degrees, based on wind direction and nacelle angle, are removed. Sensor stuck faults, indicated by zero variance in 10-minute intervals of wind direction and wind speed measurements or multiple consecutive identical readings, are also excluded. Alarm annotations from turbine status logs are filtered out. Turbines producing less than 100kW of active power are conisdered inactive and are removed from the cost function and wake simulation, assuming the wake effect is limited. This is limited to half of each wind farm. Timestamps are not considered if less than half of the turbines in a wind farm are active, or if any of the above criteria are not met. For more details on the filtering procedure, Binsbergen et al. (2023) [24] can be consulted.

### 3.2. Optimization procedure

The multi-level optimization procedure, as outlined in van Binsbergen et al. (2023) [24] and based on the Optuna framework [26], is structured on the following 3 stages for each calibrated timestamp:

- Stage 1: Optimization of wind speed.
- Stage 2: Optimization of wind speed and wind direction.
- Stage 3: Optimization of wind speed, wind direction and wake parameters.

Each stage of the process involves the use of specific samplers. During the first and second stages, the Quasi Monte Carlo (QMC) Sampler [27] is utilized to explore the search space for wind speed and wind direction. Initially, a precise estimate of the wind speed and wind direction is not required, and instead an approximation is preferred. For the third stage the Multivariate Tree-Parzen Estimator is used, which fits a Gaussian Mixture Model (GMM). This stage is combined with 50 random start-up trials, to explore the search space. An example of this is depicted in Figure 3, where $A$ represents the tuning parameter of the TurbOPark model with Gaussian wake profile [14]. This figure highlights the importance of incorporating freeflow wind conditions in the calibration framework. While these conditions are not the primary target of the calibration process, neglecting them can lead to inaccurate results, as relying solely on the SCADA data for wind speed and wind direction might not always reflect true freeflow conditions.

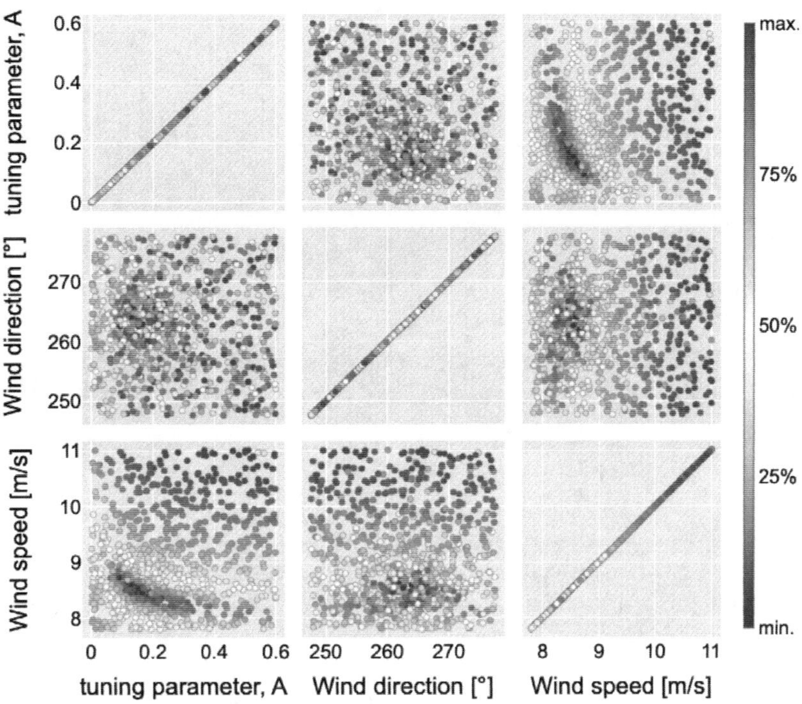


**Figure 3.** Optimization history of the TPE sampler applied to the cost-function over a 10-minute average of the entire Belgian-Dutch concession zone. Lower percentage errors indicate a better match between SCADA data and wake model.

## 4. Results

The results section is divided into two subsections. Initially the distribution of the tuning parameter across different modeling levels is compared. These scales include internal wake effects, combined internal-external wake effects, and cluster-scale wake effects with and without blockage.

Subsequently, the residual power error of the cost-function is analyzed in the context of cluster-wake effects, blockage and inflow heterogeneity for wind coming from 315 and 225 degrees, respectively.

### *4.1. Model calibration on different optimization scales*

The calibrated values of the tuning parameter as function of wind speed are visualized in Figures 4, 5, and 6. Figure 4 illustrates the tuning parameter for wind directions ranging between 200 and 250 degrees, ensuring the wind farms primarily operate under freeflow conditions. In Figure

5, the calibration takes into account the wakes from surrounding wind turbines. Figure 6 shows the expected value of the tuning parameter for the median of wind farms 1-5 (shown in blue), the entire cluster (in orange), and the entire cluster including a blockage model (in green). It is noticeable that the tuning parameters in Figures 4 and 5 are approximately of the same magnitude. However, for Figure 6 a significant difference is observed between the farm-based and cluster-based simulation, where the tuning parameter is notably lower when the entire cluster is incorporated into the cost-function of the calibration.

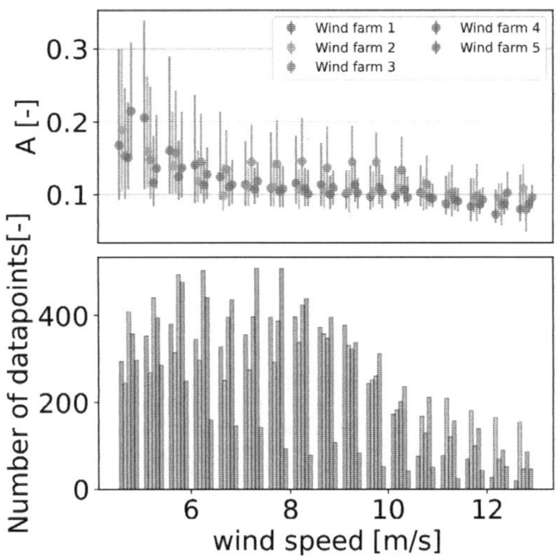


**Figure 4.** Expected value of the tuning parameter, $A$, based on the internal wake calculation for wind direction ranging between 200 and 250°.

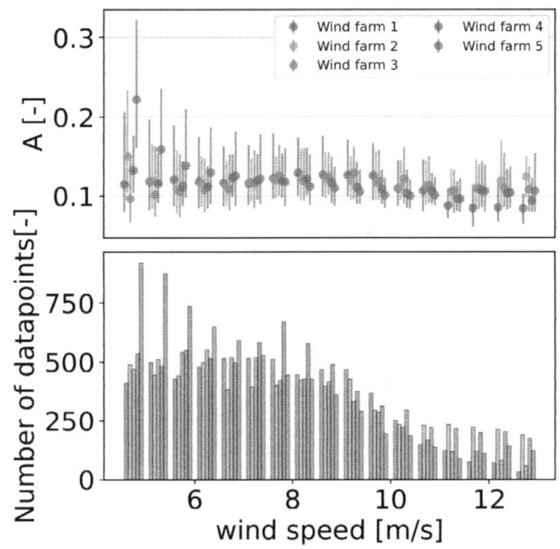


**Figure 5.** Expected value of the tuning parameter, $A$, based on the internal and external wake calculations considering all wind directions.

The expected value of the tuning parameter as function of wind direction is illustrated in Figures 7, 8, and 9 for wind speeds ranging between 6.0 and 12.0 $m/s$. Here, Figure 7 represents the calibration based on internal wakes only. Figure 8 displays the calibration considering both internal and external wakes for each wind farm, and Figure 9 compares the calibration results between the median over all wind farms for the internal-external wake scenario, the cluster calibration case, and the cluster calibration including blockage.

It is evident from Figure 7 how the tuning parameter is significantly overestimated when external wakes are not accounted for, particularly from wind directions that are affected by external wakes. This overestimation does vary notably across different wind farms. The calibration involving both internal and external wake effects, as shown in Figure 8, still shows some variation, especially between wind farms, but the overall magnitude of variation is more contained. Similar trends are observed in Figure 9 as to Figure 6. The calibration involving internal and external wake effects leads to an overestimation of the tuning parameter if used as reference to the entire cluster, and subsequently results in an underestimation of wake effects. This is consistent with observations made in recent years [11, 14, 16, 25].

### *4.2. Cluster wake effects, blockage, and inflow heterogeinity*

The performance of the wake model calibration in a cluster wake scenario is evaluated by examining data from north-west wind directions, specifically between 300 and 330 degrees and wind speeds between 7.0 and 9.0 $m/s$. The results are depicted in Figures 10 and 11. Figure

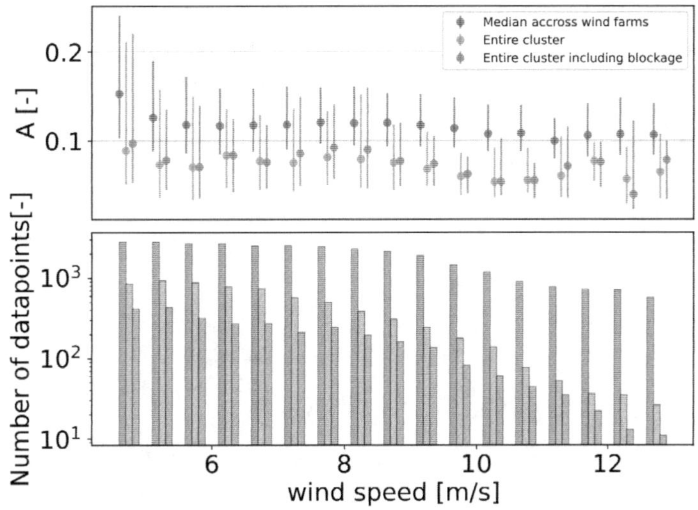


**Figure 6.** Comparison between the expected value of the tuning parameter, $A$, for the median across each wind farm for the combination of internal and external wake effects, the entire cluster, and the entire cluster including blockage.

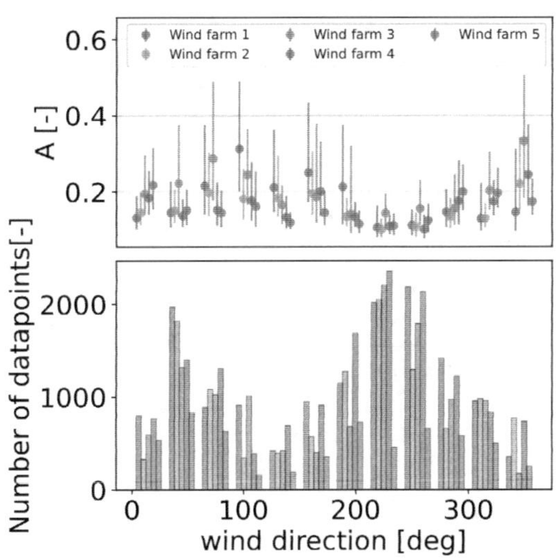


**Figure 7.** Expected value of the tuning parameter, $A$, based on the internal wake calculations considering wind speeds between 6.0 and 12 $m/s$.

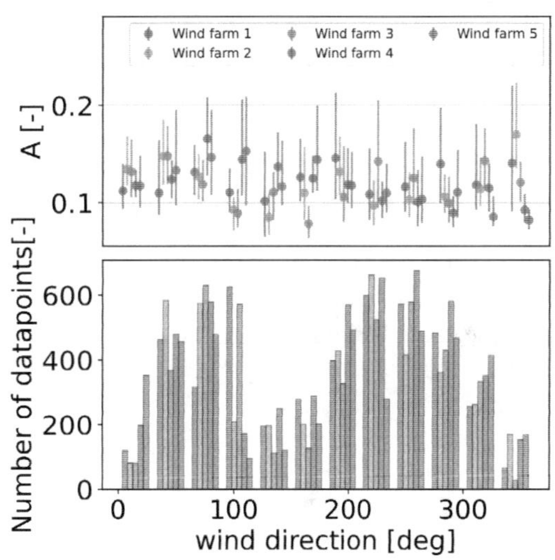


**Figure 8.** Expected value of the tuning parameter, $A$, based on the internal and external wake calculations considering wind speeds between 6.0 and 12 $m/s$.

10 shows the relative error between the calibrated wake model and SCADA data, while Figure 11 presents a scatter plot comparing active power from SCADA with the wake model predicted power. Here, the predictions from the wake model are based on the tuning parameter, A, that have been individually calibrated for each considered timestamp on the entire cluster. The $R^2$ of the data is equal to 0.90, indicating a strong correlation for the cluster wake case and suggesting that the model accurately represents cluster wake effects in this scenario.

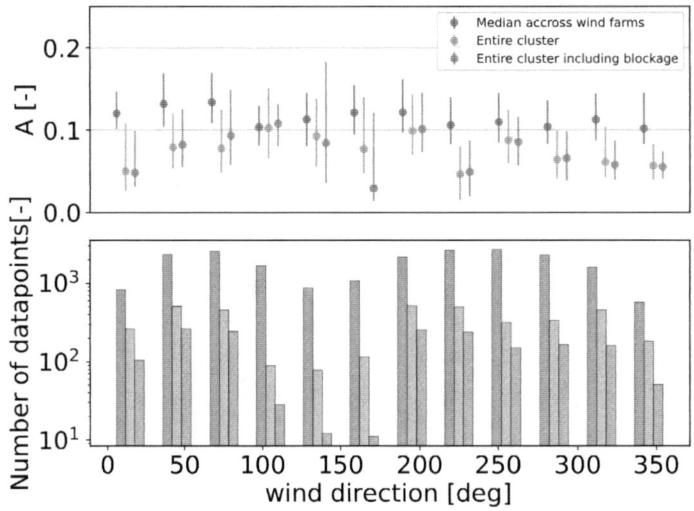


**Figure 9.** Comparison between the expected value of the tuning parameter, $A$, for the median across each wind farm for the combination of internal and external wake effects, the entire cluster, and the entire cluster including blockage. for wind speeds between 6.0 and 12 $m/s$.

Results quantifying inflow heterogeneity are visualized in Figures 12 and 13. Figure 12 reveals a consistent trend for the turbines operating in the freeflow, with turbines parallel to the wind direction showing a pattern where the calibration framework underestimates active power from north-west and overestimates the active power from the south-east. This suggests that the wind farm experiences significant and consistent heterogeneous inflow. This is further supported by Figure 13, which shows that under homogeneous inflow conditions the model fails to capture these inflow characteristics, with an $R^2$ value equal to 0.40, compared to 0.90 in the cluster wake case. The increasing overestimation between the model and SCADA data from south-east to north-west indicates that this discrepancy is not solely due to blockage. Typically, blockage effects would show forms of symmetry across both sides of the concession area, but the current pattern shows asymmetry. This observation complicates accurately quantifying the extent of blockage observed.

## 5. Conclusion

A multi-level hyperparameter optimization framework for calibrating analytical wake models using SCADA data is applied to multiple wind farms within the Belgian-Dutch offshore cluster. The calibration is performed at different scales using the TurbOPark model with Gaussian wake profile by Pedersen et al. (2022) [14]. Initially, the calibration focused on internal wake effects only, followed by individual wind farm calibration considering both internal and external wake effects. Finally, the hyperparameter optimization was performed using SCADA data from multiple wind farms in the offshore concession zone, with and without blockage considerations.

Comparing the calibration results at different scales, it was observed that calibrating wakes for freeflow wind directions using internal wake data shows similar trends to the calibration involving both internal and external wakes. Some minor variations in the predicted tuning parameter were observed between wind farms, and wind speed and wind direction dependent trends were evident across all cases. However, when calibrating the tuning parameter on the entire cluster, as opposed to individual wind farms, a significant reduction in the tuning parameter was observed. This reduction has a direct inverse effect on wake losses, suggesting that losses for large wind

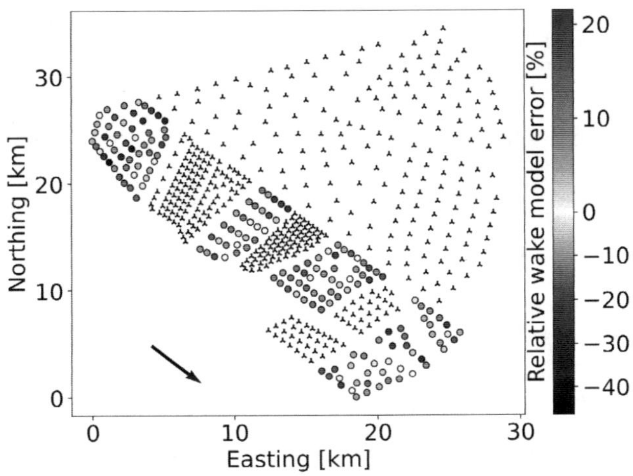


**Figure 10.** Relative wake model error $(100 \cdot \frac{P_{model}-P_{SCADA}}{(P_{model}+P_{SCADA})/2})$ in % for wind coming from north-west (300-330 degrees).

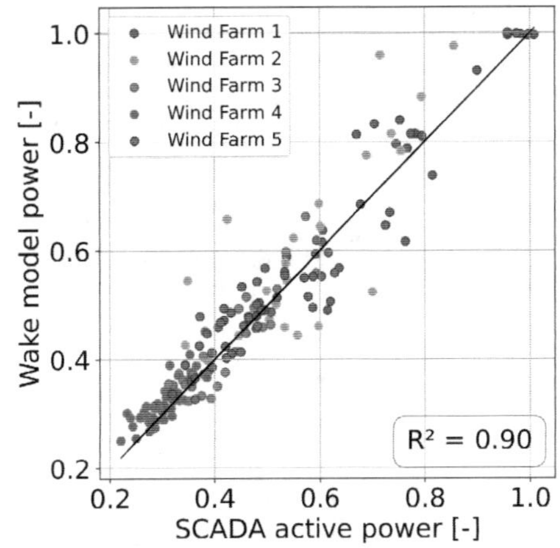


**Figure 11.** SCADA data power versus wake model predicted power for wind coming from north-west (300-330 degrees).

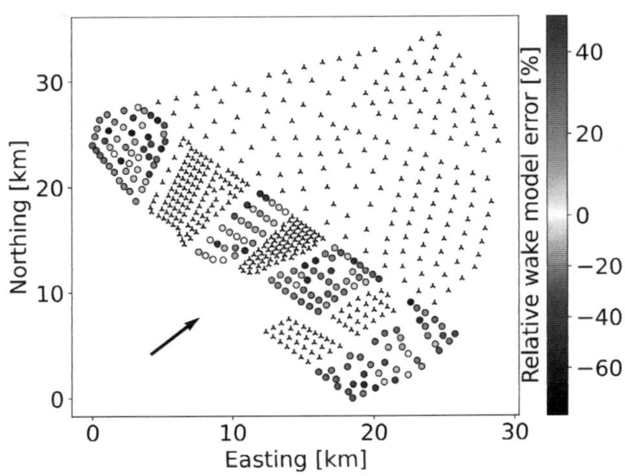


**Figure 12.** Relative wake model error $(100 \cdot \frac{P_{model}-P_{SCADA}}{(P_{model}+P_{SCADA})/2})$ in % for wind coming from south-west (210-240 degrees).

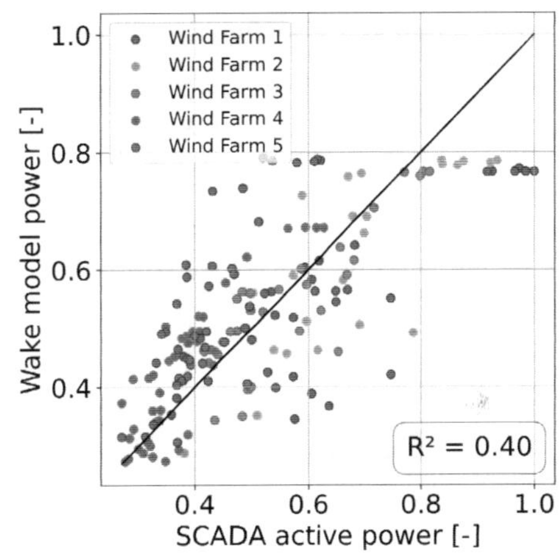


**Figure 13.** SCADA data power versus wake model predicted power for wind coming from south-west (210-240 degrees).

farms might be underestimated if the calibration of tuning parameters is based on data from smaller wind farms.

The analysis of the error between the analytical wake model and SCADA data from the north-west showed that the calibrated wake model accurately accounts for cluster wake effects. Additionally, it revealed a consistent and significant heterogeneous inflow perpendicular to the wind direction for wind originating from south-west. The model with homogeneous inflow failed to capture these effects.

In future research, the aim is to analyze the impact of heterogeneous inflow using different levels of inflow complexity (linear, quadratic, cubic, or a combination thereof). Moreover, turbulence intensity measurements will be incorporated as input to the model to assess their influence on the tuning parameter and the error between the wake model and data. Additionally, future analysis will extend to evaluating the error not only for active power but also for wind direction, potentially providing insights for quantifying blockage of the concession zone. Stability-based ad hoc analysis of the results will be conducted, and the performance of different wake models will be compared.

**Acknowledgments**
The authors would like to acknowledge the support of VLAIO in the context of the Blauwe Cluster project Cloud4Wake. The authors would also like to acknowledge the Energy Transition Funds for their support of the Poseidon project. This research was supported by funding from the Flemish Government under the "Onderzoeksprogramma Artificiële Intelligentie (AI) Vlaanderen" programme.

## References

[1] IEA 2022 Wind electricity https://www.iea.org/reports/wind-electricity licence: CC BY 4.0
[2] Veers P, Dykes K, Lantz E, Barth S, Bottasso C L, Carlson O, Clifton A, Green J, Green P, Holttinen H, Laird D, Lehtomäki V, Lundquist J K, Manwell J, Marquis M, Meneveau C, Moriarty P, Munduate X, Muskulus M, Naughton J, Pao L, Paquette J, Peinke J, Robertson A, Sanz Rodrigo J, Sempreviva A M, Smith J C, Tuohy A and Wiser R 2019 *Science* **366** ISSN 1095-9203
[3] Porté-Agel F, Bastankhah M and Shamsoddin S 2019 *Boundary-Layer Meteorology* **174** 1–59 ISSN 1573-1472
[4] Meyers J, Bottasso C, Dykes K, Fleming P, Gebraad P, Giebel G, Göçmen T and van Wingerden J W 2022 *Wind Energy Science* **7** 2271–2306 ISSN 2366-7451
[5] Veers P, Dykes K, Basu S, Bianchini A, Clifton A, Green P, Holttinen H, Kitzing L, Kosovic B, Lundquist J K, Meyers J, O'Malley M, Shaw W J and Straw B 2022 *Wind Energy Science* **7** 2491–2496 ISSN 2366-7451
[6] Jensen N 1983 A note on wind generator interaction
[7] Katić I, Højstrup J and Jensen N 1987 A simple model for cluster efficiency
[8] Bastankhah M and Porté-Agel F 2014 *Renewable Energy* **70** 116–123 ISSN 0960-1481
[9] Niayifar A and Porté-Agel F 2016 *Energies* **9** 741 ISSN 1996-1073
[10] Zong H and Porté-Agel F 2020 *Journal of Fluid Mechanics* **889**
[11] Nygaard N G, Steen S T, Poulsen L and Pedersen J G 2020 *Journal of Physics: Conference Series* **1618** 062072
[12] King J, Fleming P, King R, Martínez-Tossas L A, Bay C J, Mudafort R and Simley E 2021 *Wind Energy Science* **6** 701–714 ISSN 2366-7451
[13] Bastankhah M, Welch B L, Martínez-Tossas L A, King J and Fleming P 2021 *Journal of Fluid Mechanics* **911** ISSN 1469-7645
[14] J G Pedersen E Svensson L P and Nygaard N G 2022 *Journal of Physics: Conference Series* **2265** 022063
[15] Bay C J, Fleming P, Doekemeijer B, King J, Churchfield M and Mudafort R 2023 *Wind Energy Science* **8** 401–419 ISSN 2366-7451
[16] Ørsted 2019 Ørsted presents update on its long-term financial targets
[17] Branlard E and Gaunaa M 2014 *Wind Energy* **18** 1973–1987 ISSN 1099-1824
[18] Troldborg N and Meyer Forsting A R 2017 *Wind Energy* **20** 2011–2020 ISSN 1099-1824
[19] Bleeg J 2020 *Journal of Physics: Conference Series* **1618** 062054 ISSN 1742-6596
[20] Braunbehrens R, Vad A and Bottasso C L 2023 *Wind Energy Science* **8** 691–723 ISSN 2366-7451
[21] Bleeg J, Purcell M, Ruisi R and Traiger E 2018 *Energies* **11** 1609 ISSN 1996-1073
[22] Bleeg J 2019
[23] Munters W, Adiloglu B, Buckingham S and Beeck J v 2022 *Journal of Physics: Conference Series* **2265** 022049 ISSN 1742-6596
[24] van Binsbergen D, Daems P J, Verstraeten T, Nejad A and Helsen J 2023
[25] Nygaard N G, Poulsen L, Svensson E and Pedersen J G 2022 *Journal of Physics: Conference Series* **2265** 022008
[26] Akiba T, Sano S, Yanase T, Ohta T and Koyama M 2019 Optuna: A next-generation hyperparameter optimization framework
[27] Bergstra J and Bengio Y 2012 *Journal of Machine Learning Research* **13** 281–305 URL `http://jmlr.org/papers/v13/bergstra12a.html`

Journal of Physics: Conference Series **2745** (2024) 012015 doi:10.1088/1742-6596/2745/1/012015

# Wind Speed Ramp Rate Predictions Using Wind Farm SCADA Data Assimilation and a WRF Ensemble

**M Diallo[1], L Letang[2], B Totel[2], P Poncet[3], P Y Andri P[1].**

[1] TRACTEBEL-ENGIE, boulevard Simon Bolivar 34, 1000 Bruxelles - BELGIUM
[2] ENGIE GEMS, boulevard Simon Bolivar 34, 1000 Bruxelles – BELGIUM
[3] ENGIE DIGITAL, 1 Pl. Samuel de Champlain, 92400 Courbevoie, FRANCE

Corresponding author: mouhamet.diallo@tractebel.engie.com

**Abstract.** This study presents a novel method for improving wind power ramp events forecasts up to six hours ahead by utilizing data assimilation of SCADA measurements with an ensemble of Weather Research and Forecasting (WRF) models estimates. Leveraging data from nine wind farms in France and Belgium, the approach aims to improve WRF model predictions for wind speed and ramp event timing. The methodology employs grid and observational nudging techniques, enhancing model accuracy by incorporating real-time observational data. Key findings demonstrate that nudging significantly reduces Mean Absolute Error (MAE), decreases the Time Distortion Index (TDI), and increases the Probability of Detection (POD) of ramp events. Nudged ensemble members outperform non-nudged counterparts, exhibiting better accuracy in identifying true ramp events and reducing false alarms. MAE, TDI and POD improvements are as high as 3.7%, 8.5% and 37%, respectively. The study also explores the benefits of an ensemble approach, highlighting improved accuracy in predicting ramp rate magnitudes and providing valuable insights for grid stability management. This research contributes to wind power forecasting, showcasing the importance of integrating SCADA data into predictive models.

## 1. Introduction

In recent times, there has been a significant rise in global wind and solar energy production, and it is anticipated that a considerable share of the overall energy output will be derived from these sources [1]. Nonetheless, both wind and solar energy exhibit an inherent variability, rendering them less dependable as sources for electrical power generation. This inherent inconsistency adds a layer of complexity to the task of managing power systems, as grid operators must rapidly and effectively harmonizing supply and demand in real-time. Power system operations have implemented strategies to manage demand variability and uncertainty, such as regulation reserves, load-following reserves, and sub-hourly economic dispatch. In spite of the implemented measures, the September 2021 winter storms resulted in significant power losses affecting over five million residents in the US. The challenge of upholding a proper supply-demand equilibrium was amplified by the abrupt reduction in electricity generation, triggering the activation of grid protection mechanisms that ultimately culminated in a widespread blackout. These incidents highlight the utmost importance of accurate forecasting for such exceptional events [2] [3]. Therefore, accurate predictions of sudden surges or drops in wind and or solar electricity generation, referred to as ramp events, are essential for integrating grids, planning systems, and trading electricity in specific electricity markets [4].

Despite being crucial for the industry, ramp events are infamously challenging to forecast [5]. This challenge stems from the accurate assessment of the timing and magnitude of the ramp event. Errors in ramp magnitude occur when a ramp is predicted, but the actual value fluctuates significantly more or less than anticipated. Similarly, errors in ramp timing occur when the actual change in power precedes or follows the predicted time by a significant margin [6].

Early attempt of forecasting ramp rate date back on the early 2000 [7]. Nowadays studies attempting to forecast wind speed ramp rates can be classified in two categories: deterministic and probabilistic forecast. Deterministic forecasting provides a single predicted value for wind speed and direction at a specific location and time in the future, typically derived from physical models of the atmosphere. This method yields a single, definite output of wind speed and direction for a specified time and location. However, deterministic forecasts do not account for the inherent uncertainties in atmospheric conditions, which can lead to decreased reliability for longer time horizons. Consequently, they are not suitable for risk assessment as they do not provide a range of possible outcomes, which is crucial for applications that require evaluating different scenarios and their associated risks. Studies such as those referenced in [8], [9] and [10] employed the mesoscale weather prediction model "WRF" (Weather Research and Forecasting) [11] to forecast wind ramp rates. The results from these studies suggest that the WRF model can generally predict the timing of a ramping event reasonably well, but it struggles with accurately assessing the magnitude of the event. This limitation highlights the challenges associated with deterministic forecasting, particularly in applications that require precise quantification of event magnitudes for effective planning and decision-making. Probabilistic forecasts on the other hand provides a range of possible wind speeds and directions along with the associated probabilities. These forecasts are usually obtained using ensemble models that run multiple simulations with different initial conditions or model parameters. The output is a probability distribution of wind speeds and directions. Probabilistic forecasts are suitable for risk assessment; yet their output can be more complex to interpret as it provides a range of possible values with associated probabilities. Additionally, they are computation intensive: running multiple simulations for ensemble modelling requires more computational resources compared to deterministic models. Example of such approach are [12], [13], and [14] where an ensemble of WRF forecasts were used to forecast ramp rates. The findings from these investigations indicate that employing an ensemble approach enhances the accuracy of the magnitude of the ramp rate forecast. However, they also underscore the challenge associated with weighting the ensemble members, a critical step in ensemble forecasting. Often, a simple averaging of the ensemble members is used, but this approach may not always yield the most accurate results.

A recent trend involves the use of data assimilation (DA) to address the shortcomings of weather forecast models. This technique combines the physical information provided by numerical models with onsite observations, to enhance the forecast accuracy [15]. This combination can be categorized as either empirical, statistical or hybrid depending on how the weighting function is computed to produce an accurate estimate of atmosphere true state: the “analysis”. Statistical DA [16] [15] while robust, requires extensive computational power to identify the most accurate analysis, making it less feasible for operational and time-sensitive applications. On the other hand, empirical DA [17] [18] [19] [20] also known as four-dimensional DA (FDDA), offers a faster, more straightforward and user-friendly approach. However, it is important to note that for empirical DA a notable challenge arises in adjusting the analysis toward an arbitrary state, especially when model predictions greatly diverge from sparse and low-confidence observations in data-deficient areas. In their studies, researchers in references [17] [18] [19] and [20] employed techniques of observation and analysis nudging. Across these studies, it was consistently found that data assimilation improved the accuracy of wind speed and direction forecasts. A critical factor identified was the spatial extent over which an observation was assimilated and the degree of influence each observation exerted on the model. Moreover, the study in reference (17) highlighted that denser observational networks enhanced the benefits of nudging, particularly in accurately forecasting wind speed and direction. However, while these studies concentrated on evaluating the general accuracy of wind speed metrics, they did not address the quantification of the magnitude and timing errors associated with ramp events. Additionally, there was no effort made to measure how nudging techniques affected estimates of wind power production.

The purpose of this research is to introduce a robust approach for accurately forecasting wind power real-time ramp events six hours ahead. Our method's novelty lies in employing an ensemble approach along with data assimilation of Supervisory Control and Data Acquisition (SCADA) measurements from surrounding wind farms relative to the target wind farm. To the best of our knowledge a similar approach to the one proposed in this study has not been conducted yet. The structure of this article is organized as follows: Section 2 introduces the input data and models, which include SCADA data, the WRF model, and the implementation of the nudging technique. Section 3 outlines the workflow of the nudging method the validation metrics employed in the study. Finally, Section 4 and 5 present the results and conclusions drawn from this research.

## 2. Data and models

This section presents the data and model used for conducting the nudging and forecasting ramp rate events.

### 2.1. SCADA data

This research harnessed data from 9 wind farms across France and Belgium. Data were sourced from Darwin, the Engie group's specialized digital platform for data storage and power asset optimization. For confidentiality reasons, specific details such as the exact locations, turbine types, and power curves of these wind farms are not disclosed in this study. The dataset from each wind farm encompassed measurements of wind speed, direction, and air temperature, recorded at 15-minute intervals from 2023-05-14 to 2023-05-16. These data underwent comprehensive automated quality checks within Darwin to rectify common issues, including:

1. Correction of frozen values resulting from communication losses;
2. Identification and removal of outliers using the Local Outlier Factor statistical algorithms;
3. Assessment and correction of abnormal values by comparison with data from neighbouring assets
4. Elimination of duplicate entries;
5. Addressing missing values through techniques like linear and spline interpolation, forward/backward filling, or referencing data from proximate wind turbines within the same farm.

The compiled database was divided into two segments: one for data assimilation, 8 wind farms, and the other for validation purposes, 1 wind farm (figure 1). The validation data were exclusively derived from a wind farm composed of a single wind turbine which is a specificity of Belgium market. This particular wind farm is situated with no other turbines within a radius of five rotor diameters, ensuring that it is unaffected by inter-farm wake effects and have minimal or no wake interference from neighbouring wind farms as concluded in [21]. Furthermore, validation was carried out from 08:00 to 18:00 due to local noise restriction curtailments [22].

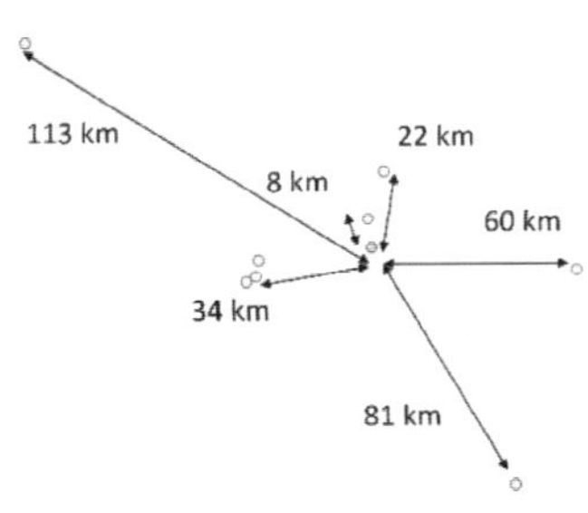


**Figure 1.** Relative distance between the validation site (green) and nudging sites (yellow)

### 2.2. Synthetic SCADA data

Increased observations significantly enhance the advantages of data assimilation, as evidenced in Reference (17). However, wind farm operators typically do not share data from their assets for competitive reasons. Global Wind Power Tracker database [23] illustrates that there are currently numerous wind farms in operation globally. These farms collectively offer an abundant source of measurements, including wind speed, wind direction, and temperature. To replicate the accessibility of these real-world data, we utilized data from the NCEP global surface [24] and upper air observational weather data [25] while manually removing measurements of wind speed, wind direction, and temperature at heights greater than 200 meters above ground level. The database obtained and used contains 2202 stations, aligning with the estimated number of wind farms in the study area, which is around 2500 as per the Global Wind Power Tracker data. This database is available with a one day delay and was quality checked using the Obsgrid module of WRF.

### 2.3. WRF model

WRF [11] is a numerical weather prediction model that solves numerically a set of non-linear equations over a Cartesian grid. These equations model the physical process that occur in the atmosphere and its interaction with the ocean and land surface. WRF takes global model forecasts as initials and boundary conditions inputs to account for large scale. The effects of Physical phenomena with length smaller than the grid size are approximated using parameterization scheme (PS). Table 1 gives WRF model setup.

Table 1: WRF model configuration

| | Configuration specifics |
|---|---|
| version | V 3.8 |
| Initialization | Global Forecast System (GFS) every 6 hours [26] |
| Domain extent | 1°W-9°E, 47°N-55°N |
| Horizontal resolution | 9km and 3km for parent and child domain |
| Vertical resolution | User specified: 64 levels |
| Cumulus PS | Kain Fritsch [27] |
| Radiation PS | RRTMG [28] and Dudhia [29] |
| Microphysics PS | WSM6 [30] |
| Planetary boundary layer (PBL) PS | MYNN 2.5 [31], YSU [32], MYJ [33], Shin Hong [34] |
| Land surface PS | Noah [35] |

To combine WRF with SCADA observations an empirical DA method is used: the "nudging". A short description of the nudging and its implementation in WRF is given in the subsections below.

#### 2.3.1. Observational nudging

Observational nudging involves continuously relaxing the model state, typically hourly to sub-hourly, towards the observed state, as detailed in reference [36]. This approach is particularly advantageous when there is a scarcity of observational data, as it facilitates dynamic and frequent adjustments to the model. Observational nudging is executed by adding to the prognostic equations of the variable to be nudged, an artificial tendency term, which is based on the difference between the two states as illustrated in equation (1).

$$\frac{\partial p\alpha_m}{\partial t}(\vec{x},t) = F_\alpha(\vec{x},t) + G_\alpha p \frac{\sum_{i=1}^{N} W_{\alpha,i}^2(\vec{x},t) \times \left(\alpha_0(\vec{x}_i,t') - \alpha_m(\vec{x}_i,t)\right)}{\sum_{i=1}^{N} W_{\alpha,i}(\vec{x},t)} \quad (1)$$

where, p is the hydrostatic pressure; α is the nudged variable: either the components of the wind speed vector or the temperature; $F_\alpha$ is The model's physical forcing terms for the variable α; $G_\alpha$ is the nudging strength for the variable $\alpha$; $W_{\alpha,I}$ is the weighting function of the observation number $i$ at a location $\vec{x}$ and time t of the WRF computation grid; $\alpha_0$ is the value of observation number $i$ located at $\vec{x}_i$ and measured at time t'; $\alpha_m$ is the WRF value of variable $\alpha$ either at location $\vec{x}$ (native domain grid point) or spatially interpolated to the observation location.

Table 2 presents the WRF input parameters chosen empirically to compute the weight $W_{\alpha,i}$ and define the nudging strength $G_\alpha$.

**Table 2.** Observational nudging settings. Nudging is done only on the parent domain

| | Value | Description |
|---|---|---|
| obs_rinxy | 60 | Radius of influence when computing analysis weights (in km). |
| rinblw | 60 | Radius of influence when computing analysis confidence (in km). |
| obs_coef_wind | 0.0006 | Nudging coefficient. |
| obs_twindo | 0.07 | Half of the length of the time period over which observations are applied (in hours). |
| obs_ionf | 2 | Interval between recalculations of innovations (in units of coarse grid time steps). |
| obs_rinsig | 0.1 | Vertical radius of influence in terms of sigma. |
| auxinput11_interval | 1 | How frequently the model should check to see if additional observations are available (in units of coarse grid time steps). |
| auxinput11_end_h | 99999 | End time for reading observations (in hours). |

In this study, the observation nudging make use of SCADA data of wind speed, wind direction, and temperature, with a frequency of 15 minutes as detailed in section 2.1 of the paper.

*2.3.2. Analysis nudging*

Analysis nudging, as referenced in [36], is an intermittent technique for initializing the model. This approach employs objective analysis techniques, where observed values are interpolated onto the model's grid points to refine the atmospheric state estimate at regular intervals, typically every 6 hours. In this process, Newtonian relaxation terms are integrated into the prognostic equations for wind and temperature. These relaxation terms dictate the rate at which the initial conditions, derived from the first guess, converge towards the gridded analysis. This approach is based on the expectation that enhanced initial conditions will lead to a reduction in forecast errors. However, it is crucial to acknowledge that the production of this gridded analysis demands a greater volume of data compared to observational nudging. Therefore, analysis nudging is more commonly applied on larger scales with coarser grid resolutions.

Equation (2) details the objective analysis employed in this study, while Equation (3) illustrates how the gridded analysis is integrated into the prognostic equation..

$$\alpha_a(\vec{x},t) = \alpha_m(\vec{x},t) + \frac{\sum_{i=1}^{N} W_{\alpha,i}(\vec{x},t) \times \left(\alpha_0(\vec{x}_i,t') - \alpha_m(\vec{x}_i,t)\right)}{\sum_{i=1}^{N} W_{\alpha,i}(\vec{x},t)} \quad (2)$$

$$\frac{\partial p\alpha_m}{\partial t}(\vec{x},t) = F_\alpha(\vec{x},t) + G_\alpha\, p\, W_\alpha(\vec{x},t)\, \varepsilon_\alpha(\vec{x}) \times \left(\alpha_a(\vec{x},t) - \alpha_m(\vec{x},t)\right) \quad (3)$$

where $W_\alpha$ is the weighting function at a location $\vec{x}$ and time t of the WRF computation grid; $\alpha_a$ is the value of the objective analysis of variable $\alpha$ at time t; $\varepsilon_\alpha$ is the analysis quality factor which is based on

Journal of Physics: Conference Series **2745** (2024) 012015 doi:10.1088/1742-6596/2745/1/012015

the quality and distribution of the data used to produce the gridded analysis. Other variables are identical to those presented Equation (1).

Table 3 presents the WRF input parameters chosen empirically to compute the weight $W_\alpha$ and define the nudging strength $G_\alpha$.

**Table 3.** Analysis nudging settings. Nudging is done only on the parent domain

| | Value | Description |
|---|---|---|
| oa_type | 'Cressman' | Observation analysis methodology |
| radius_influence | 30 | Radius of influence when computing analysis weights (in km). |
| fdda_start | 0 | Specifies analysis nudging start (beginning of forecast). |
| fdda_end | 24 | Specifies analysis nudging end (in hours). |
| guv_sfc | 0.0002 | Nudging strength for surface wind speed analysis. |
| gt_sfc | 0.0002 | Nudging strength for surface temperature analysis. |
| gq_sfc | 0.0000 | Nudging strength for surface mixing ratio analysis (deactivated). |
| guv | 0.0002 | Nudging strength for upper air wind speed analysis. |
| gt | 0.0002 | Nudging strength for upper air temperature analysis. |
| gq | 0.0000 | Nudging strength for upper air mixing ratio analysis (deactivated). |

In this study, the analysis nudging make use of synthetic SCADA data of wind speed, wind direction, and temperature, with the same frequency than the initial and boundary conditions of GFS: 6 hours. Synthetic SCADA data are further detailed in section 2.2 of the paper.

## 3. Method and validation metrics

This section presents the methodology used for accurately forecasting wind power ramp events as well as the validation carried out

### *3.1. Methodology workflow*

Our methodology incorporates two types of nudging techniques: grid nudging 2.3.2 and observational nudging 2.3.1. In the lead-up to the day of interest (2023-05-14 00:00:00 to 2023-05-15 00:00:00), grid nudging is employed to refine our estimate of the atmosphere true state. This enhancement is achieved by adjusting the large-scale initial and boundary conditions from GFS to align with the gridded analysis. Consequently, at the start of the focus period (2023-05-15 00:00:00), we possess an improved understanding derived from the data of the preceding day. Subsequently, we employ observational nudging to integrate new, incoming SCADA observations. Every 6 hours, new SCADA observations are assimilated by initiating a fresh forecast from a designated restart point. These restart points are automatically generated at 6-hour intervals and define the span of our forecast range. The workflow of the methodology is shown in figure 2.

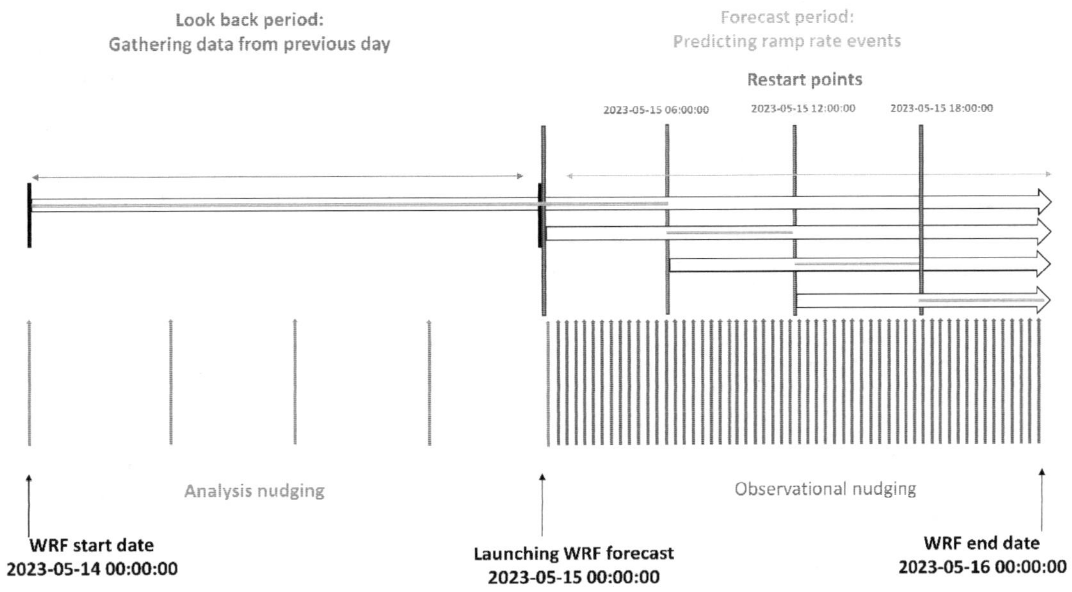


**Figure 2.** Methodology workflow

### *3.2. Ensemble approach*

Various sensitivity studies were carried in WRF by Carvalho and al [37]; Draxl and al [38]; Krogsaeter and Reuder [39] and Muñoz Esparza and al [40] for a total of 4 different Planetary boundary layer (PBL) schemes. An intercomparison of these studies showed that the accuracy of PBL parameterization schemes is site and stability condition dependent. The choice of the best model setup for a forecasting system for a particular region being dependant on the typical distribution of atmospheric stability conditions at the site, we decided to use a multi ensemble approach to benefit from the strength of each individual ensemble members.

**Table 4.** Ensemble member description. Other physical and numerical parameters are identical

| Ensemble | Nudging | PBL scheme |
|---|---|---|
| e1/e1n | No/Yes | MYNN 2.5 [31] |
| e2/e2n | No/Yes | YSU [32] |
| e3/e3n | No/Yes | MYJ [33] |
| e4/e4n | No/Yes | Shin Hong [34] |

Carvalho and al (35), Krogsaeter and Reuder (37) and Shin and Hong (39) indicate that local closure schemes, such as MYNN2.5 and MYJ, are better suited for atmospheres with stable stability. These schemes are designed to assume that fluxes are dependent only on the values and local gradients at nearby grid points. Conversely, for unstable atmospheric conditions, non-local closure schemes like YSU and Shin Hong are often preferred. These non-local schemes account for turbulent fluxes influenced by large eddies, which transport variables over more considerable distances. However, it is important to note that non-local closure schemes can struggle in replicating accurate stable wind speed vertical profiles, particularly in stable and stratified atmospheric conditions, as they tend to cause excessive turbulent mixing near the surface (35). This overestimation of mixing can lead to inaccuracies in modelling wind profiles under such conditions. The choice between local and non-local closure schemes, therefore, depends on the specific atmospheric stability being modelled, with each having its strengths and limitations in different meteorological scenarios. The selection of these schemes, table 4, was strategic to encompass turbulence modelling across various terrain types, complexities, and meteorological conditions.

### *3.3. Ramp rate detection*

To identify ramp events, we employ the swinging door algorithm [6]. This algorithm facilitates the extraction of ramps from signals, employing a piecewise linear method, while accommodating a threshold parameter that influences its sensitivity to variations in ramps. The implementation of Aleksandr F. Mikhaylov was used for this study [42]. The threshold parameter selected is 0.01 which highlighted thirty-six power ramp events during the validation period.

### *3.4. Validation procedure*

The validation of our approach hinges on several metrics, including the Probability of Detection (POD), False Alarm Rate (FAR), Mean Absolute Error (MAE), and Time Distortion Index (TDI) [6] [12]. While POD and FAR gauge the ensembl''s capacity to detect ramp occurrences, MAE and TDI quantify the accuracy of ramp magnitude and temporal alignment using a dynamic time warping algorithm [43].

These metrics are described as follows:

$$POD = \frac{Total\ number\ of\ correct\ ramp\ event\ forecasts}{Total\ number\ of\ ramp\ event\ observed} \times 100 \tag{4}$$

$$FAR = \frac{Total\ number\ of\ false\ alarms}{Total\ number\ of\ ramp\ event\ forecasted} \times 100 \tag{5}$$

$$MAE = \frac{\sum_i |P_{WRF}(\mathrm{i}) - P_{SCADA}(i)|}{\sum_i P_{SCADA}(i)} \times 100 \tag{6}$$

$$TDI = \frac{\text{area under optimal warping path}}{area\ under\ identity\ path} \tag{7}$$

where $P_{SCADA}$ and $P_{WRF}$ are the power measured by SCADA and computed from WRF estimated wind speed; the warping and identity path are obtained by computing the cost matrix: a measure of the amplitude difference between a sample from the SCADA and a sample from WRF. Further information on the TDI can be found in [6] and its implementation in [44].

The MAE, POD, and FAR are metrics that range from 0% to 100%. A 0% MAE or FAR indicates a perfect forecast, where the former denotes no deviation from observed values and the latter reflects no false alarms. Conversely, a 100% POD signifies a perfect model in terms of correctly identifying ramp events. The TDI is a dimensionless number varying in the interval [0,1], where 0 corresponds with the null temporal distortion and 1 with the maximum temporal distortion.

## 4. Results

This section presents the power production data measured at the validation wind farm via SCADA, alongside various estimates derived from WRF ensemble members.

It is observed from the data analysis that the impact of nudging dissipates after 2023-05-15 13:30:00 and that the forecast accuracy decreases (figure 3). This phenomenon can be explained as follows: from 2023-05-14 00:00:00 to 2023-05-15 00:00:00, the abundance of synthetic SCADA data contributes to improved initial conditions for the model. However, from 2023-05-15 00:00:00 to 2023-05-15 13:00:00, only observational nudging from 8 stations is applied, and this is updated every 6 hours. After the second update cycle, the domain's inherent momentum likely becomes dominant, diminishing the influence of the limited observational data points. Consequently, we infer that our methodology is capable of forecasting ramp rates up to 6 hours ahead but loses efficacy beyond 13 hours from the initial observation nudging cycle when limited SCADA data are available.

To evaluate the effectiveness of the proposed nudging methodology, key performance metrics such as the Probability of Detection (POD), False Alarm Ratio (FAR), Mean Absolute Error (MAE), and Time Displacement Index (TDI) were calculated for each ensemble member (table 5). These metrics, detailed in 3.4, cover the time period from 2023-05-15 08:00:00 to 2023-05-15 13:00:00. The results indicate that ramp events, identified using the Swinging Door Algorithm, are consistently captured by the nudged members, with a POD of 100%. With the exception of member 1n, the nudged members generally exhibit equal or lower FAR compared to their non-nudged counterparts, suggesting enhanced accuracy in detecting true events and reducing false alarms.

In terms of the magnitude of ramp events, the MAE values demonstrate notable improvements. For instance, member 1n exhibited a 37.4% improvement in MAE over member 1 (table 5). The smallest improvement was observed in members 3n, with a 3.1% enhancement in MAE. These findings highlight the nudging methodology's efficacy in refining the accuracy of ramp event magnitude predictions. Regarding the time alignment of ramp events, the TDI scores of the nudged members, with the exception of member 1n, are lower, indicating superior accuracy in predicting the timing of ramping events.

It should be noted that the accuracy of the forecasting could be further enhanced. Both observation and analysis nudging techniques are significantly influenced by factors such as the radius of influence and the specific parameters of the nudging process. While a sensitivity study to explore these aspects could provide valuable insights and improved metrics, it was beyond the scope of this current research.

Figure 3 provides a probabilistic perspective on the ramp rate forecast, illustrating the median, maximum, and minimum power production values for both nudged and non-nudged members. The median value offers an estimate of the most probable ramp magnitude, while the ensemble spread reflects the associated uncertainty. For example, during the interval between 2023-05-15 12:00:00 and 2023-05-15 13:00:00, when multiple ramp events are observed, the nudged members predict an increase of 800kW, albeit with considerable uncertainty as indicated by the ensemble spread. However, this range of uncertainty still encompasses the actual value recorded by the SCADA system: 910 kW power increase. Grid operators can leverage this information to establish threshold values for activating grid protection mechanisms, considering various cost factors. These include the cost incurred for responding to a forecasted event, the cost arising from unnecessary purchases of ancillary services, and the cost associated with unanticipated events that were not forecasted and thus not prepared for. By balancing these costs, operators can more effectively manage grid stability and efficiency.

Table 5. Statistical metrics. The data considered for obtaining the statistic below range from 2023-05-15 08:00:00 to 2023-05-15 13:00:00.

| Ensemble | POD(%) | FAR(%) | MAE(%) | TDI(-) |
|---|---|---|---|---|
| e1/(**e1n**) | 94% (**100%**) | 6% (**19%**) | 10.21% (**6.47%**) | 0.803 (**0.872**) |
| e2/(**e2n**) | 81% (**100%**) | 19% (**19%**) | 12.09% (**11.78%**) | 0.747 (**0.721**) |
| e3/(**e3n**) | 100% (**100%**) | 25% (**25%**) | 12.47% (**9.71%**) | 0.761 (**0.662**) |
| e4/(**e4n**) | 63% (**100%**) | 44% (**19%**) | 10.15% (**8.88%**) | 0.798 (**0.779**) |

Journal of Physics: Conference Series **2745** (2024) 012015 doi:10.1088/1742-6596/2745/1/012015

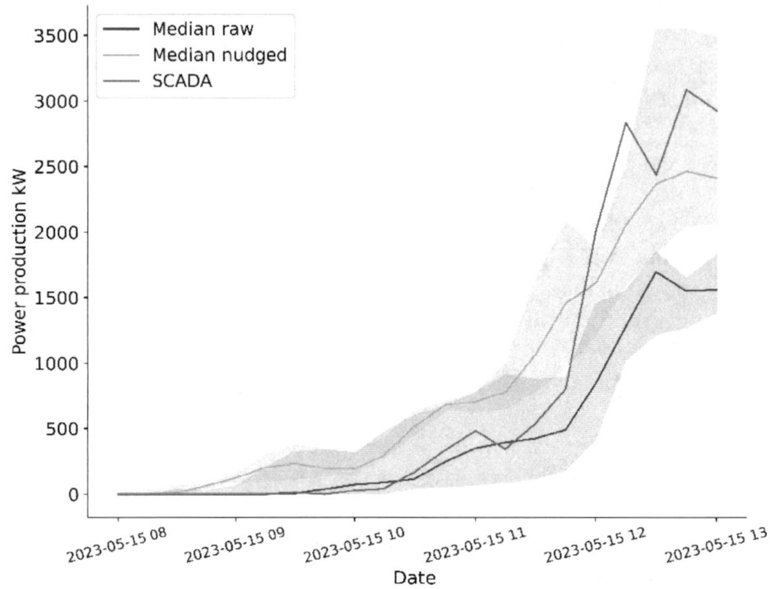


**Figure 3.** Power production probabilistic forecast. The band thickness is given by the ensemble member with lowest and highest value for each time steps

## 5. Conclusions

This study presents a methodological advancement in the forecasting of wind power ramp events, utilizing data assimilation of SCADA measurements within an ensemble of Weather Research and Forecasting (WRF) models. The approach leverages observational data from multiple wind farms in France and Belgium, aiming to refine the predictive capabilities of the WRF models, particularly for wind speed and ramp event timing.

The findings of this research indicate that incorporating grid and observational nudging techniques within the data assimilation process results in a reduction of Mean Absolute Error (MAE) and an improvement in the Probability of Detection (POD) of ramp events. The analysis demonstrates that the ensemble members enhanced with nudging techniques generally exhibit superior performance compared to their non-nudged counterparts, particularly in terms of accurately detecting true ramp events and minimizing false alarms. The improvement is highly dependent of the availability of measurements which dictates the forecast horizon over witch improvements are observed. Moreover, the study highlights the utility of an ensemble-based approach in forecasting. This approach provides essential information for grid operators, facilitating more effective grid stability management.

Overall, this research contributes to the field of wind power forecasting by illustrating the potential benefits of integrating SCADA measurements with ensemble-based WRF models. This advancement is particularly relevant for the effective integration of renewable energy sources into power grids, optimizing system planning, and enhancing the operation of electricity markets.

**Acknowledgments**
Authors wishing to acknowledge assistance from Engie Darwin and TRACTEBEL's green fuels, thermal and nuclear teams for their support in allocating computing hours.

## Bibliography

[1] Renewables 2022-Analysis 2022 (Paris, FR: IEA).

[2] Andersson G and al., 2005 Causes of the 2003 major grid blackouts in North America and Europe, and recommended means to improve system dynamic performance *IEEE Transactions on Power Systems* **20** pp 1922-1928.

[3] Wikipedia 2023 List of major power outages [Online] Available: https://en.wikipedia.org/wiki/List_of_major_power_outages. [Accessed on 01 09 2023].

[4] Wang Q, Wu H, Florita A, Anido C and Hodge B 2016 The value of improved wind power forecasting: Grid flexibility quantification, ramp capability analysis, and impacts of electricity market operation timescales *Applied Energy* **184** pp 693-713.

[5] Bessa R, Browell J and Pinson P 2019 The future of forecasting for renewable energy *WIREs Energy and Environment* **9** e365.

[6] Florita A, Hodge B and K. Orwig K 2013 *Identifying Wind and Solar Ramping Events* (Denver, CO, USA: IEEE Xplore) pp 147-152.

[7] Cristobal Gallego-Castillo and Alvaro Cuerva-Tejero 2015 Oscar Lopez-Garcia A review on the recent history of wind power ramp forecasting *Renewable and Sustainable Energy Reviews* **52** pp 1148-1157.

[8] Bradford K, Carpenter R and Shaw B 2010 *Forecasting Southern Plains wind ramp events using the WRF model at 3km* (Atlanta, GA: American Meteorological Society) pp 1-10.

[9] Haupt S and al 2011 *Forecasting wind power ramps* (Amsterdam, NL: ICWE).

[10] Yang Q, Berg L, Pekour M, Fast J, Newsom R, Stoelinga M and Finley C 2013 Evaluation of WRF-Predicted Near-Hub-Height Winds and Ramp Events over a Pacific Northwest Site with Complex Terrain *Journal of Applied Meteorology and Climatology* **52** pp 1753-1763.

[11] Skamarock B 2008 *A description of the Advanced Research WRF version 3* opensky.ucar.edu.

[12] Gallus W, Ames 2011 *Wind ramp events at an Iowa wind farm: a climatology and evaluation of WRF ensemble forecast skill* (Seattle, WAS: American Meteorological Society).

[13] Deppe W, Gallus W and Takle E 2013 A WRF Ensemble for Improved Wind Speed Forecasts at Turbine Height *Weather and Forecasting* **28** pp 212-228.

[14] D. M. S. a. R. B. Stull 2017 Benefits of a multimodel ensemble for hub-height wind prediction in mountainous terrain *Wind Energy* **21** pp 783-800.

[15] Diallo. M. 2018 Solar irradiance forecast and assesment in the intertropical zone, PhD Thesis.

[16] James Wilczak and al 2015 The Wind Forecast Improvement Project (WFIP): A Public–Private Partnership Addressing Wind Energy Forecast Needs *Bulletin of the American Meteorological Society* **96** pp 1699-1718.

[17] Reen B and Dumais R 2014 *Assimilating Tropospheric Airborne Meteorological Data Reporting (TAMDAR) Observations and the Relative Value of Other Observation Types* (Adelphi, MD: Army Research Laboratory).

[18] Jonassen O, Ólafsson H, Ágústsson H, Rögnvaldsson Ó and Reuder J 2012 Improving High-Resolution Numerical Weather Simulations by Assimilating Data from an Unmanned Aerial System *Monthly Weather Review* **140** pp 3734-3756.

[19] Tran T, Tran H, Mansfield M, Lyman S and Crosman E 2018 Four dimensional data assimilation (FDDA) impacts on WRF performance in simulating inversion layer structure and distributions of CMAQ-simulated winter ozone concentrations in Uintah Basin *Atmospheric Environment* **177** pp 75-92.

[20] Kanda I and Nayak S 2023 Examining the Effectiveness of Doppler Lidar-Based Observation Nudging in WRF Simulation for Wind Field: A Case Study over Osaka, Japan *atmosphere* **14** pp 1-17.

[21] Porté-Agel F and Akbar M 2015 Influence of atmospheric stability on wind-turbine wakes: A large-eddy simulation study *physics of fluid* **27** 035104.

[22] Vlaanderen VLAREM II - bijlagen: Bijlage 5.20.6.1. Richtwaarden voor windturbinegeluid [Onlinee]. Available: https://navigator.emis.vito.be/detail?woId=42564. [Accessed on 15 01 2024].

[23] Global wind power tracker 2024. [Online]. Available: https://globalenergymonitor.org/projects/global-wind-power-tracker/tracker-map/. [Accessed on 01 11 2024].

[24] National Centers for Environmental Prediction/National Weather Service/NOAA/U.S. Department of Commerce updated daily. NCEP ADP Global Surface Observational Weather Data, October 1999 - continuing. Research Data Archive at the National Center for Atmospheric Research, Computational and Information Systems Laboratory 2024. [Online]. Available: https://doi.org/10.5065/4F4P-E398. [Accessed on 15 01 2024].

[25] Satellite Services Division/Office of Satellite Data Processing and Distribution/NESDIS/NOAA/U.S. Department of Commerce, and National Centers for Environmental Prediction/National Weather Service/NOAA/U.S. Department of Commerce, updated daily. NCEP ADP Global Upper Air Observational Weather Data, October 1999 - continuing. Research Data Archive at the National Center for Atmospheric Research, Computational and Information Systems Laboratory2004. [En ligne]. Available: https://doi.org/10.5065/39C5-Z211. [Accessed on 15 01 2024].

[26] National Centers for Environmental Prediction/National Weather Service/NOAA/U.S. Department of Commerce CEP GDAS/FNL 0.25 Degree Global Tropospheric Analyses and Forecast Grids (Updated daily) 2015. [Online]. Available: https://doi.org/10.5065/D65Q4T4Z. [Accessed on 15 01 2024].

[27] J. S. Kain 2004 The Kain–Fritsch Convective Parameterization: An Update *Journal of Applied Meteorology and Climatology* **43** pp 170-181.

[28] Iacono J, Delamere J, Mlawer E, Shepard M, Clough S and Collins W 2008 Radiative forcing by long-lived greenhouse gases: Calculations with the AER radiative transfer models *Journal of geophysical research* **113** D13103.

[29] J. Dudhia 1989 Numerical Study of Convection Observed during the Winter Monsoon Experiment Using a Mesoscale Two-Dimensional Model *Journal of the Atmospheric Sciences* **46** pp 3077-3107.

[30] Hong S and Lim J 2006 The WRF Single-Moment 6-Class Microphysics Scheme (WSM6) *Asia-pacific Journal of Atmospheric Sciences* **42** pp 129-151.

[31] Nakanishi M and Niino H An Improved Mellor-Yamada Level 3 Model: Its Numerical Stability and Application to a Regional Prediction of Advection Fog *Boundary Layer Meteorology* **119**.

[32] Hong S, Noh Y and Dudhia J 2006 A new vertical diffusion package with an explicit treatment of entrainment processes **134** pp 2318-2341.

[33] Z. Janjic 1994 The Step-Mountain Eta Coordinate Model: Further Developments of the Convection, Viscous Sublayer, and Turbulence Closure Schemes *Monthly Weather Review* **122** pp 927-945.

[34] Shin H, Hong S 2015 Representation of the subgrid-scale turbulent transport in convective boundary layers at gray-zone resolutions *Monthly Weather Review* **143** pp 250-271.

[35] Ek M, Mitchell K, Lin Y, Rogers E, Grunmann, Koren V, Gayno G and Tarpley J 2003 Implementation of NOAH land surface model advances in the NCEP operational mesoscale Eta model *Journal of geophysical research* **108** GCP12.

[36] Aijun D, Stauffer D, Dudhia J, Otte T and Hunter G 2007 *Update On Analysis Nudging FDDA In WRF-ARW* (Boulder, CO: NCAR).

[37] Carvalho D, Rocha A, Gomez Gesteira M and silva Santos C 2014 Sensitivity of the WRF model wind simulation and wind energy production estimates to planetary boundary layer parameterizations for onshore and offshore areas in the Iberian Peninsula *Applied energy* **135** pp 234-246.

[38] Draxl C, Hahmann A, Peña A and Giebel G 2014 Evaluating winds and vertical wind shear from Weather Research and Forecasting model forecasts using seven planetary boundary layer schemes: Evaluation of wind shear in the WRF model *Wind energy* **17** pp 39-55.

[39] Krogsaeter O 2015 Validation of boundary layer parameterization schemes in the weather research and forecasting model under the aspect of offshore wind energy applications- Part I **18** pp 769-782.

[40] Canadillas B, Muñoz Esparza and Neumann T 2012 *Forecasting the Diabatic Offshore Wind Profile at FINO1 with the WRF Mesoscale Model* (Amsterdam, NL: EWEA Offshore).

[41] Shin H an Hong S 2011 Intercomparison of Planetary Boundary-Layer Parametrizations in the WRF Model for a Single Day from CASES-99 *Boundary Layer Meteorology* **139** pp 261-281.

[42] A. F. Mikhaylov, «Swinging Door,» [Online]. Available: https://pypi.org/project/swinging-door/. [Accessed on 15 01 2024].

[43] H. Sakoe and S. Chiba 1978 Dynamic programming algorithm optimization for spoken word recognition, Acoustics, Speech, and Signal processing *IEEE Transactions on Acoustics, Speech, and Signal Processing* **26 pp** 43-49.

[44] R. Tavenard, «tslearn's documentation,» 2017. [En ligne]. Available: https://tslearn.readthedocs.io/en/stable/. [Accessed on 15 01 2024].

[45] Tegen I, Hollrig P, Chin M, Fung I, Jacob D and Penner J Contribution of different aerosol species to the global aerosol extinction optical thickness: Estimates from model results *Journal of geophysical research* **102** pp 23895-23915.

[46] Cutululis N, Litong Palima M and Sorensen P 2012 *North Sea Offshore Wind Power Variability in 2020 and 2030* (Lisbon, POR: Wind and Solar Integration Workshop).

Journal of Physics: Conference Series **2745** (2024) 012016 doi:10.1088/1742-6596/2745/1/012016

# Enhancing standard CFD based practices for site assessment through polynomial surrogates for estimating uncertainty in wind speed

**Zahra Lakdawala[1], Hassan Kassem[1] and Jonas Schulte[1]**

[1]Fraunhofer Institute for Wind Energy Systems, Küpkersweg 70, 26129 Oldenburg

E-mail: zahra.lakdawala@iwes.fraunhofer.de

**Abstract.** One of the challenges in digitizing the wind energy sector is to have reliable, accurate and efficient data assessment and prediction tools. We believe that accurate surrogates are key towards improving efficiency and reliability for informed decision making. For a quantitative understanding of wind and energy in complex terrains, a detailed site assessment is done, most often using mesoscale and/or microscale computational fluid dynamics (CFD) based approach. Typically, one would require multiple scenarios/case for a CFD simulator to incorporate the influence of a parameter on variability of wind speed and energy production. These simulations are computationally expensive, and the industry is constantly on a look out for more efficient methods. Towards this end, we propose using a polynomial surrogate based on Polynomial Chaos Expansion (PCE) for uncertainty quantification that requires only a few CFD simulations to assess the variability of wind speed with respect to variability in an input parameter. The variability estimates from PCE were first validated against the standard Monte Carlo method on simple benchmarks and then tested for complex terrains. In this work, we present results that help us understand how canopy foliage, particularly its height, impacts the variability in wind speed in complex terrains. Since the polynomial surrogate relies only on a limited set of input variations (forest height) to estimate the variability distributions for wind speed locally (for specified points of interests), it has proven to be a low-cost and accurate method for estimating uncertainty. In this paper, we discuss the results for wind speed variability for five sites with varying complexity using the polynomial surrogate model. Furthermore, the singular value decomposition (SVD) entropy value is used to quantify terrain complexity for the sites considered. We show how the variability in wind speed can potentially be correlated to terrain complexity. This is our first attempt in providing an empirically (including multiple sites) derived relationship that can be used as a basis for fast and reliable variability estimator for wind resource planners.

## 1 Introduction

The uncertain behavior of wind poses a challenge towards accurately determining the wind speed, the wind energy yield and predicting the overall performance of wind farms [1]. Information on the complex terrain, site quality and energy yields of a wind farm (in planning) can only be ascertained with the aid of simulations. Multiple model scenarios are usually considered for the complex terrains where the performance is predicted using computational fluid dynamics (CFD) simulations with varying grid resolutions and input parameters such as forest tree height, roughness length, wind direction, and/or thermal stability parameters. Most of these parameters vary over a wide range of time and length scales. To incorporate the impact of parameter variation, a typical site assessment workflow looks as follows:

- a domain of possible inputs (multiple scenarios) is defined
- a deterministic computation on the inputs using a physics-based CFD simulator is performed
- the results are aggregated for further analysis (risk assessment and decision making)

The uncertainty is conventionally quantified using a Monte-Carlo (MC) approach that would require a substantial number of cases/CFD-simulation runs. However, the CFD-based simulations are computationally expensive and only a small subset of realizations can be realistically selected to do a site assessment of wind farms. We explore the use of surrogate models to obtain accurate aggregates to quantify uncertainty using only a subset of simulations. In this work we employ a polynomial surrogate for estimating uncertainty w.r.t. wind speed for variations in canopy foliage height. We then correlate the variability estimate derived from the polynomial surrogate to terrain complexity. One of the challenges was to find a way that correctly quantifies the terrain complexity. We employed the Singular Value Decomposition (SVD) entropy value as a measure for quantifying terrain complexity. To find a correlation between a parameter variability and terrain complexity, a normalized (0-1) ranged scale was used whereby the terrains are ranked w.r.t. their physical complexity on a 10x10m sized domain. We discuss the following topics in this work:

(i) uncertainty is quantified in terms of variability of a parameter, such as wind speed, using the small subset of CFD simulations via polynomial surrogates (to replace the expensive Monte Carlo approach);

(ii) terrain complexity is quantified in terms of the SVD entropy using the special factorization of geometric property of the terrain, such as height, gradients, directional gradients;

(iii) the SVD entropy for the mean of directional gradients of sites is correlated with the maximum variability of the wind speed to provide an insight on how the two are linked to each other. The polynomial surrogate, its application, its associated results and a way to link the variability of a parameter to terrain complexity is discussed in this work.

This work is organized in the following way. Section 2 lays out the essentials on how we (statistically) tackle the topic of input variability in our modeling setup and gives an overview of the polynomial surrogate model using PCE. Section 3 gives an overview of the SVD approach used to calculate the entropy value w.r.t a geometric property of the terrain. The results of the surrogate and the SVD are provided in Section 4. We show how wind speed variability estimates can be derived for varying forest heights in complex terrains for a real site. Furthermore, the sites are ranked according to their complexity and the a linear correlation plot is shown of wind speed variability against terrain complexity.

## 2 The problem of estimating wind speed variability

Let us first discuss how the forest height distribution is interpreted in our CFD setup. Let $v$ denote the forest height (in principle, it can be any single scalar parameter of interest) that enters the setup of the CFD simulations. Let's assume we wish to model a scenario where $v$ is distributed according to a normal distribution with mean $v_0$ and standard deviation $\sigma$,

$$p(v) = \frac{1}{\sigma\sqrt{2\pi}} \exp\left[-\frac{1}{2}\left(\frac{v - v_0}{\sigma}\right)^2\right] . \tag{1}$$

The resulting stochastic effects of $v$ on the CFD results are denoted by $\mathbf{u} = (u_1, \ldots, u_m)$, based on a small number of required simulations $m$ for individual values of $v$.

Then the variable transformation

$$\xi = \frac{v - v_0}{\sigma} \tag{2}$$

maps the random variable $v$ onto another random variable $\xi$ which follows a normal distribution centered at the origin with unit standard deviation, $\hat{p}(\xi) = \frac{1}{\sqrt{2\pi}} w(\xi)$, where $w(\xi)$ is the weight function $w(\xi) = e^{-\frac{1}{2}\xi^2}$. The expectation value of a function $U(\xi)$ is then given by

$$E[U] = \frac{1}{\sqrt{2\pi}} \int_{-\infty}^{\infty} U(\xi)\; w(\xi)\; \mathrm{d}\xi\ . \tag{3}$$

Interpreting $v$ as a random variable that follows a certain distribution $p(v)$, the non-intrusive PCE described in Section 2.1 provides a method for uncertainty propagation that approximates the resulting stochastic effects of $v$ on the CFD results, denoted by $\mathbf{u} = (u_1, \ldots, u_m)$, based on a small number of required simulations for individual values of $v$. PCE is the polynomial surrogate that enables us to understand how uncertainty in forest height is propagated into wind speed calculations.

### *2.1 Polynomial chaos expansion surrogate*

The aim of uncertainty quantification is to describe the effect of uncertain inputs to an output parameter. Most often this is done using several statistical methods such as Monte Carlo, direct (numerical) integration methods or stochastic expansion methods. We describe the stochastic polynomial chaos expansion method to estimate the variability in wind speeds for complex sites in this Section.

A polynomial chaos expansion surrogate models the response of a system under stochastic effects of a random input parameter and can be represented via some (properly constructed, orthogonal) polynomial basis functions [2, 3]. We are mainly interested in the effects of the the random input (forest height) on the output of the function (variability of wind speed).

A stochastic process $U(\boldsymbol{\xi})$, depending on a random variable $\boldsymbol{\xi}$, is defined as a linear combination of basis functions $\boldsymbol{\phi}$ and coefficients of expansion denoted by $\boldsymbol{\alpha}$,

$$U(\boldsymbol{\xi}) = \sum_{n=0}^{\infty} \alpha_n \phi_n(\boldsymbol{\xi}) \approx \sum_{i=0}^{n} \alpha_n \phi_n(\boldsymbol{\xi}). \tag{4}$$

$U(\boldsymbol{\xi})$ is the estimate or the response described as a polynomial surrogate of order $n$. The choice of the polynomial basis $\phi$ (Hermite, Legendre, Jacobi,...) depends on the choice of the underlying input distribution(normal, uniform, beta) [4]. The orthogonal polynomials enables us to express the moments (mean and variance) of the polynomial surrogate, with the mean $\mu[U]$ and variance $\sigma^2[U]$ expressed as

$$\mu[U] := E[U] = \alpha_0; \quad \sigma^2[U] := E[U^2] - E^2[U] = \sum_{n=1}^{\infty} \alpha_n^2 < \phi_i^2 > . \tag{5}$$

The mean is the zeroth coefficient of expansion and the variance is the sum of product of the square of coefficients and polynomial basis inner product. For example, in the case of Hermite polynomial basis

$$\langle \phi_n, \phi_m \rangle = \frac{1}{\sqrt{2\pi}} \int_{-\infty}^{\infty} \phi_n(\xi)\phi_m(\xi) w(\xi) d\xi = \Psi_n^2\ \delta_{nm}\ , \tag{6}$$

| Order | Hermite $\phi$ | Legendre $\phi$ |
|---|---|---|
| 0 | 1 | 1 |
| 1 | $\xi$ | $\xi$ |
| 2 | $\xi^2 - 1$ | $\frac{1}{2}(3\xi^2 - 1)$ |
| 3 | $\xi^3 - 3\xi$ | $\frac{1}{2}(5\xi^3 - 3\xi)$ |

**Table 1.** First five Hermite and Legendre basis polynomials.

where $\Psi_n^2 = \langle\phi_n^2\rangle = n!$ and $\delta_{nm}$ is the Kronecker delta. Table 1 describes the first four components of $\phi$ for the Hermite polynomials (and Legendre for when $\xi$ has an underlying uniform distribution).

A linear system of the form $\boldsymbol{\Phi\alpha} = \boldsymbol{U}$ is constructed. Here, $\boldsymbol{U}$ is a $k$-sample set corresponding to random variables $\boldsymbol{\xi} = (\xi_1, \ldots, \xi_k)$. To incorporate the $k$ simulation response into $U(\boldsymbol{\xi})$, one would need $n + 1$ to be equal to $k$. If number of $k > n + 1$, the system $\boldsymbol{\Phi\alpha} = \boldsymbol{U}$ is over-determined. A least squares approach using linear regression is then used to estimate the coefficients of expansion $\boldsymbol{\alpha}$, such that it minimizes the residual sum of squares problem (7)

$$\boldsymbol{\alpha} = \arg\min||\Phi\boldsymbol{\alpha} - \boldsymbol{U}||_2^2. \tag{7}$$

The least squares enables use to pick the best response in terms of a polynomial surrogate of order $n$ for any given number of samples in the linear system. The response distribution can then be easily constructed using the mean and variance (Equation (5)) for any point of interest.

## 3 The problem of quantifying terrain complexity

Once the wind variability is understood and quantified, we'd like to define a quantifiable measure for terrain complexity. Quantifying the complexity of terrains is still an elusive topic of research. Most definitions describing terrain complexity, such as the Ruggedness Index (RIX) [5] or Terrain Ruggedness Index (TRI) values [6], express it in terms of a fraction of terrain surface that is steeper than a critical slope or as a measure of total number of contour intercepts or fluctuations (ups and downs) along the transect. Some parameterizations are also statistically derived with further correction terms.

For our work, we employ a factorization technique, an idea borrowed from the field of image processing, called singular value decomposition (SVD). The geometrical property of the terrain is represented as an image/matrix $A$ with $m \times n$ real values. Each pixel/component of $A$ is the (color) intensity value at its spatial coordinate. Interpreting $A$ as the field map of a geometric property of the terrain, the factorization using SVD transforms the field into another (basis) representation thereby making it easier to quantify its complexity. If a lot of information in incorporated in the factored form, it is deduced that the data is highly complex. The entropy of the singular values enables us to quantify the terrain complexity with respect to a geometric property and rank them on a 0-1 scale.

### *3.1 Singular value decomposition*

We present a way of estimating terrain complexity using the mean of directional slopes with SVD Entropy. SVD is the factorisation of a matrix of values A expressed as

$$A = U\Sigma V^T$$

where $U$ and $V$ are $m \times m$ and $n \times n$ orthogonal matrices, respectively. One of the many ways to achieve this factorization is using a least squares approach [7]. The columns in these matrices are called the left- and right-singular vectors of $A$, were $U = AA^T$ and $V = A^TA$. $\Sigma$ is a diagonal matrix that only contains non-negative values (referred to as singular values of A) along its diagonal Let $\vec{\sigma}$ denote the diagonal values of $\Sigma$ where $\sigma_i = \Sigma_{ii}$. The singular values are unique, though the singular vectors (in $U$ and $V$) may not necessarily be unique. Conventionally, the singular values are in descending order.

The number of non-zero entries (after rounding of small values if required due to numerical precision issues in computing the factorisation) in $\sigma$ is the rank of matrix $A$, denoted as $k = rank(A)$. Because only the first $k$ elements of $\sigma$ are non-zero, and that the result of the SVD is a simple matrix multiplication, one can define a truncated SVD containing only the first $k$ singular values. Small $k$ implies the data is can be represented with less information. The magnitude of $\sigma$ further becomes a quantitative measure of complexity/stretch in principal axes in the data.

Intuitively, the singular value $\sigma_i$ measures how much of the data set is (proportionally) explained by each vector and this leads us to measure the entropy of $\sigma$ [8]. High values of SVD entropy reflects that all vectors are equally important, i.e., that the terrain structure is not easily compressible, indicating high complexity [7]. Since the terrains usually have different dimensions and inherent complexity, the entropy values are normalized. The entropy for $A$, used as a complexity measure is defined as

$$E(A) = -\frac{1}{\ln(k)} \sum_{i=1}^{k} \bar{\sigma_i} \ln(\bar{\sigma_i})$$

where $\bar{\sigma_i} = \frac{\sigma_i}{\mathrm{sum}(\vec{\sigma})}$

## 4 Results and discussions

We discuss the modeling setup here and provide more insights on the sites under consideration. Figure 1 shows the five sites with varying complexity with elevation variation of about 0-1000m, and varying density of forest coverage across sites. For the sites considered: 1) We studied the

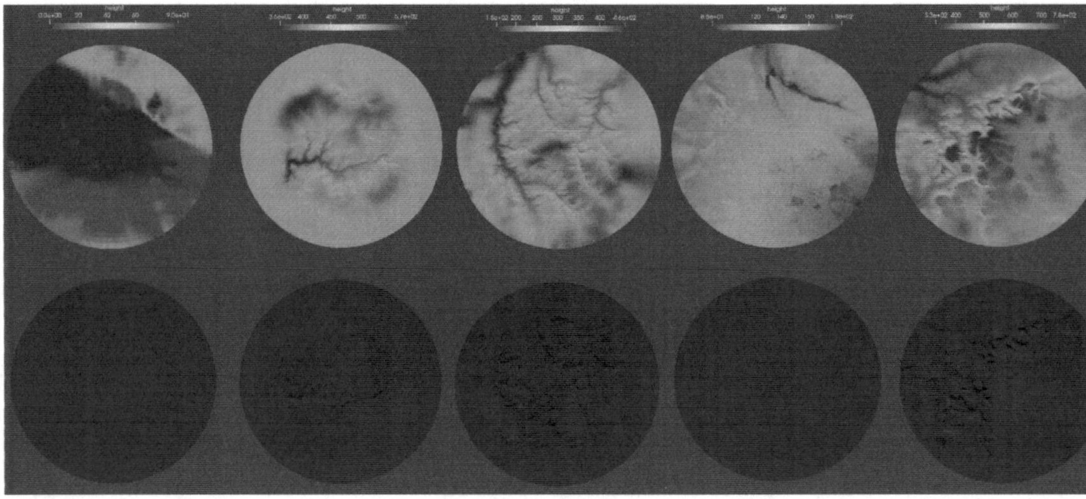

**Figure 1.** Top: Elevation maps varying from about 0-1000m across sites. Bottom: Forest coverage with varying density across sites marked by red regions.

impact of varying height of the canopy foliage on the wind speed in complex terrains. A normal distribution of mean of 15m and standard deviation of 10m was considered for the forest height.

2) We linearly correlate wind speed variability and terrain complexity, for which quantify terrain complexity using the mean of directional gradients (of 36 sectors) for each site.

### *4.1 Wind speed variability response as a function of varying forest height*

The wind speed over the entire site is predicted using a CFD simulation software on a fixed grid resolution and varying forest tree height values. For this work, we have used the software FIWind-OpenFOAM, an in-house CFD simulator which is an OpenFOAM based tool-chain for micro-siting by Fraunhofer IWES [9, 10]. Five sets of CFD simulations were done for each site with forest height (5, 10, . . . , 25m) to estimate the wind speed in each site.

The wind speeds together were provided into the polynomial surrogate as described in Equation (4) to compute the corresponding expansion of coefficients for every grid point. The mean and variance of wind speed is consequently calculated locally for every grid point. The plots in Figure 2 show the column profiles of the CFD simulations corresponding to the five forest heights for two selected points of interests in Site C. The shaded grey area is the variability (variance) computed using the polynomial surrogate. The plots in Figure 3 show the probability density function (pdf) of the wind speed for four points of interest for State 13 and State 22. The pdfs can be reconstructed for any point for any state using the coefficients of expansions (mean and variance) of the polynomial surrogate. It is also interesting to observe that the pdfs are different for every location implying that the variation of wind speed indeed depends on local conditions.

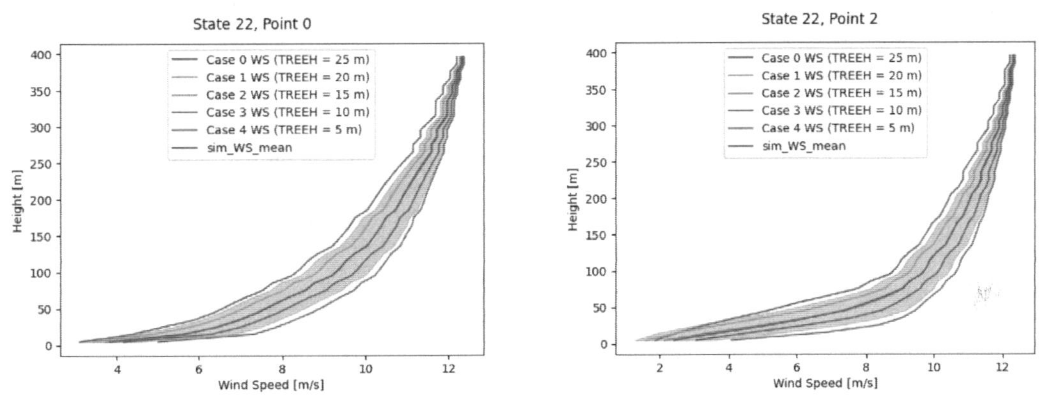


**Figure 2.** The results in this figure correspond to state 22, for Site C. The five curves are the CFD-computed wind speed values along the height for different values of forest height. The grey shaded area is the variance computed using PCE along the height. Right-most plot: The distributions (with mean and variance from PCE) reconstructed for every point of interest (POI).

Figure 4 (Left and Middle) shows a 2D slice view of the mean and variance of wind speed at 100m hub height for Site C. It provides a better insight towards identifying high variability regions. As described above, a pdf can be reconstructed for every point individually. We believe a location-accurate evaluation of the probability densities with parameter variations is a new and useful feature for CFD-based site assessment. The information generated can provide valuable insights for wind farm planning, depending on the application. We observed a consistent trend where the turbulent intensity correlated rather well with the variance of wind speed (sector-wise) (as shown in Figure 4) (Right). The same trend was observed for all sites, all sectors. This motivated us to investigate and correlate the wind speed variability to terrain complexity.

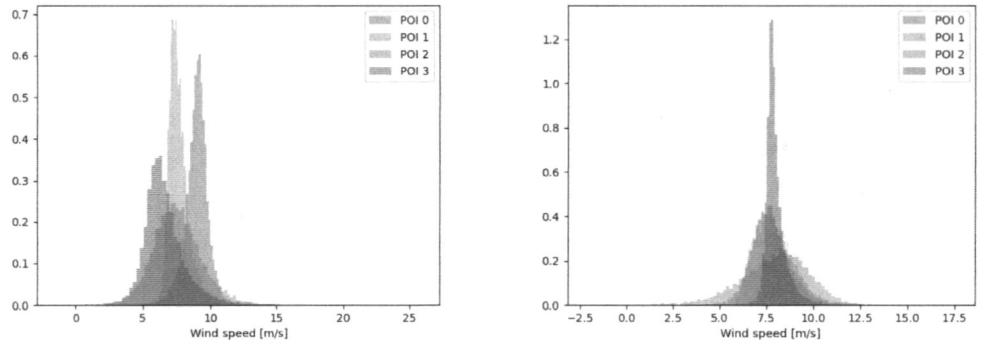


**Figure 3.** The distributions (with mean and variance from PCE) reconstructed for four selected points of interest (POI) for state 13 (left) and state 22 (right).

*4.2 Terrain complexity quantification, classification and correlation*

We apply SVD, as described in Section 3, on a data image/map of a geometrical property (such as height, gradient, directional gradient based on wind directional sector) of the terrain. To define a scale of complexity across sites, we analyze a 10x10km extent in x-y direction for each of the considered sites. The data maps for elevation and the mean of directional gradients are shown in Figures 5 and 6.

Using the same five sites (as above) of varying complexity, we use the height and the mean of directional slopes (over 36 sectors) to compute the SVD entropy. Table 2 shows the entropy values for height and mean directional slopes. We observe that all sites exhibit a very high ($> 0.86$), almost maximal, SVD entropy values illustrating high physical complexity (with degradations ofcourse).

Figure 7 (Left) shows the scale that maps and ranks each site with respect to SVD entropy of mean directional slopes. Figure 7 (Right) shows how the terrain complexity correlates with variability in wind speed due to forest height variations.

## 5 Conclusions

The polynomial surrogate models the response of a system under stochastic effects of a random input parameter. The proposed surrogate is used to estimate the variability in wind speed for complex terrains linked to canopy foliage. The variability is realized as probability density

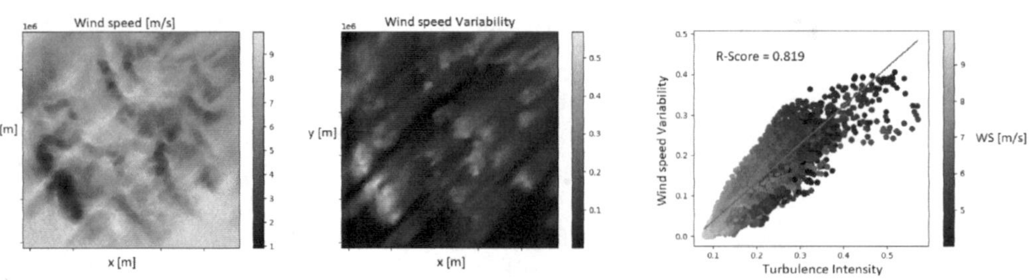


**Figure 4.** Left, Middle: The mean and variance computed for a selected region of interest at hub height 100m. Right: Turbulence intensity correlated with variability in wind speed. All results in this figure correspond to state 22, for Site C.

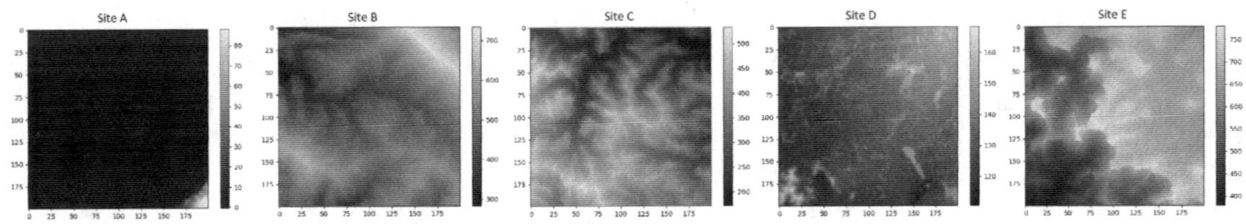


**Figure 5.** Elevation for sites on a 10 x 10km range.

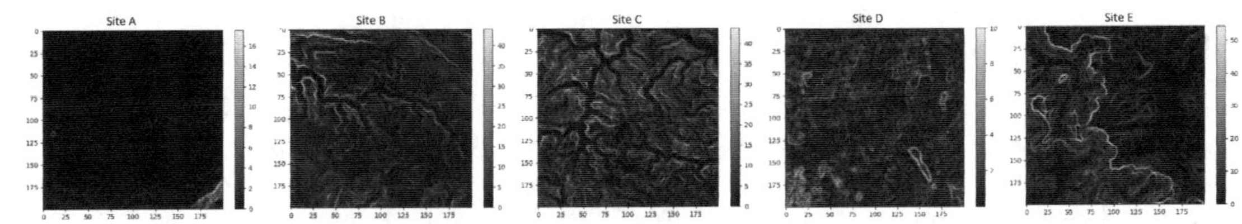


**Figure 6.** Mean of all sector-directional slope on a 10x10km range.

| Site | SVD Entropy Height | SVD Entropy Mean directional slopes | PCE Variability Wind Speed |
|---|---|---|---|
| A | 0.6005 | 0.8640 | 0.1402 |
| B | 0.3872 | 0.9093 | 0.5632 |
| C | 0.4421 | 0.9219 | 1.1850 |
| D | 0.2148 | 0.9186 | 0.1314 |
| E | 0.3467 | 0.9380 | 4.3441 |

**Table 2.** The sites ranked by SVD Entropy scale ranging from 0-1 using elevation and directional slopes The maximum variability at a hub height of 100m is extracted from the polynomial surrogate model.

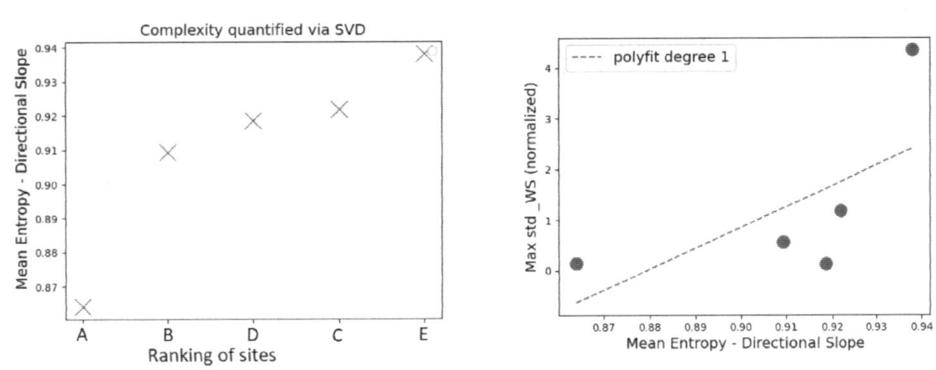


**Figure 7.** Left: Ranking of sites with respect to SVD entropy using mean directional slopes. Right: Correlation plot of SVD entropy and maximum variability at hub height 100m for each site.

functions for velocity for selected observation points. The surrogate described in our work offers a cheap and reliable alternative to the conventional Monte Carlo approach for estimating the uncertainty of wind speed (and other fields of interest) with respect to model and terrain specific input parameters.

We believe that incorporating well-derived surrogates, such as the polynomial surrogate, for quantifying uncertainty for faster and reliable results for site assessment is a valuable resource for industry. Surrogates offer a promising way to enhance the workflow involving computationally expensive CFD simulations. With the proposed surrogate the following is possible

- Fewer scenarios are required for reliable assessment of variability
- Variability in a parameter field can be constructed efficiently for selected region or points of interest.
- Parameter variations can be analyzed using the PCE surrogate, both along vertical and horizontal wind profiles (column and and slice views through the point/domain of interest)

The idea of correlating variability (of a parameter) to the terrain complexity is motivated by the fact that reliable site assessment (using modeling and CFD based simulation workflow) is a time-consuming and costly process. Our attempt is to map out a scientifically driven way to have a reference correlation graph that could potentially provide a smart way of predicting variability with few or no simulations involved. There is room for including more sites on the complexity scale, making the complexity measure multi-dimensional and/or and deriving a better (non-linear) correlation estimate for a faster estimation of parameter variability. This work lays out a motivation for investigating such correlations in a rigorous scientifically-driven way.

**Acknowledgments**
The authors would like to thank and acknowledge the support of Federal Ministry for Economic Affairs and Climate Action (BMWK). This work is done as part of the "Simulation uncertainties for the detailed assessment of wind energy yield (SUnDAY)" project funded by BMWK (grant no. 03EE2010). The simulations were partly performed at the HPC Cluster MOUSE, located at the University of Oldenburg (Germany) funded by BMWK (grant no. 03EE3067A-B).

## References

[1] S. Barber, A. Schubiger, S. Koller, D. Eggli, A. Radi, A. Rumpf, and H. Knaus. The wide range of factors contributing to wind resource assessment accuracy in complex terrain. *Wind Energy Science*, 7(4):1503–1525, 2022.

[2] Serhat Hosder, Robert Walters, and Rafael Perez. A non-intrusive polynomial chaos method for uncertainty propagation in cfd simulations. *44th AIAA Aerospace Sciences Meeting and Exhibit*, 2006.

[3] Andrés Santiago Padrón, Jared Thomas, Andrew P. J. Stanley, Juan J. Alonso, and Andrew Ning. Polynomial chaos to efficiently compute the annual energy production in wind farm layout optimization. *Wind Energy Science*, 4(2):211–231, May 2019.

[4] M. S. Eldred, C. G. Webster, and P.G. Constantine. Evaluation of non-intrusive approaches for wiener-askey generalized polynomial chaos. *49th AIAA/ASME/ASCE/AHS/ASC Struct., Struct. Dyn. Mater. Conf.*, pages 1–22, 2008.

[5] Niels Gylling Mortensen, Ole Rathmann, Andrew Tindal, and Lars Landberg. Field validation of the $\delta$rix performance indicator for flow in complex terrain. 2008.

[6] Shawn Riley, Stephen Degloria, and S.D. Elliot. A terrain ruggedness index that quantifies topographic heterogeneity. *Internation Journal of Science*, 5:23–27, 01 1999.

[7] Gene H. Golub and Christian Reinsch. Singular value decomposition and least squares solutions. *Linear Algebra*, pages 134–151, 1971.

[8] C.E. Shannon. A mathematical theory of communication. *Bell System Technical Journal*, 27(3):379 – 423, 1948. Cited by: 31222.

[9] S. Akbarzadeh, H. Kassem, R. Buhr, G. Steinfeld, and B. Stoevesandt. Adjoint-based calibration of inlet boundary condition for atmospheric computational fluid dynamics solvers. *Wind Energy Science*, 4(4):619–632, 2019.

[10] Chi-Yao Chang, Jonas Schmidt, Martin Dörenkämper, and Bernhard Stoevesandt. A consistent steady state cfd simulation method for stratified atmospheric boundary layer flows. *Journal of Wind Engineering and Industrial Aerodynamics*, 172:55–67, 2018.

Journal of Physics: Conference Series **2745** (2024) 012017 doi:10.1088/1742-6596/2745/1/012017

# Reconstructing turbulent wind-fields using inverse-distance-weighting interpolation and measurements from a pulsed mounted-hub lidar

**Esperanza Soto Sagredo, Ásta Hannesdóttir, Jennifer M. Rinker, Michael Courtney**

Department of Wind and Energy Systems, Technical University of Denmark, Denmark

E-mail: espa@dtu.dk

**Abstract.** This study evaluates a numerical multi-beam pulsed lidar mounted on the hub of the NREL 5MW reference wind turbine using the HAWC2 v13.1 numerical sensor for synthetic lidar measurement generation. While initially designed for single-beam operations, it facilitates multi-beam configuration simulations. We conducted an analysis of full-rotor longitudinal wind speed reconstruction by combining inverse-distance-weighting with synthetic sensor data from HAWC2. Utilizing a Mann-generated turbulence box for wind input at U = 11.4 m/s, we examined three lidar configurations for efficacy. The hub-mounted lidar proved efficient in capturing the incoming flow towards the turbine, showing a 60% improvement in overall reconstruction accuracy across the plane compared to the baseline, where hub-lidar measurements are simple average across the rotor plane. The rotor average wind speed showed a 30% enhancement compared to the baseline. Crucially, the lidar configuration, which impacted the spatial distribution across the rotor plane, emerged as a pivotal factor for effective reconstruction. Proper configuration assessment is essential, especially given the implications of rotational sampling and its impact under various wind conditions, for optimal performance. The proposed method, combining inverse-distance weighting with hub-lidar data for high spatial resolution measurements across the rotor plane, shows significant potential for real-time wind-flow estimation and lidar-assisted control applications.

## 1. Introduction

Nacelle-mounted lidars are crucial for wind measurements in wind turbine applications, but their placement behind the rotor leads to blade blockage, significantly reducing data availability and increasing uncertainty in wind field reconstruction [1, 2].

We propose a spinner-mounted multi-beam pulsed lidar (hereinafter referred to as hub-lidar), which enables spatial measurements across the rotor plane and may enhance wind-field estimation accuracy.

Previous research has explored the concept of spinner-mounted single-beam lidars. In [3], a single-beam continuous wave (CW) lidar (ZephIR) on a 2.3 MW turbine (Vestas NM80) successfully measured upwind wind and turbulence structures in real-time. In [4] multiple configuration scenarios for single-beam CW and pulsed lidars on the NREL 5MW turbine for feed-forward control were examined, finding optimal error minimization with a lidar scan radius of 75% and preview distances between 100 and 160 m. More recently, [5] implemented a

Journal of Physics: Conference Series **2745** (2024) 012017 doi:10.1088/1742-6596/2745/1/012017

spinner-mounted single-beam lidar, both CW and pulsed, to assess its advantages in feed-forward blade pitch control compared to a nacelle-mounted lidar, revealing greater control benefits with the spinner-based single-beam lidar, akin to a four-beam nacelle lidar in feed-forward control performance.

This study utilizes a multi-beam pulsed lidar on the hub of the NREL 5MW [6] with HAWC2 v13.1's hub-lidar numerical sensor [7]. We focus on two main objectives: (1) reconstructing the longitudinal wind component across the full rotor plane using inverse-distance-weighting (IDW) interpolation with synthetic hub-lidar measurements from HAWC2 v13.1, and (2) evaluating the effect of three different hub-lidar configurations on our method's efficacy.

This paper is structured as follows: Section 2 details the specifications of the wind-field turbulence simulation used as the reference case, the chosen lidar configurations, and a description of the implemented reconstruction method. Section 3 presents the results for wind-field reconstruction and rotor average wind speed estimation. A discussion on the study's assumptions is provided in Sec. 4. Finally, Sec. 5 covers the key conclusions.

## 2. Methodology

This section outlines the parameters for the reference turbulence box (Sec. 2.1), details the generation of numerical hub-lidar measurements (Sec. 2.2), and describes the wind-field reconstruction method using inverse distance weighting interpolation (Sec. 2.3).

### *2.1. Mann-generated turbulence box*

In this study, we use the Mann model [8] to simulate a turbulent wind for a wind speed of $U = 11.4$ m/s. The wind condition was defined using IEC-recommended [9] parameters for the Mann model (L=33.6 m, Γ=3.9), with the parameter $\alpha\epsilon^{2/3}$ adjusted for a target turbulence intensity of 10%.

The turbulence box dimensions are $4096 \times 64 \times 64$ (x × y × z), with a grid spatial resolution of 2.0625 m in the rotor plane, resulting in box dimensions of 132 m × 132 m (y × z). The longitudinal dimension extends to 6840 m over 600 s. The inflow wind profile was represented using a power-law model with a shear exponent of 0.2. This turbulence box serves as our reference case for extracting the synthetic lidar measurements.

### *2.2. Synthetic lidar measurements*

In HAWC2 version 13.1 [7], a novel sensor simulating a hub-mounted single-beam wind lidar with a pulsed laser system, known as hub-lidar, has been implemented. The hub-lidar rotates with the rotor and is constrained to the azimuthal angle of blade 1 body in the HAWC2 wind turbine model [10].

We utilized this sensor to generate synthetic lidar measurements from a wind-field created as outlined in Sec. 2.1, conducting aeroelastic simulations on a rigid tower without tilt. This will neglect the effect of the tilt angle and the relative tower moment on the line-of-sight measurements from the hub-lidar.

The sensor computes line-of-sight (LOS) velocities, $V_{LOS}$, at each scanning location by converting the $u$, $v$, and $w$ components of the synthetic turbulent field into the LOS coordinate system. To mimic the lidar's probe volume, the sensor employs the weighting function described in [11].

Users can simulate various lidar configurations by incorporating multiple beams with distinct parameters in the HAWC2 output channels. Additionally, the sampling frequency can be adjusted by post-processing the HAWC2 output results.

In this study, we explored three six-beam lidar configurations, with each beam sampled consecutively at 200 ms intervals without any switching delay. The six-beam hub-lidar settings in HAWC2 included range lengths from 50 to 300 m at each 10 m (26 ranges), a range-gate length of $\Delta P = 38.4$ m, a beam width of $\Delta L = 24.75$ m, normalized integration volume half-width of $dv = 3.0$, and integration points equal to $N_{lidar} = 200$. These parameters were chosen based on a convergence study of the HAWC2 hub-lidar sensor detailed in [10].

The half-cone and azimuthal angles for each configuration are specified in Table 1, with the initial azimuthal angle determined by the starting position of blade 1, measured clockwise from downwind perspective, as detailed in [10]. These configuration were determined via a preliminary investigation into good beam settings, not detailed here for brevity.

**Table 1.** Lidar configurations: Half-cone ($\theta_i$) and azimuthal ($\phi_i$) angles

| | **Beam Number** | | | | | | | | | | | |
|---|---|---|---|---|---|---|---|---|---|---|---|---|
| | **1** | | **2** | | **3** | | **4** | | **5** | | **6** | |
| **Configuration** | $\theta_1$ | $\phi_1$ | $\theta_2$ | $\phi_2$ | $\theta_3$ | $\phi_3$ | $\theta_4$ | $\phi_4$ | $\theta_5$ | $\phi_5$ | $\theta_6$ | $\phi_6$ |
| 1 | 20° | 0° | 20° | 120° | 20° | 240° | 20° | 0° | 20° | 0° | 20° | 120° |
| 2 | 20° | 0° | 20° | 120° | 20° | 240° | 20° | 0° | 20° | 120° | 20° | 240° |
| 3 | 0° | 0° | 20° | 72° | 20° | 144° | 20° | 216° | 20° | 288° | 20° | 0° |

HAWC2 generates outputs for each measurement, including the location (x, y, z), nominal and volume-averaged LOS, and free wind components u, v, w.

The $u_{lidar}$ velocity is obtained by projecting $V_{LOS}$ onto the longitudinal axis, disregarding the v- and w-components, and is defined as $u_{lidar} = V_{LOS}/\cos(\theta)$, where $\theta$ represents the half-cone angle of each beam. To ensure accuracy, measurements are filtered to fit within the dimensions of the turbulence box.

Figure 1 presents the selected lidar configurations in two perspectives: a rotating frame showing beams in their initial positions on the spinner, and a non-rotating frame demonstrating beam placements after a complete scan, considering a rotational speed of $\omega = 12.1$ rpm for the NREL 5MW reference wind turbine at $t = 380$ s. The appearance of the non-rotating frame changes with time due to variations in rotational speed. Although configurations 1 and 2 appear similar in the rotating frame, their primary distinction is found in the sequence of beam scanning, attributed to differences in their azimuthal angle definitions.

### *2.3. Inverse-distance-weighting interpolation*

IDW interpolation is a technique used to estimate values at unmeasured locations using known values from nearby points. It is based on the concept that a known point's influence diminishes with increasing distance, meaning points closer to the target location have a larger impact on the interpolated value.

Let's define $d(x, x_i)$ as the distance between the target value $u(\mathbf{x})$ (wind speed value that we want to estimate) located at $\mathbf{x}$ and a measurement point $u_i$ located at $\mathbf{x_i}$. In this method, if the target value is located at the same location of a measurements point ($\mathbf{x} = \mathbf{x_i}$), then our target value $u(\mathbf{x})$ will be equal to the wind speed at the measurement point, $u_i$.

Mathematically, IDW is expressed as:

$$u(\mathbf{x}) = \frac{\sum_{i=1}^{n} w_i(\mathbf{x}) \cdot u_i}{\sum_{i=1}^{n} w_i(\mathbf{x})}, \tag{1}$$

if $d(x, x_i) \neq 0$ for all i, or equal to $u_i$ if $d(x, x_i) = 0$ for some i, where $u(\mathbf{x})$ is the estimated value at location $\mathbf{x}$, $u_i$ are the known values at surrounding points, $w_i$ are the weights assigned

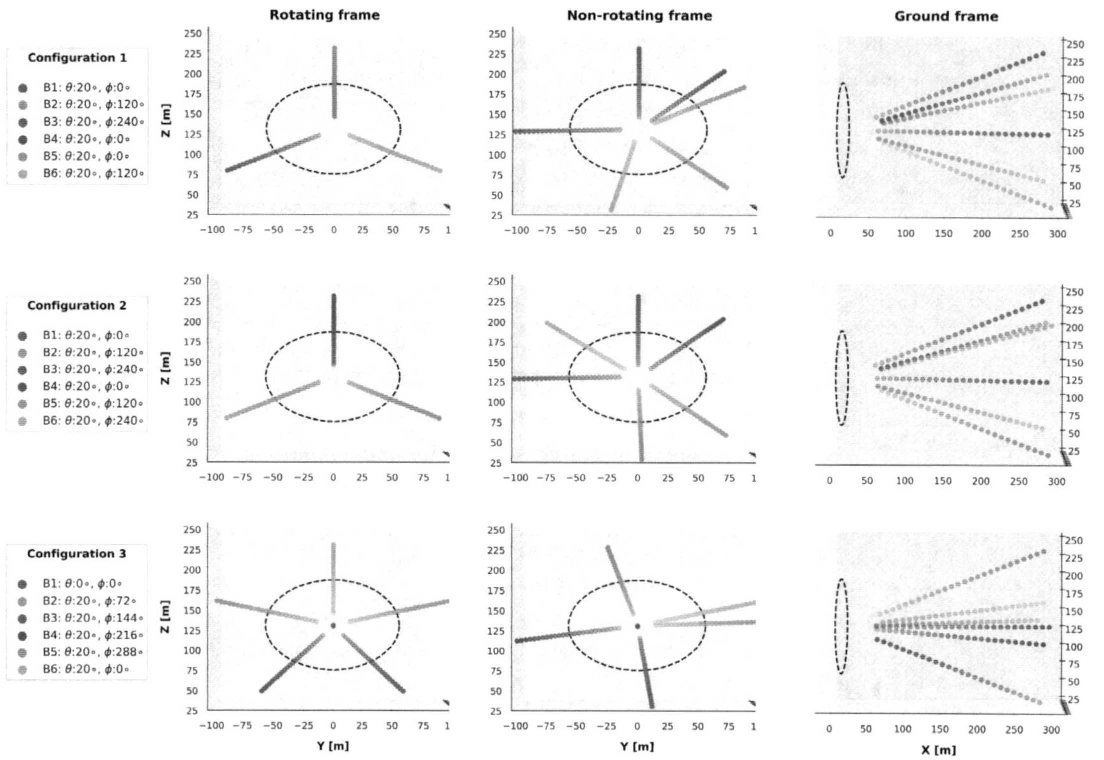


**Figure 1.** For the study's selected lidar configurations, the rotating frame is depicted on the left, illustrating beam positions in a fixed plane, while the non-rotating frame on the right shows beam positions after one complete scan at $\omega = 12.1$ rpm for the NREL 5MW reference wind turbine, starting at $t = 380$ s. It is important to note that the key distinction between configurations 1 and 2 lies in the sequence of azimuthal angles, thereby affecting the order in which the lidar beams conduct scanning.

to each known value, and $n$ is the number of known points used in the estimation. The weight $w_i$ is calculated as:

$$w_i(\mathbf{x}) = \frac{1}{d(x, x_i)^p}, \tag{2}$$

where $p$ is a positive exponent that controls the rate at which the weight decreases with distance. A higher value of $p$ results in a steeper decrease in influence with distance. In this study, we chose $p = 3$, after it showed better performance in a preliminary evaluation.

To calculate longitudinal wind speeds across the rotor plane (YZ) at each time step $t$, we adopt Taylor's frozen hypothesis [12]. For the IDW reconstruction, we select hub-lidar measurements from a specific longitudinal range, including all measurements within this span. This range is determined by the duration of a number of complete six-beam lidar scan, which is $t_{lidar} = 1.2$ s, given that each beam scans at 5 Hz, equating to 0.2 s per scan.

An illustration of data selection for a specific time step, say $t_{target} = 200$ s, considering two full scans ($n_{scans} = 2$) and a wind speed of $U = 11.4$ m/s, is provided in Fig. 2. Here, the YZ plane reconstruction occurs at $x_{target} = 2280$ m (calculated as $U \times t_{target}$), with selected hub-lidar measurements contained within a defined cuboid. This cuboid extends (in the x-axis) from $x_{initial} = 2266.32$ m (calculated as $U \times (t_{target} - \frac{t_{lidar}}{2} \times n_{scans})$) to $x_{end} = 2293.68$

m ($U \times (t_{target} + \frac{t_{lidar}}{2} \times n_{scans})$). Measurements within this cuboid (highlighted in blue) are utilized in the IDW reconstruction.

In this study, we evaluated from one to five full-scans (1.2 to 6 s), corresponding to longitudinal ranges of length, $x = x_{initial} - x_{end}$, from 13.6 m up to 68.4 m.

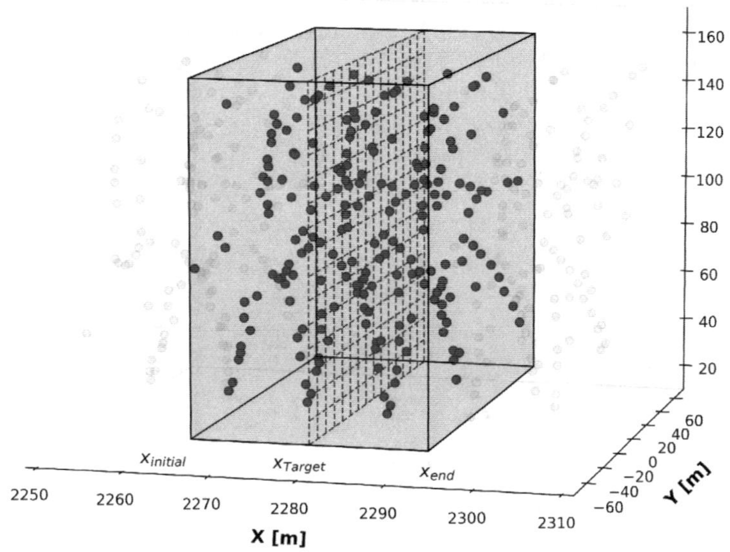


**Figure 2.** Illustration of hub-lidar measurement selection based on the number of scan considered. The hub-lidar measurements inside the limits of the blue cuboid are the ones considered for the reconstruction with inverse-distance-weighting interpolation. The length of the cuboid will be determined by the number of scans selected, in this case $n_{scans} = 2$, for a time step $t_{target} = 200$ s.

It's important to note that $x_{target}$ represents a longitudinal distance across the entire length of the turbulence box where the reconstruction is executed. In a real-time case scenario, the nearest distance from the wind turbine where the reconstruction can be performed would be approximately 70 m, determined by the hub-lidar's minimum measurement range and the number of scans under consideration.

Using selected data (all measurement points situated across the selected range), IDW calculates longitudinal wind speeds at $x_{target}$, weighting available $V_{LOS}/\cos(\theta)$ values. This results in longitudinal wind speeds at each grid point on the YZ plane ($N_y = 64, N_z = 64$).

This process is replicated at every time step to produce reconstructed turbulence boxes for the longitudinal component. The finalized longitudinal dimension of these turbulence boxes is $Nt = 3413$, after excluding the initial 80 s and final 20 s from the aeroelastic HAWC2 simulation. This exclusion ensures that lidar measurements are encompassed within the generated box, culminating in a 500 s turbulence box.

For comparison, a baseline is established by averaging lidar measurements across the YZ plane at each $t_{target}$ time step, considering a range for one full scan (13.68 m). This is to mimic the standard practice of estimating the rotor-average wind speed from lidar measurements.

## 3. Results

### *3.1. Wind-field reconstruction*

To assess the accuracy of IDW-reconstructed wind-fields against reference fields, we calculated the Root Mean Square Error (RMSE) as $= (\sqrt{1/n \sum_i^n (y_i - \hat{y}_i)^2})$, comparing the reconstructed ($y_i$) to the reference field ($\hat{y}_i$), with $n = 3413$ representing the grid size longitudinally over 500 s.

Table 2 outlines the RMSE for the longitudinal component of the wind-field across varying ranges (or number of scans), for both IDW reconstructions and baseline cases. This RMSE spans the entire reconstructed turbulence box ($3413 \times 64 \times 64$).

The optimal case for each configuration varies. For instance, configuration 1 yields the best results at three scans, showing a 60% improvement over the baseline. For configuration 2, the highest performance is at four scans, marking a 57% enhancement compared to the baseline. Configuration 3's optimal result is at five scans, achieving a 43% improvement relative to the baseline.

**Table 2.** Root-mean-square error (RMSE) [m/s] for the longitudinal wind-field reconstruction across varying ranges (or number of scans), for three different lidar configurations.

| Number of scans | Length | Configuration 1 | Configuration 2 | Configuration 3 |
|---|---|---|---|---|
| 1 | 13.68 m | 0.595 | 0.641 | 0.848 |
| 2 | 27.36 m | 0.578 | 0.627 | 0.838 |
| 3 | 41.04 m | 0.572 | 0.621 | 0.829 |
| 4 | 54.72 m | 0.573 | 0.620 | 0.821 |
| 5 | 68.40 m | 0.579 | 0.622 | 0.814 |
| Baseline | 13.68 m | 1.437 | 1.437 | 1.454 |

Additionally, it's evident that configuration 1 achieves the best performance, while configuration 3 shows the least effectiveness among the chosen setups. Notably, less optimal configurations benefit from a higher number of scans, enhancing performance. In contrast, for an optimal configuration, extending the range length beyond a certain point begins to diminish the method's overall effectiveness.

Lastly, the benefit of adding an extra scan (range) for reconstruction is marginal. For configuration 1, an improvement of 4% is observed when increasing from one to three scans. In configuration 2, the enhancement is 3% moving from one to four scans. For configuration 3, the improvement is 4% when expanding from one to five scans.

Figure 3 illustrates the IDW reconstruction across the YZ plane for a specific time step ($t = 315.95$ s), showcasing the best performance ranges (number of full-scans) for each of the three evaluated configurations. The sequence, from left to right, includes available hub-lidar measurements for reconstruction, followed by the original and reconstructed longitudinal wind speed at the selected time step, and concludes with a comparison of the reference and reconstructed cases, by calculating the difference at each grid point.

The configuration's influence is evident, as the spatial distribution of measurements directly affects method performance. In configuration three (Fig. 3 c), despite having a similar number of points, the beams' overlapping leads to a limited distribution across the rotor plane, consequently reducing the reconstruction's effectiveness.

The comparison reveals that the most significant differences in interpolation occur in areas where no lidar measurement points are available within specific grids of the YZ plane. Additionally, it's important to note that even though the configurations were estimated for a constant rotational speed, since the wind speed selected represents the rated wind speed, the

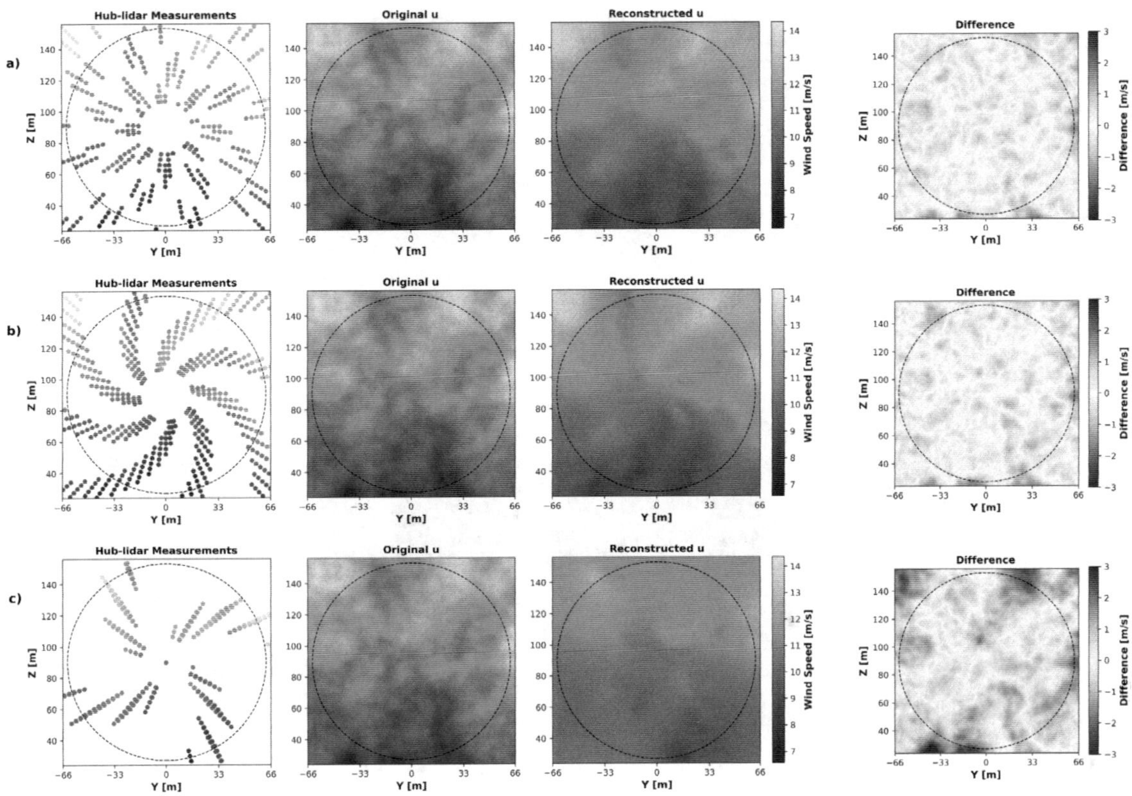


**Figure 3.** Wind-field reconstruction comparison at $t = 315.95$ s, at slice 2157, for (a) Configuration 1 with three scans, (b) Configuration 2 with four scans, and (c) Configuration 3 with five scans. Sequence from left to right: Available hub-lidar measurements for reconstruction, reference and reconstructed wind-fields, and the differences between them across the YZ plane.

wind turbine is operating at different regimes. This variation in wind speed leads to fluctuations in the rotational speed, and therefore to variations in the spatial distribution for the lidar measurements.

### *3.2. Rotor average wind speed*

To evaluate the rotor average wind speed (RAWS), the aeroelastic response of the NREL 5MW reference wind turbine in HAWC2 was evaluated.

In HAWC2, the RAWS is calculated as the rotor average free wind speed excluding tower top motion. In this study, only the longitudinal wind speed was included for the RAWS calculation.

Figure 4 illustrates the RMSE for the RAWS over a 400-second period, showing that all RAWS estimates from the hub-lidar outperformed the baseline, with improvements of 30%, 33%, and 28% for configurations one, two, and three, respectively, compared to their baselines.

This mirrors trends seen in Sec. 3.1, where the lowest overall RMSE occurs at three, four, and five scans for configurations one, two, and three, respectively. The optimal reconstruction was achieved by configuration one at three scans, with a $RMSE = 0.589$ [m/s].

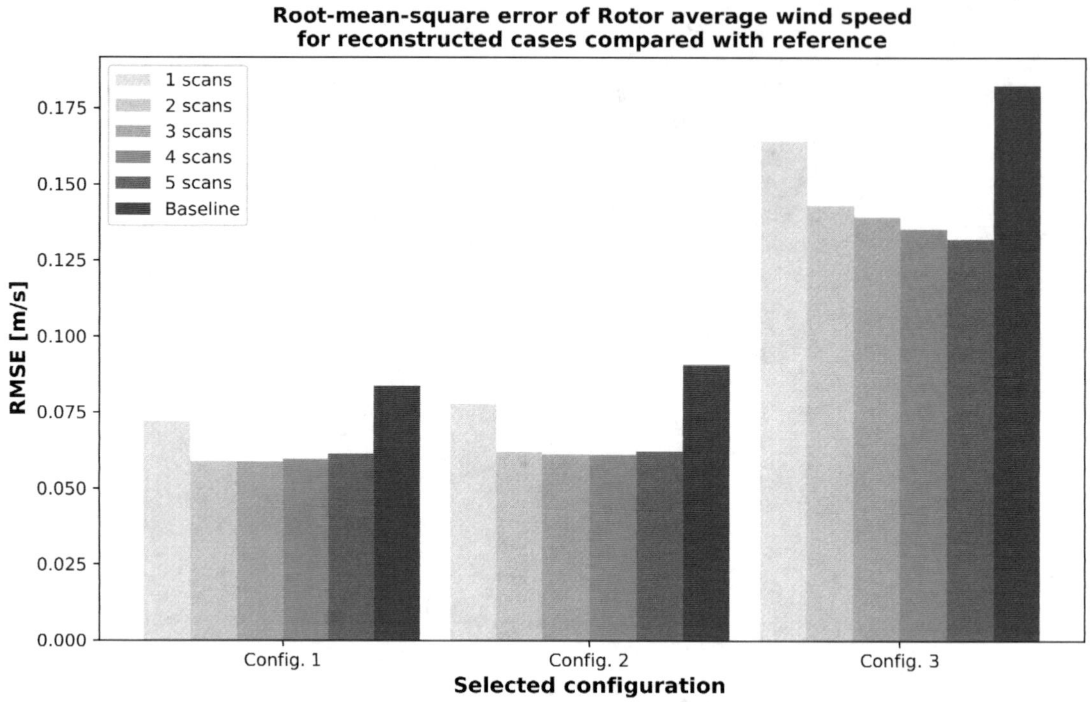


**Figure 4.** Root-mean-square error (RMSE) for the RAWS, calculated between each reconstructed case with the reference one. The selected cases are the best-performing range for each configuration. Additionally, the baseline cases are also presented.

## 4. Discussion

Our study introduces a straightforward IDW interpolation method using hub-mounted lidar measurements for full-rotor flow estimation. However, some assumptions made under this study merit further discussion.

In our initial study, the LOS wind speed is scaled by $\cos(\theta)$, omitting $v$ and $w$ components. This approach may need adjustments for more detailed analyses. Additionally, errors in LOS measurements due to a flexible tower and tilt could impact the accuracy of longitudinal wind speed estimates used in reconstruction. These elements will be examined in subsequent phases of our research.

The concept of perfectly known shear wind speed profiles is overly idealistic. While lidar measurements allow reasonable shear profile estimation, real-world scenarios with more complex shear patterns may pose greater challenges. Future phases of this study will involve performance evaluation using actual hub-lidar data to ensure comprehensive analysis of this phenomenon.

Furthermore, the assumption of a constant wind-flow rate may not hold true in practice (Taylor's frozen hypothesis). Additional research is required to verify the hypothesis's applicability, especially considering factors like rotor induction.

Finally, the method's dependency on measurement spatial distribution presents challenges for turbines with variable rotor speeds. One potential solution could be selecting scans based on data availability to fill in measurement gaps.

## 5. Conclusions

Our research focuses on reconstructing longitudinal wind speed across the full-rotor plane at each time step, employing inverse distance weighting (IDW) interpolation with data from an innovative hub-mounted lidar that overcomes blade blockage issues.

We used synthetic measurements from HAWC2 v13.1's numerical hub-lidar sensor, simulating aeroelastic responses with a Mann-generated turbulence box [8] at 11.4 m/s wind speed and 10% turbulence intensity. Three lidar configurations were assessed to gauge the IDW method's effectiveness, alongside a baseline case determined by calculating rotor average wind speed within data a 13.68 m range.

When comparing full-rotor reconstruction, the study revealed that our method significantly enhances accuracy compared to the baseline, with the best configuration yielding a 60% improvement. For rotor average wind speed, we observed a 30% improvement over the baseline, achieving an RMSE of 0.589 m/s.

We observed the hub-lidar configuration's substantial impact on both baseline and IDW reconstruction performance, emphasizing the need for spatially distributed measurements across the rotor plane. Poor configurations led to error increases up to 42%, underscoring the disparity between the worst (configuration 3) and best (configuration 1) reconstructions.

Expanding the measurement range (i.e. including high number of scans in the reconstruction) yielded modest improvements of about 3%, benefiting configurations with fewer data points. However, for data-rich configurations, extending the range eventually results in diminishing returns.

The hub-mounted lidar demonstrates significant promise for real-time wind-flow estimation, with the proposed method averaging a computation time of 0.15 s per time step on a standard personal computer. Its high spatial resolution provides detailed measurements across the rotor plane, facilitating real-time wind-field reconstruction when paired with an appropriate reconstruction technique. This capability is vital for the development of a flow-aware controller, aligning with the objectives of the CONTINUE project.

Ongoing research is exploring further techniques, such as Proper Orthogonal Decomposition (POD) coupled with Inverse Distance Weighting (IDW), to effectively bridge data gaps.

## Acknowledgments

This work is part of the CONTINUE project, which has received funding from the Danish Energy Technology Development and Demonstration Programme (EUDP), under grant agreement 640222-496980.

## References

[1] Schlipf, D., et al. (2018). IEA Wind Task 32: Best Practices for the Certification of Lidar-Assisted Control Applications. https://doi.org/10.1088/1742-6596/1102/1/012010
[2] Angelou, N., & Sjöholm, M. (2022). Data Reliability Enhancement for Wind-Turbine-Mounted Lidars. Remote Sensing, 14(13). https://doi.org/10.3390/rs14133225
[3] Mikkelsen, T., Angelou, N., Hansen, K., Sjöholm, M., Harris, M., Slinger, C., Hadley, P., Scullion, R., Ellis, G., & Vives, G. (2013). A spinner-integrated wind lidar for enhanced wind turbine control. Wind Energy, 16(4), 625–643. https://doi.org/10.1002/we.1564
[4] Simley, E., Pao, L. Y., Frehlich, R., Jonkman, B., & Kelley, N. (2014). Analysis of light detection and ranging wind speed measurements for wind turbine control. Wind Energy, 17(3), 413–433. https://doi.org/10.1002/we.1584
[5] Fu, W., Guo, F., Schlipf, D., & Peña, A. (2023). Feedforward pitch control for a 15 MW wind turbine using a spinner-mounted single-beam lidar. Wind Energy Science, 8(12), 1893–1907. https://doi.org/10.5194/wes-8-1893-2023
[6] Jonkman, J., Butterfield, S., Musial, W., & Scott, G. (2009). Definition of a 5-MW Reference Wind Turbine for Offshore System Development. http://www.osti.gov/bridge

[7] DTU (2019). HAWC2 Horizontal Axis Wind turbine simulation Code 2nd generation, Official Web-page. http://www.hawc2.dk/hawc2-info/.

[8] Mann J. (1998). Wind field simulation. Probabilistic engineering mechanics, 13(4):269–282. https://doi.org/10.1016/S0266-8920(97)00036-2

[9] International Electrotechnical Commission et al. IEC 61400-1 (2019): Wind energy generation systems - Part 1: Design requirements. International Electrotechnical Commission.

[10] Soto Sagredo, E., Rinker J.M and Lund, R.S. (2023). Verification of numerical lidars in HAWC2: Analysis of nacelle- and hub-mounted lidars, Technical report, DTU Wind Energy, Roskilde, Denmark.

[11] Meyer Forsting, A. R., Troldborg, N., & Borraccino, A. (2017). Modelling lidar volume-averaging and its significance to wind turbine wake measurements. Journal of Physics: Conference Series, 854(1). https://doi.org/10.1088/1742-6596/854/1/012014

[12] Taylor G.I. (1938). The spectrum of turbulence, Proceedings of the Royal Society of London. Series A-Mathematical and Physical Sciences 164 (919) 476-490. https://doi.org/10.1098/rspa.1938.0032

Journal of Physics: Conference Series **2745** (2024) 012018 doi:10.1088/1742-6596/2745/1/012018

# Incorporating Electricity Prices in Wind Turbine Design: Introducing the AEV Metric

**A Bechmann[1], T Barlas[1], and H A Madsen[1]**

[1] Department of Wind and Energy Systems, Technical University of Denmark (DTU)

E-mail: andh@dtu.dk

**Abstract.** This paper addresses the challenge of incorporating electricity prices into wind turbine design methods and shows how price volatility drives wind turbines towards larger rotors and lower specific power. Since wind speed and electricity prices fluctuate, current efforts to estimate a wind turbine's revenue are based on time-series approaches. However, this paper presents a new way of accounting for price volatility based on wind distributions, which is computationally cheap and easily integrates with current wind turbine and farm design methods and tools. The new method demonstrates that a traditional wind turbine can lose more than 15% of its revenue in open energy markets like Denmark due to price volatility. Designing turbines with lower specific power can substantially increase revenue by producing more energy at low wind speeds with higher energy demand and electricity prices.

## 1. Introduction

As we introduce more wind energy to open-market grids, the electricity prices tend to decline by the "merit-order effect" and correlate negatively with production [1, 2]. Therefore, while a wind farm owner will have high electricity production on windy days, the production revenue can be practically zero. Loth *et al.* [3] projects that the revenue for open-market grids with high wind energy penetration can be halved due to this price volatility.

Traditional wind turbines (WTs) reach their rated power at around 12 m/s and cut out at 25 m/s, i.e. they produce the most energy during high wind speeds. This leads to an imbalance between power production and consumption in regions with considerable wind and solar energy and can result in low or even negative electricity prices in periods with high wind speeds. On this background, the LowWind turbine concept [4] was proposed with the objective of making a major contribution to the system integration on the production side, where traditionally, the integration has only been on the consumption side, like battery storage and adaptive consumption. Low-wind WTs [4] have larger rotors and lower Specific Power (SP), making them more expensive. However, they produce more power during low-wind periods when electricity is expensive.

Swisher *et al.* [5] investigate the price point at which the LowWind WT concept, with an SP of only 100 W/m$^2$, becomes competitive in the energy system of Northern and Central Europe. They simulate the energy system (electricity, heat, and transport) until 2050 using the dynamic Balmorel model [6], giving it the objective of minimising the total system cost. Results show that the SP100 WT becomes disruptive if the cost is not more than 39% higher than a conventional WT. While the SP100 WT will find a lot of investments in low-wind regions, high-wind regions with transmission constraints, like Denmark, also see considerable investments.

Current WTs and WFs (wind farms) are designed to maximise the Annual Energy Production ($AEP$), a simple design metric to calculate and use. On the other hand, optimising for revenue requires dynamic information on the hourly energy price and production of the future electricity market, making it an impractical metric for WT and WF design. This paper defines a new design metric called the Annual Energy Value ($AEV$) that accounts for the negative production-price correlation. $AEV$ is based on wind distributions instead of time-series and is therefore computationally cheap, can be implemented in current design methods, and can be used as a new "objective function" when optimising WTs and WFs. $AEV$ can, therefore, be used to design a new generation of LWTs that better integrate with market demands.

### 1.1. Research questions

This paper investigates how we can design WTs that better align with the fluctuating energy demand. We first derive a mathematical model of calculating the WT revenue based on wind distributions instead of time-series. The mathematical model leads to the new AEV metric that can be seen as a price volatility correction to the well-known AEP metric. Next, we present an example of the new price curve for a site in Denmark. Finally, a parametric study is conducted of how the SP of a WT impacts the revenue when the electricity price is fixed and fluctuating. The paper tries to answer the following central questions:

(i) Can we calculate the annual revenue of a WT statistically instead of using time-series?
(ii) Does price volatility change the boundary conditions for WT design?
(iii) How does changing the SP of a WT affect its revenue?

## 2. A New Design Paradigm

Current objective functions for WT and WF design aim to minimise the cost of the energy produced. However, this design philosophy assumes that the value of energy is constant. In open energy markets with high shares of wind energy, wind power does not perfectly align with market demands and the energy price is reduced. Production needs to better align with the fluctuating energy demand to get higher prices and make future wind energy systems profitable. We, therefore, want a new objective function that includes the volatility of the electricity prices. This section derives such a metric that we call the Annual Energy Value (AEV).

### 2.1. A stationary revenue model

The revenue of a WT depends on the price of electricity and power production, which depends on wind speed. Since spot prices change each hour and there are distinct yearly weather patterns, we use these time scales to calculate the Annual Energy Revenue (AER). We can calculate AER using time-series [3] or with a statistical approach [7]:

$$\begin{aligned} AER &= 8766\,\mathbf{E}(r) \\ &= \sum_{i=1}^{8766} v_i\, p(u_i) \\ &= 8766 \int_0^{\infty} \int_{-\infty}^{\infty} v\, p(u)\, f(v,u)\, dv\, du \end{aligned} \tag{1}$$

$\mathbf{E}$ denotes the expected value, $r$ is the mean hourly revenue, $u_i$ and $v_i$ are the mean wind speed and spot price values at a particular hour, $p(u) = \mathbf{E}(p|u)$ is the power curve, and the 8766 factor is the number of hours per year. Since the joint probability density function, $f(v,u)$, is hard to model, the current efforts use time-series of electricity prices and wind speed (Figure 1 left)

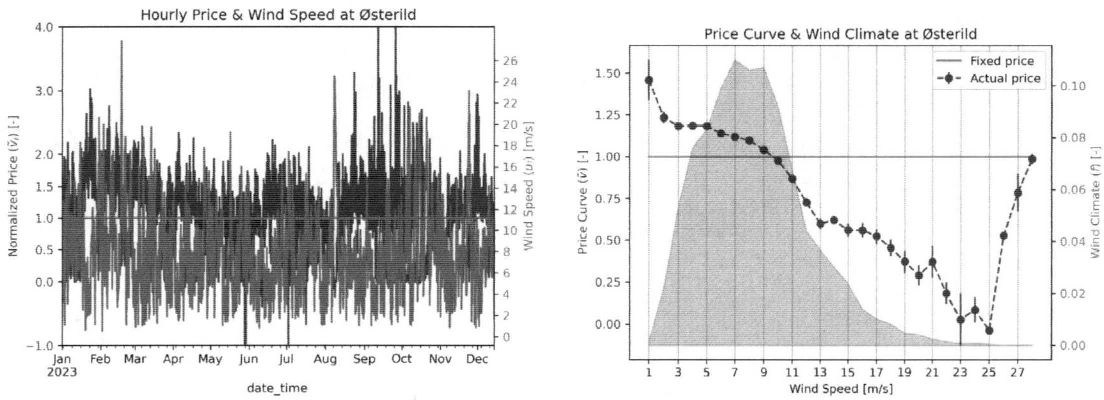


Figure 1: The same data from Østerild for 2023 is presented as time-series and statistically. Left: The hourly electricity price (blue) and wind speed (green). Right: The price curve (blue) and the wind climate (green). The annual mean price is shown in red.

to calculate the revenue. However, even though it is possible to dynamically simulate the spot price and wind speed at a wind farm location, these models are computationally expensive and have coarse spatial resolution.

To predict the revenue of each individual WT for layout optimisation or WT design, we want a stationary modelling approach that can quickly update the revenue estimation for each design iteration. We can do this by introducing the spot price's dependency on the wind speed:

$$\begin{aligned}\mathbf{E}(r) &= \int_0^\infty \int_{-\infty}^\infty v\, p(u)\, f(v,u)\, dv\, du \\ &= \int_0^\infty \int_{-\infty}^\infty v\, p(u)\, f(v|u)\, f(u)\, dv\, du \\ &= \int_0^\infty \left( \int_{-\infty}^\infty v\, f(v|u)\, dv \right) p(u)\, f(u)\, du \\ &= \int_0^\infty \mathbf{E}(v|u)\, p(u)\, f(u)\, du \\ &= \int_0^\infty v(u)\, p(u)\, f(u)\, du\end{aligned} \tag{2}$$

$f(u)$ is the marginal probability density function of wind speed, often referred to as the wind climate, and $v(u) = \mathbf{E}(v|u)$ is a "price curve" that expresses how the mean spot price depends on the site-specific wind speed. For a given wind climate and price curve, we can revenue-optimise the WT as the wind climate and price curve are independent of the WT power curve. The price curve and wind climate can be described with only a few parameters instead of 8766 time-steps (Figure 1 right). Suppose we know the wind climate and price curve at one position in the wind farm (e.g. from measurements during the wind farm prospecting), then we can quickly extrapolate them to all WT positions using well-known stationary methods like the Wind Atlas methodology [7].

### *2.2. The Annual Energy Value*

The IEC 61400-12-1 standard [8] evaluates the performance of wind turbines using the Annual Energy production (AEP). Therefore, most WF assessments and WT design tools calculate the AEP without considering the hourly price volatility. To make it straightforward to include price volatility, we define a new metric called the Annual Energy Value (AEV) that can directly replace AEP. AEV has the same units (Wh) as AEP but includes the energy devaluation when production and market demand misalign. Like Loth *et al.* [3], we use the average annual price ($\overline{v}$) to define the hourly normalised price ($\tilde{v} = v/\overline{v}$). If we combine this with Equation (2), we can write the AER as a function of the non-dimensional price:

$$\begin{aligned} AER &= 8766 \int_0^{\infty} v(u)\, p(u)\, f(u)\, du \\ &= \overline{v} \left( 8766 \int_0^{\infty} \tilde{v}(u)\, p(u)\, f(u)\, du \right) \\ &= \overline{v}\, AEV = \overline{v}\, (\phi_V\, AEP) \end{aligned} \tag{3}$$

where $\phi_V = AEV/AEP$ is a price volatility correction factor that is the ratio of actual revenue to expected revenue when discounting price volatility. For markets with fixed electricity prices, $AEV$ equals $AEP$ and $\phi_V = 1$. As demonstrated in the Results section, the volatility loss can be more than 15% in markets with fluctuating prices, so it is important to consider this during WF development and WT design.

## 3. Method

Traditional AEP-optimisation of WT designs requires models of the power curve, $p(u)$, and the wind climate, $f(u)$. We need to add the price curve model, $\tilde{v}(u)$, to demonstrate how the inclusion of price volatility affects WT design. The models used in this paper are shown in Figure 2 and further described in the following section.

### *3.1. Wind climate $f(u)$ and price curve $\tilde{v}(u)$*

Figure 1 left shows the time-series of the hourly wind speed and the normalised electricity for 2023. The wind speed is taken at a height of 106 meters from a met mast at Test Centre Østerild, located on the west coast of Northern Jutland, and the spot price is taken from the same price area (west Denmark). Figure 2 uses the same data set but presents the wind speed as a probability density function (left) and the spot price as a price curve (middle); the error bars represent the standard error of the mean.

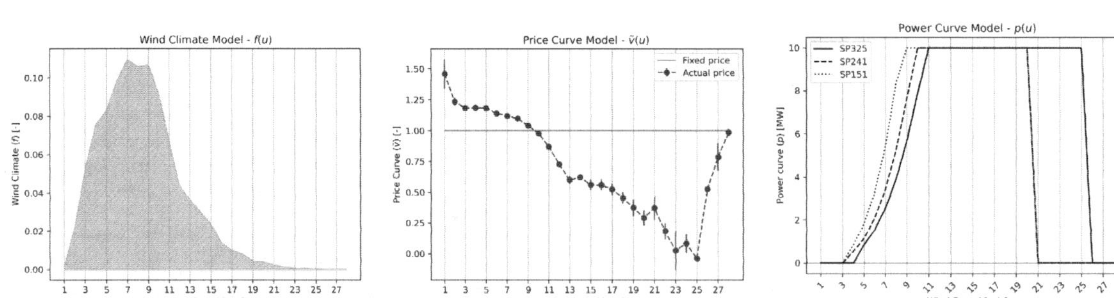


Figure 2: The three model types needed to calculate the Annual Energy Revenue (AER). Left: $f(u)$ - The wind climate (PDF). Middle: $\tilde{v}(u)$ - The price curves for Østerild/Denmark ("Actual price") and a fixed price market ("Fixed price"). Right: $p(u)$ - The WT power curve models. SP indicates the specific power (the ratio of the rated power to the swept area).

As seen in the middle of Figure 2 ("Actual price"), there is a clear trend of decreasing electricity prices at increasing wind speeds. The supply of wind-produced electricity is low at low wind speeds, and electricity prices increase. In contrast, the energy supply increases at higher speeds and prices decrease. At wind speeds above 24 m/s, many wind farms reach their cut-out wind speed, reflected in increasing electricity prices.

Traditionally, the wind climate is represented in compact form by the Weibull distribution:

$$f(u) = \frac{k}{A}\left(\frac{u}{A}\right)^{k-1} \exp\left(-\left(\frac{u}{A}\right)^{k}\right)$$

where $A$ and $k$ are the all-sector Weibull scale and shape parameters. Similarly, the price curve can be represented as a simple linear function of $u$:

$$\tilde{v}(u) = \alpha u + \beta$$

where $\alpha$ and $\beta$ are also site-specific constants. In this work, however, we do not use the compact forms but apply the wind speed binned values (Figure 2) directly. As Denmark has a high share of wind energy, the price curve ("Actual price") represents a market with high price volatility. To study how price volatility influences wind turbine design across markets, we will also consider a market with fixed pricing, $\tilde{v}(u) = 1$ ("Fixed price" on Figure 2). Comparing these two price curves can indicate how wind turbine design will develop for markets approaching Denmark's wind energy penetration. The wind climate is fixed for all result comparisons presented in the paper.

### *3.2. Power curve models $p(u)$*

We want to evaluate the effects of changing the WT's specific power (SP)– the ratio of the rated power to the swept area. The WT is modelled using a simple parametric power curve. The power in the full load region is equal to the rated power (10 MW), while we use the power equation in the partial load region, assuming the optimal Power Coefficient ($C_p$=0.49):

$$p(u) = \frac{1}{2}\rho u^3 C_p \pi (D/2)^2$$

$D$ is the rotor diameter, and the air density is $\rho = 1.225\,\mathrm{kg/m^3}$. We ignore wake effects since this work only analyses a single WT. The IEA 10-MW off-shore WT [9] is used as reference WT. It has cut-in and cut-out wind speeds of $u_{in} = 4\,\mathrm{m/s}$ and $u_{out} = 25\,\mathrm{m/s}$ and a rotor diameter of $D = 198\,\mathrm{m}$, giving it a $SP = 325\,\mathrm{W/m^2}$ (SP325). In addition to the reference SP325 WT, we define two LWT with rotor diameters of $D = 230\,\mathrm{m}$ (SP241) and $D = 290\,\mathrm{m}$ (SP151), with cut-in and cut-out wind speed of $u_{in} = 3\,\mathrm{m/s}$ and $u_{out} = 20\,\mathrm{m/s}$ (see Figure 2 right).

## 4. Results

### *4.1. The energy value density*

Evaluating the revenue and production generation potential at varying wind speed ranges is important for designing WTs, as we can shift the production to lower or higher wind speeds by lowering or increasing the WTs' SP. We can assess the average value of diverse wind speeds by defining the average energy value density $\left(\mathrm{Wm^{-2}}\right)$ as follows:

$$E_V(u) = \frac{1}{2}\rho\, u^3\, \tilde{v}(u)\, f(u)$$

where $f(u)$ is the wind climate, and $\tilde{v}(u)$ is the price curve at the specific location. A graph of this function shows which wind speeds are essential for energy revenue generation. Figure 3 left shows the energy density value for Østerild with and without the inclusion of price volatility.

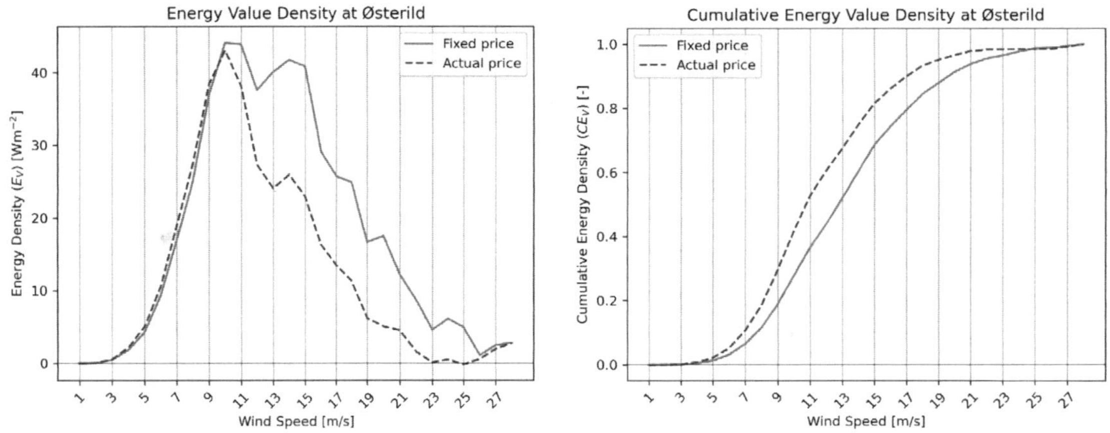


Figure 3: This figure shows the energy value density in order to evaluate the revenue generation potential at varying wind speeds. Left: Comparison of the energy density value with and without price volatility. Right: Comparison of the cumulative energy density value with and without price volatility.

When price volatility is included ("Actual price"), the energy value at higher wind speeds is reduced and even negative. The revenue of the produced electricity is markedly reduced at wind speeds higher than 10 m/s. Therefore, the effect of price volatility on the AER is similar to reducing the mean wind speed of the wind climate. LWTs may, therefore, be preferred in volatile markets despite good wind conditions.

### *4.2. The cumulative energy value density*

To further illustrate how the main part of the AER is shifted to lower wind speeds in markets with price volatility, we define the cumulative energy value density:

$$CE_V(u) = \frac{\int_0^u E_V(u)}{\int_0^\infty E_v(u)}$$

The cumulative energy value density of Østerild is shown in Figure 3 right. The figure shows that 90% of the potential revenue is generated at wind speeds of 17 m/s and below ("Actual price"), despite only 80% of the energy can potentially be produced in this wind speed range ("Fixed price"). 20 m/s is a point of diminishing returns; almost no revenue can be created at higher wind speeds. Only 3% of the potential AER comes from the 20-25 m/s wind speed range, while 9% of the potential AEP can be found here. These numbers illustrate why WTs that operate in volatile markets should be designed for lower SP than traditional WT.

### *4.3. Price volatility loss*

Now, we will look at how the SP325 reference WT performs in the fixed pricing market compared to the volatile one (Figure 2 middle). It is worth remembering that the power curve, wind climate, and mean electricity price are identical in this example; only the price volatility is changed.

Table 1 shows the AEP and AEV for the reference WT (SP325) in the two electricity markets. It is seen that the revenue in the market with price volatility is only 84.5% ($AEV/AEP$) of a market with fixed prices. In other words, there is no demand for 15.5% of the electricity produced, which might as well be curtailed. WF developers that aim to maximise AEP (or

Table 1: The AEP and AEV for the SP325 WT in a volatile and fixed-price market

| | Fixed price | Volatile price |
|---|---|---|
| AEP [GWh] | 41.6 | 41.6 |
| AEV [GWh] | 41.6 | 35.2 |
| AEV/AEP [%] | 100 | 84.6 |

Table 2: The AEP and AEV for the SP325, SP241, and SP151 WT in a volatile price market.

| | SP325 | SP241 | SP151 |
|---|---|---|---|
| Diameter [m] | 198 | 230 | 290 |
| AEP [GWh] | 41.6 | 47.7 | 56.5 |
| AEV [GWh] | 35.2 | 42.2 | 51.9 |
| AEV/AEP [%] | 84.6 | 88.5 | 91.9 |

minimise the LCOE) are unaware of this volatility loss; only by calculating AEV can we see this loss.

### *4.4. Marked-aligned wind turbines*
In Table 2, we compare the performance of WT SP325, SP241, and SP151 in the volatile price market. The wind climate and price curve are kept fixed; only the power curve is changed. As expected, the revenue (AEV) increases with lower SP-values. WT SP241 and SP151 have 19.8% and 47.4% higher AEV than SP325, respectively. The LWTs not only produce more electricity (higher capacity factor), but a larger fraction of the electricity is in demand ($AEV/AEP$). Only 8.1% of the energy produced by SP151 is not in demand and therefore curtailed.

Swisher *et al.* [5] showed that the LowWind SP100 WT would be disruptive in many European high-wind speed regions towards 2050 if the WT cost is not more than 39% higher than a conventional WT. The results presented here show that an SP151 WT is already competitive in Denmark if its cost is not more than 47.4% higher than a conventional WT.

## 5. Conclusions
This work explored the concept of incorporating electricity prices into wind turbine design. A new design metric, Annual Energy Value, was proposed to address the issue of price volatility and energy supply/demand mismatch in open energy markets. The paper also delved into the potential changes in wind turbine design under fluctuating electricity prices and examined how the specific power of a wind turbine influences its revenue. Based on these investigations, we can draw the following conclusions:

(i) The proposed AEV metric provides a feasible method for calculating the annual revenue of a wind turbine in dynamic pricing scenarios. It successfully accounts for price volatility using a new "price curve" that integrates with the traditional AEP calculation. The price curve is the only new "ingredient" compared to AEP calculations, and it can be determined by simply logging the electricity price during the wind measurement campaign. Keeping to the traditional wind distribution approach rather than time series offers a computationally efficient method of optimising wind turbines and farms that integrates with existing design methods and tools. As such, the AEV can be a one-to-one replacement for the IEC standard AEP performance metric.

(ii) Electricity price volatility significantly alters the boundary conditions for wind turbine design. In markets with high price volatility, there is a clear shift towards valuing energy produced at lower wind speeds more highly. This underscores the importance of re-evaluating traditional wind turbine designs, prioritising high-wind speed energy production. Especially in open markets like Denmark, which has high wind energy penetration, it is important to reconsider the design objectives; in the example presented, the traditional wind turbine (SP325) lost 15.5% of its revenue due to price volatility.

(iii) Changing a wind turbine's specific power significantly impacts its revenue in price-volatile markets. Lower specific power generates higher revenues in these markets, implying that larger rotor designs are more profitable despite increased costs. The increased revenue is caused by more low-wind speed energy production that better aligns with the fluctuating energy demand, leading to higher electricity prices.

## References

[1] Matsuo Y 2022 *Energies* **15** 6845 ISSN 1996-1073 URL `https://www.mdpi.com/1996-1073/15/18/6845`

[2] Pearre N S and Swan L G 2023 *Wind Energy* **26** 446–457 ISSN 1095-4244, 1099-1824 URL `https://onlinelibrary.wiley.com/doi/10.1002/we.2809`

[3] Loth E, Qin C, Simpson J G and Dykes K 2022 *Advances in Applied Energy* **8** 100112 ISSN 26667924 URL `https://linkinghub.elsevier.com/retrieve/pii/S2666792422000300`

[4] Madsen H A, Zahle F, Meng F, Barlas T, Rasmussen F and Rudolf R T 2020 *Journal of Physics: Conference Series* **1618** 032011 ISSN 17426596, 17426588

[5] Swisher P, Murcia Leon J P, Gea-Bermúdez J, Koivisto M, Madsen H A and Münster M 2022 *Applied Energy* **306** 118043 ISSN 03062619 URL `https://linkinghub.elsevier.com/retrieve/pii/S0306261921013337`

[6] Wiese F, Bramstoft R, Koduvere H, Pizarro Alonso A, Balyk O, Kirkerud J G, Tveten G, Bolkesjø T F, Münster M and Ravn H 2018 *Energy Strategy Reviews* **20** 26–34 ISSN 2211467X URL `https://linkinghub.elsevier.com/retrieve/pii/S2211467X18300038`

[7] Troen I and Petersen E L 1989 *European wind atlas* (Roskilde, Denmark: Risø National Laboratory) ISBN 87-550-1482-8 URL `http://orbit.dtu.dk/en/publications/european-wind-atlas(335e86f2-6d21-4191-8304-0b0a105089be).html`

[8] IEC 61400 2017 *Wind energy generation systems Part 12-1: Power performance measurements of electricity producing wind turbines* edition 2.0 ed (British Standard) ISBN 978-2-8322-3823-3 oCLC: 1040063022

[9] Bortolotti P, Tarres H C, Dykes K, Merz K, Sethuraman L, Verelst D and Zahle F 2019 IEA wind task 37 on systems engineering in wind energy – WP2.1 reference wind turbines Tech. rep. International Energy Agency tex.howpublished: NREL/TP-73492 URL `https://www.nrel.gov/docs/fy19osti/73492.pdf`

# Influence of wind gusts on aeroelastic loads of NREL 5MW wind turbine

**Neeraj Paul Manelil, Johannes Nicolaas Theron, Leo Höning, Bernhard Stoevesandt**

Fraunhofer Institute for Wind Energy Systems, Department Aerodynamics and Numerical Energy Meteorology, Küpkersweg 70, 26129 Oldenburg, Germany

E-mail: neeraj.paul.manelil@iwes.fraunhofer.de

**Abstract.**
This study employs blade-resolved computational fluid dynamics simulations to investigate the aero-elastic loads on a yawed NREL 5MW wind turbine subject to full and partial wind gusts. A novel spatial gust representation, tailored towards application to large eddy simulations, is introduced. This methodology permits the injection of gusts with customizable spatial dimensions inside the flow domain. Simulations are conducted with gusts having the same diameter as the turbine rotor, half rotor diameter, and asymmetric impact with respect to the center of rotation of the turbine. Simulations of the turbine under yawed conditions are also considered. The flow characteristics are reported using contours and iso-surfaces of velocity magnitude and Q-criteria. Insights into the load characteristics of the turbine are described using torque, thrust, and root bending moments. This research highlights that partial gust impact contributes to noteworthy load fluctuations and blade tip deformations, stemming from the imbalance in force distribution across the rotor plane.

## 1. Introduction

Wind gusts exert a substantial impact on wind turbine structural loads, encompassing critical parameters such as root bending moments and tower base bending moments. Conforming to the stipulations of the International Electrotechnical Commission (IEC), wind turbines included in its certification program are mandated to substantiate their resilience against a 50-year gust event [1]. Prevailing certification calculations predominantly employ lower-order engineering tools rooted in the Blade Element Momentum theory (BEM), a framework that fails to account for the intricate three-dimensional repercussions stemming from the blade geometry. In light of these considerations, this study endeavors to execute gust simulations by harnessing fluid-structure interaction capabilities within a blade-resolved 5MW NREL wind turbine model. The simulations are conducted within a Computational Fluid Dynamics (CFD) domain, facilitating a more comprehensive exploration of the three-dimensional effects that significantly influence wind turbine performance and safety.

The spatial and temporal characterization of extreme gusts presents a considerable challenge. This complexity arises from the fact that most deterministic gust descriptions, such as the extreme operating gust and extreme coherent gust outlined in the IEC standards [1], do not align with observations derived from field data [2]. Moreover, these simplified models have limitations, primarily their one-dimensional nature, assuming that gusts instantaneously affect

all rotor blades concurrently [3]. State-of-the-art numerical methods, exemplified by the high-fidelity Large Eddy Simulations (LES) conducted in this study, as well as LiDAR(light detection and ranging) measurements, offer the capability to explore fully three-dimensional wind gusts. However, it is noteworthy that studies addressing this aspect remain relatively scarce, with some notable examples including [3] and [4]. In their study, [3] conducted Large Eddy Simulation (LES) simulations of wind turbines in turbulent atmospheric boundary layers during strong gust events. The turbines were modeled using the actuator disk method, and descriptions of one-dimensional and two-dimensional gust shapes based on limited field measurements were presented. They indicated that multidimensional gust descriptions are crucial for accurately establishing loads on structures such as turbine blades, bridges, and aircraft structures.
In reality, gusts, as anticipated in [2], are expected to make partial impacts on wind turbines, potentially resulting in non-uniform loads across individual blades and consequential power fluctuations. The current study addresses this crucial aspect by conducting a comparative analysis of the aero-elastic loads and wake characteristics experienced by an NREL 5MW wind turbine under both uniform and partial gust impact conditions. The reference [4] describes Large Eddy Simulation (LES) techniques applied to wind turbine modeling, wherein the turbines are represented using actuator disks. Additionally, it describes the incorporation of a simplified wind gust within the simulations. The present study extends this work by performing blade-resolved fluid-structure interaction simulations while implementing a gust model that allows tuning of the gust dimensions.

## 2. Problem definition

Figure 1 illustrates a top-down view of the computational domain employed for this study. The wind gust is introduced within a specified volume zone, referred to as the "injection zone," positioned four rotor diameters upwind of the wind turbine (WT). To mitigate any potential complexities stemming from interactions between the turbine blade and tower at large deflections, the simulations consider blades and hub without tower. The figure illustrates the wake of a turbine that is yawed at an angle of 30 degrees relative to the incoming flow. The turbine is subject to a uniform wind speed of $10 m/s$, and rotates at the rated speed of 12.1 $rpm$. The maximum gust velocity is $14 m/s$ for all simulations in this study. Two different scenarios of gust impingement have been considered in the study. In the first scenario, the gust strikes the entire rotor plane all at once (see Fig. 2 a.), while in the second situation, the gust strikes the top section of the WT (see Fig. 2 b.). In addition to the yawed scenario, simulations are also carried out with a zero yaw angle. Unlike the standard definition, wherein the gust impacts the full rotor plane, this study also considers the impact of gusts with dimensions (diameter) half that of the rotor. The complete spectrum of simulations carried out in the present study is summarized in table 1.

**Table 1.** Problem matrix with description of different gust scenarios investigated.

| Name | Yaw angle (deg) | Gust diameter (m) | Vertical position of gust core from hub (m) |
|---|---|---|---|
| GY0D0P0 | 0 | 120 | 0 |
| GY0D1P0 | 0 | 60 | 0 |
| GY0D1P1 | 0 | 60 | 60 |
| GY30D0P0 | 30 | 120 | 0 |
| GY30D0P1 | 30 | 120 | 120 |

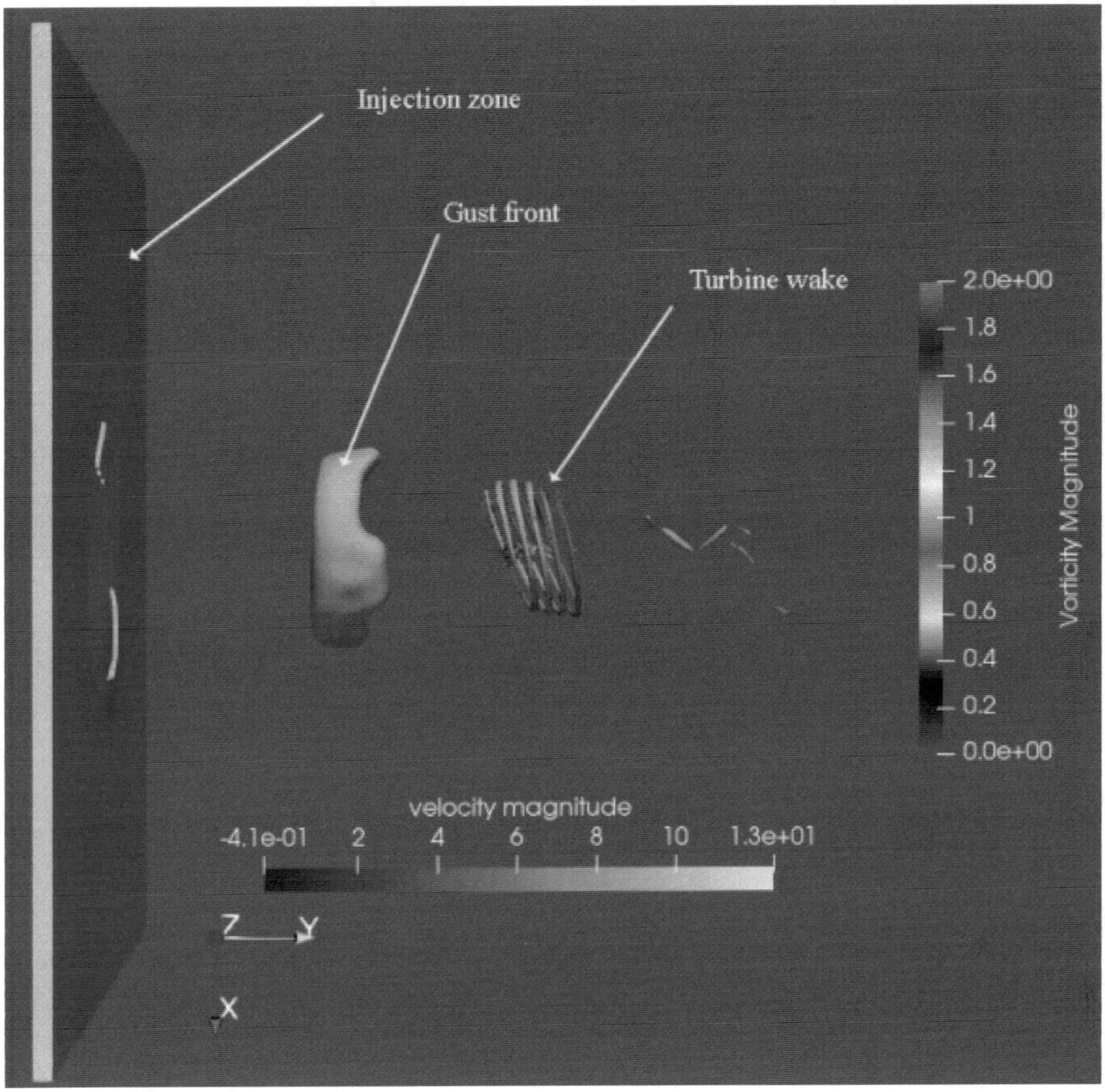


**Figure 1.** Illustration of the computational domain highlighting the injection zone, turbine orientation and wind gust

## 3. Numerical methodology

### *3.1. Gust implementation*

In order to introduce wind gusts in the computational domain, a volumetric body force source term injection method described in [5] has been employed. The body force ($F_p$) corresponding to fluid acceleration from mean velocity ($V_m$) to $V_m + V_g$, where $V_g$ is the gust velocity, is given by

$$F_p = \rho V_g \Delta n \left( V_m + \frac{1}{2} V_g \right) \tag{1}$$

here $\rho$ is the fluid density and $\Delta n$ represent the width of the injection zone in the flow direction. The injection of a gust cause a sudden change in mass flow through the CFD domain and great care must be taken to ensure that mass conservation is maintained during the injection phase

Journal of Physics: Conference Series **2745** (2024) 012019 doi:10.1088/1742-6596/2745/1/012019

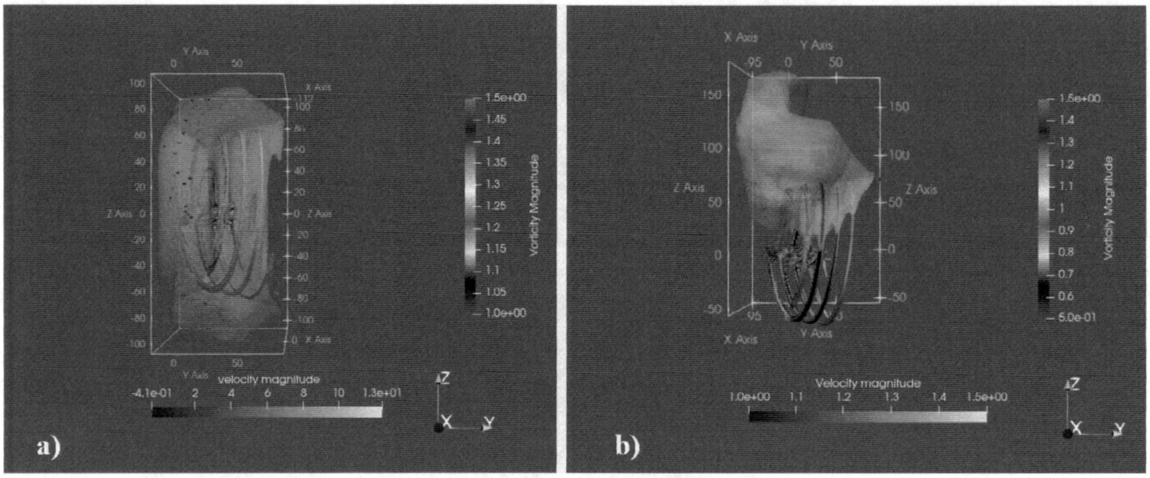


**Figure 2.** Instantaneous iso-surface of velocity-magnitude and Q value at the instance of a) full gust impact b) partial gust impact, with wind turbine rotor

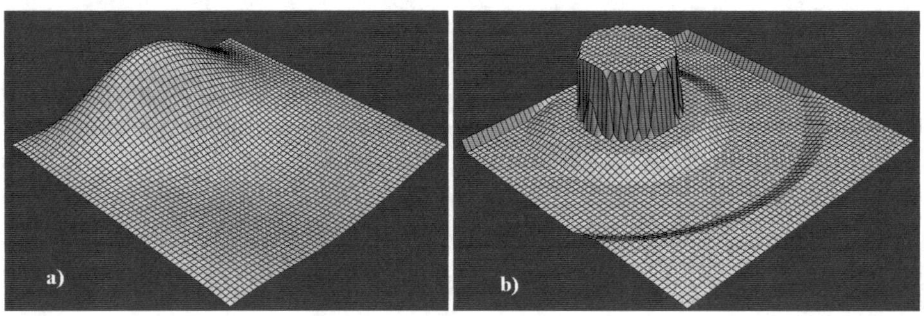


**Figure 3.** Comparison of spatial description of gust used in a) [4] with b) present study

to prevent numerical instabilities and anomalies in the LES simulation [4]. Thus it is essential to ensure that the mass is conserved in the injection zone leading to the need of describing a region of negative velocity in the injection zone to counterbalance the mass injection due to gust. Figure 3 a. and b. compares the spatial distribution of gust employed in [4] with that used in the present study. The implementation of gust in the present study describes a circular region in space defined by gust diameter ($D_G$). The velocity of in this region is described equal to the gust speed ($V_G$). $V_G = 14$ m/s for all cases in the present study. The position, $D_G$ and $V_G$ could be varied as per requirement. A sigmoid smoothing function is employed in annular region starting at $D_G$ till $D_G + 2V_G$. This is done to prevent the solution from destabilizing due to the strong gradients at the edges. Beyond this smoothing region a negative velocity region is described extending till $2D$. The negative velocity magnitude is determined iteratively, ensuring mass conservation across the injection plane. The temporal description of the gust adheres to the extreme operating gust definition in [1], and the time period of gust used in the present study id 10 seconds.

### *3.2. Flow solver*

The study utilizes an in-house framework for fluid-structure coupled computational fluid dynamics (CFD) simulations, as detailed in [6]. Within this framework, both the structural

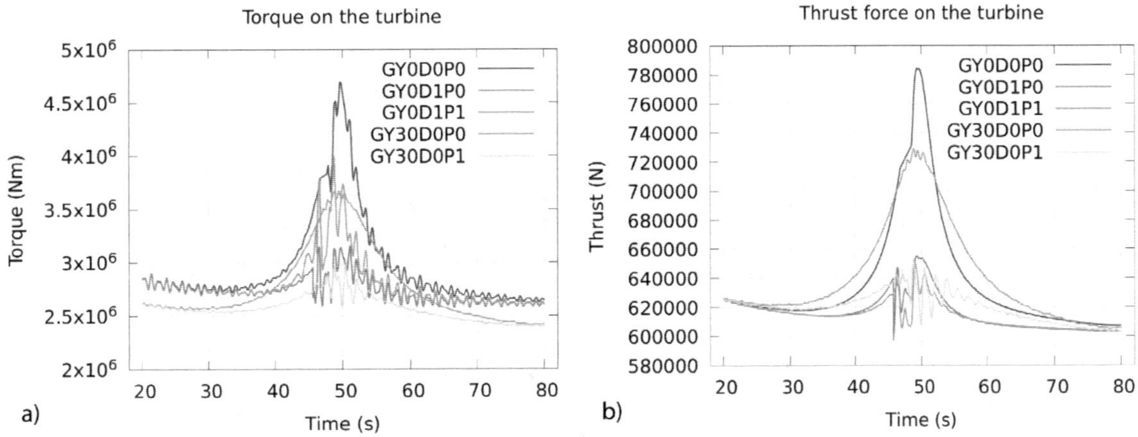


**Figure 4.** a) Comparison of torque a) and thrust b) force on the WT for different gust scenarios

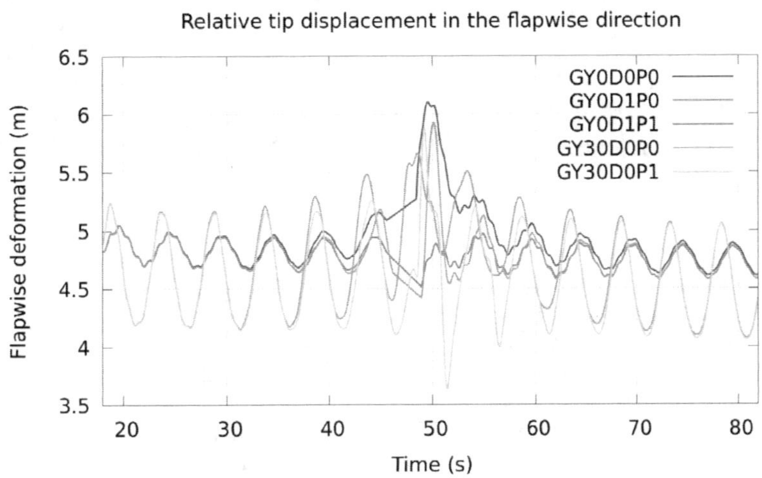


**Figure 5.** Temporal evolution of relative flap-wise tip displacement for different gust impact scenarios

solver and the mesh deformation module are integrated into the open-source CFD toolbox OpenFOAM. For the fluid domain, a time-accurate Delayed-Detached Eddy Method is employed, while the structural aspect is addressed using Geometrically Exact Beam Theory (GEBT). The robustness of this comprehensive framework for fluid-structure coupling has been previously validated in prior works, as documented in [7] and [8]. The grid distribution, time step size, and boundary treatment adopted in the present study align closely with those outlined in [6], ensuring a consistent and well-established computational approach.

## 4. Results and discussion

Figure 2 depicts the interaction between the wind gust and the wind turbine (WT) for symmetric and asymmetric gust impacts in the yawed scenario. In this study, symmetric gust impact refers to the scenario when the core of the gust aligns with the center of rotation of the turbine, while asymmetric gust impact refers to scenarios otherwise. The iso-velocity surface delineates the spatial positioning of the gust, while the WT's wake characteristics are represented by iso-Q surfaces. The distinct plug-like shape of the gust, as shown in Figure 3b, ensures that it strikes the entire rotor disk of the wind turbine uniformly. It is also noted that the gust undergoes deformation prior to impacting with the WT. This is associated with its interaction with the induction zone of the WT.

The torque and thrust forces on the WT for all the cases considered in this study are presented

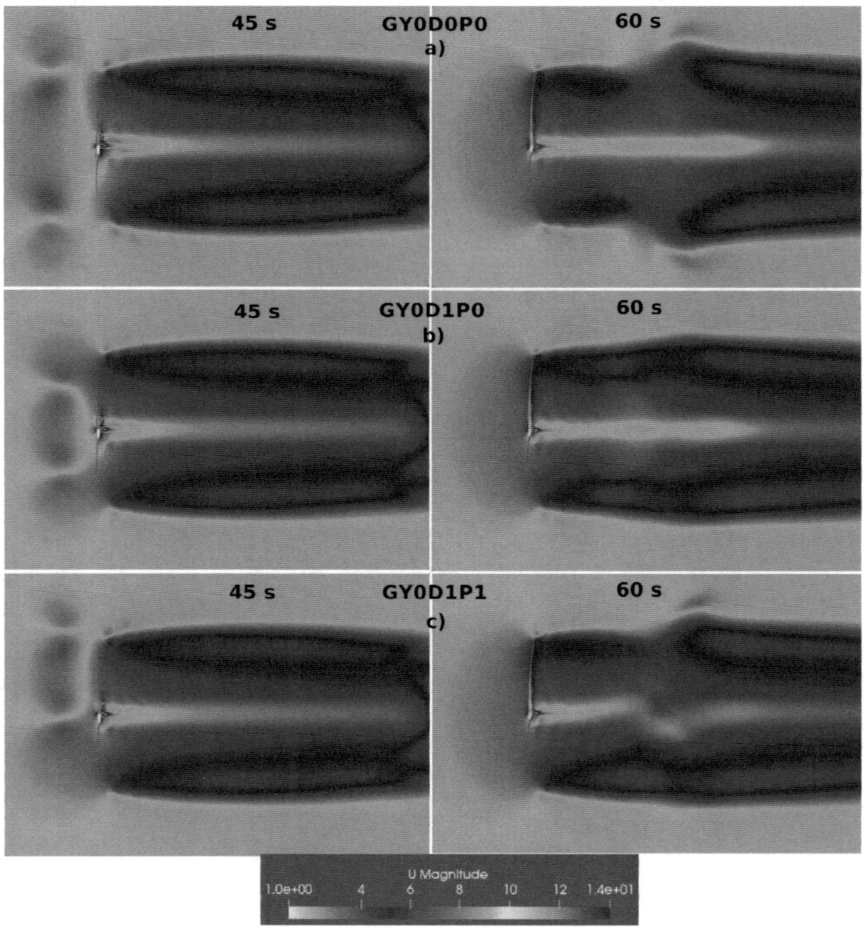


**Figure 6.** Comparison of contours of velocity magnitude before and after impact of gust on the rotor plane

in Figure 4. The gust makes full impact on the wind turbine (WT) after 50 seconds of flow simulation. Figure 4a shows that the loads on the turbine are influenced by the gust event before the gust reaches the rotor plane. This is caused by the interaction of the gust with the induction zone. The gust retards due to the relatively low-velocity region, imparting momentum to the fluid in the induction zone, thereby influencing the thrust and torque characteristics of the turbine. This is substantiated by the gradual increase in torque on the turbine from 30 seconds of simulation. This effect is prominent in the cases GY0D0P0 and GY30D0P0 when the gust impacts are symmetric. The torque curves for symmetric and asymmetric gust impacts reveal a disparity in magnitudes. Under symmetric impact, there is a significant enhancement of torque generation. Conversely, the partial gust impact results in oscillating patterns in both torque and thrust, with only marginal power increase. As expected, the torque and thrust are lower for the yawed cases. The yaw angle increases the period over which the turbine is subject to the gust.

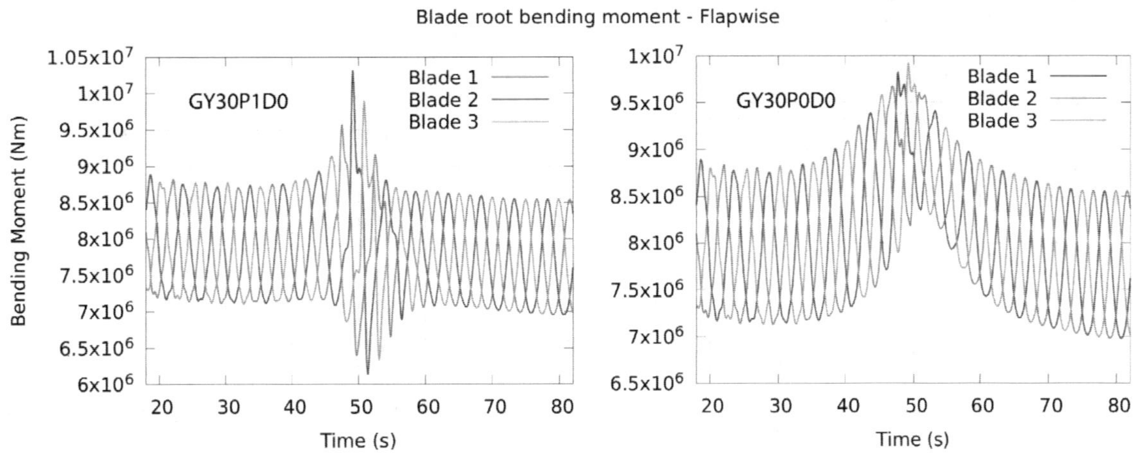


**Figure 7.** Time evolution of flap-wise bending moments at the root section of the three blades of WT for GY30P1D0 and GY30P0D0

This is the reason for the wider, more distributed curves for GY30D0P0 and GY30D0P1.
Figure 5 illustrates the flap-wise deflections at the tip of turbine blades, for all the scenarios considered in this study. It is noted that the trends (shown in Figure 5) are consistent across all three rotor blades. The edgewise deflections exhibit minimal variation between full and partial gust impacts. However, substantial flap-wise deflections are observed during partial gust impact (GY0D1P1 and GY0D0P1). Under full gust impact, the flap-wise deflection experiences a slight shift in the mean tip deflection, with nearly unchanged deflection from the mean value. In contrast, during partial gust impact, the tip undergoes rapid deflection, with the tip deflection being 2.5 times greater than that during normal yawed operation. This finding holds significant implications, as such extensive deflections could potentially lead to structural failure.

Figure 6 details the instantaneous velocity contour in the near wake of the turbine before and after gust impact for cases with yaw angle of zero degree. For GY0D0P0, the contour at 45 seconds clearly indicate a gust having the same diameter of the turbine approach the rotor plane. After impact the turbine wake is observed to expand. In the other two scenarios shown in this figure the gust has a diameter half that of the rotor plane (see Fig. 6b). The center line impact of this half diameter gust leads to a pulse like increase of velocity behind the hub of the turbine. The near wake is quiet distinct (see Fig. 6c) in the case of asymmetric gust strike. The outer region of the turbine wake is distorted outward while the wake generated by the hub distorts inward. The tip deflections in the flap-wise direction is also greater for this scenario as shown in Fig. 5.

The variation of flap-wise bending moments at the root section of the blades with respect to flow time is presented in Figure 7 for the two yawed cases investigated in this study (GY30P1D0 and GY30P0D0). When the influence of the gust is minimal (first 20-30 seconds in Fig. 7), gravity force dominates the bending moments as noted by [9]. This effect manifests as a phase difference between the flap-wise root bending moment of the turbine blades (due to the difference in azimuth angle of each blade). The root bending moment for GY30P1D0, when the gust impact is asymmetric to the rotor center, shows a short time period of approximately 10 seconds during which the amplitude of flap-wise bending moments is increased by 100% (Blade 1 in Fig. 7). This is attributed to the impact of the gust, and the duration corresponds to the gust period. It is also noted that the amplitude of fluctuation during this period is different for each blade. This is because the time period of gust duration and turbine rotation are comparable. The temporal evolution of the flap-wise bending moment for GY30P0D0 conveys a different response to the gust. The amplitude of fluctuation within a cycle of turbine rotation remains unchanged, but the bending moments of all three blades are observed to increase over a time period that is greater than the gust period. This means that the influence of the gust is longer, but the fluctuations in the loads are lower. The maximum flap-wise bending moment for this case is also observed to be lower than for GY30P1D0.

## 5. Conclusions

This study examines the relative influence of the spatial distribution of wind gusts on a 5MW NREL wind turbine operating under yawed conditions using high-fidelity fluid-structure coupled simulations. Observations from these investigations indicate that the gust undergoes deformation within the induction zone of the turbine, influencing thrust and torque even before the instance of direct impact, thereby challenging the concept of uniform gust impact on the rotor plane. This aspect is dominant when the core of the gust aligns with the center of rotation of the wind turbine (symmetric impact). On the contrary, tip deflection and root bending moments of the wind turbine blade are more pronounced when there is misalignment between the core of the gust and the center of rotation of the wind turbine (asymmetric impact), highlighting the potential for structural failure or blade-tower impact for such scenarios. This study underscores the significance of partial gust impact in relation to structural integrity. Future studies will address the effect of gusts with a sheared atmospheric boundary layer profile to improve understanding of the effects of asymmetric loading on the wind turbine.

## Acknowledgement

This work is part of the EMUwind project (FKZ 03EE2031C) funded by the German Federal Ministry of Economic Affairs and Climate Action.

## References

[1] IEC-Standard. *61400-21. Measurement and Assessment of Power Quality of Grid Connected Wind Turbines.* 2002.

[2] Wim Bierbooms and Po-Wen Cheng. Stochastic gust model for design calculations of wind turbines. *Journal of Wind Engineering and Industrial Aerodynamics*, 90(11):1237–1251, 2002.

[3] Christoph Knigge and Siegfried Raasch. Improvement and development of one-and two-dimensional discrete gust models using a large-eddy simulation model. *Journal of Wind Engineering and Industrial Aerodynamics*, 153:46–59, 2016.

[4] Stuart Edward Norris, John Edward Cater, Karl Alexander Stol, and CP Unsworth. Wind turbine wake modelling using large eddy simulation. In *Proceedings of the 17th Australasian Fluid Mechanics Conference. University of Auckland*, 2010.

[5] Niels Troldborg, Jens N Sørensen, Robert Mikkelsen, and Niels N Sørensen. A simple atmospheric boundary layer model applied to large eddy simulations of wind turbine wakes. *Wind Energy*, 17(4):657–669, 2014.

[6] B Dose, H Rahimi, B Stoevesandt, and J Peinke. Fluid-structure coupled investigations of the nrel 5 mw wind turbine for two downwind configurations. *Renewable energy*, 146:1113–1123, 2020.

[7] B Dose, H Rahimi, I Herráez, B Stoevesandt, and J Peinke. Fluid-structure coupled computations of the nrel 5mw wind turbine blade during standstill. In *Journal of Physics: Conference Series*, volume 753, page 022034. IOP Publishing, 2016.

[8] Bastian Dose, Hamid Rahimi, Iván Herráez, Bernhard Stoevesandt, and Joachim Peinke. Fluid-structure coupled computations of the nrel 5 mw wind turbine by means of cfd. *Renewable energy*, 129:591–605, 2018.

[9] Bridget Moynihan, Babak Moaveni, Sauro Liberatore, and Eric Hines. Estimation of blade forces in wind turbines using blade root strain measurements with openfast verification. *Renewable Energy*, 184:662–676, 2022.

# Methodology to assess scour development around complex structures

**J. Sarmiento[1], S. Fernández[1], R. Guanche[1]**

[1] IHCantabria – Instituto de Hidráulica Ambiental de la Universidad de Cantabria, Isabel Torres 15, 39011, Santander, Spain
sarmientoj@unican.es, feruanos@unican.es, guancher@unican.es

**Abstract**. Accurate prediction of scour evolution around fixed offshore foundations is a key parameter for selecting the most appropriate scour mitigation strategy. Various formulations can be used to estimate scour around monopiles. However, non-specific formulations have been developed for more complex foundations. In the case of jacketed offshore substations (OSS), the geometry can be one of the most complex. This paper describes a hybrid methodology (numerical model simulations, semi-empirical approaches and physical experiments) to estimate the scour evolution around OSS. The application of this methodology allows the selection of the most appropriate mitigation strategy from an early stage of the project to the final decision. The methodology is accompanied by some examples. In general, the results obtained from the semi-empirical approaches show quite good agreement with the measurements from the physical experiments. Nevertheless, considering the complex geometry of the OSS and the potential non-linearities associated with the platform geometry, physical experiments are required to ensure the correct estimation of the maximum scour (local and global).

## 1. Introduction

When structures are supported on the seabed, the presence of the foundation causes an increase in flow velocity, turbulence and vorticity around the structure. All these effects together increase the mobility of the seabed around the structure, leading to what is known as scour ( [1] and [2]). The magnitude of the scour depth depends on several parameters such as seabed characteristics, hydrodynamic conditions, water depth or platform geometry.

During the design process of a fixed offshore foundation, an accurate estimation of the scour is critical in order to select the most appropriate scour mitigation strategy [3].

1) Free scour development (the maximum scour depth is less than the maximum allowable value or the foundation is designed to allow free scour development).
2) Installation of scour protection pads to fix the seabed level to prevent lowering of the seabed around the foundation.

An accurate prediction of the scour around slender cylinders can be estimated using various methods and formulations available in the literature (e.g. [2], [4] or [5] among others). However, when the geometry of the structures is complex, the estimation of the scour along time is more difficult. In many cases reduced scale physical experiments are recommended.

At this point, the Offshore Substation (OSS) is one of the most unique, critical and complex structures to be found at an Offshore Wind Farm (OWF). Although OSS can be monopiles or even gravity foundations, they are often jacket foundations connected to the seabed by multiple piles, groups of piles or suction caissons (Figure 1). For these structures, the development of scour around this type of structure is mainly developed in two phases [1]:

1) Local Scour - Erosion around elements in contact with the seabed.
2) Global scour - general lowering of the seabed as a result of the global blocking effect created by all elements of the foundation. The time scale of global scour could take years from the installation of the platform.

In addition, seabed lowering associated with morphodynamic changes, such as those induced by the propagation of sand waves or megaripples, must also be considered in order to estimate the minimum seabed level at the foundation location.

The correct estimation of scour will provide the platform designers with the necessary information to select the most appropriate mitigation strategy. In general, the design of the scour protection pad is carried out in parallel with the selection of the most feasible mitigation strategy.

This paper focuses only on the presentation of a hybrid methodology to assess the free scour evolution around the jacket OSS, where the connection to the seabed is provided by different sets of pile clusters located around the foundation (Figure 1).

The presented methodology includes the application of semi-empirical formulations, advanced numerical model simulations (openFoam) and reduced scale physical experiments. The first two phases (numerical model simulations and semi-empirical approaches) are carried out to identify the potential scour depth using some simplifications at an early stage of the project. While, the final validation of the equilibrium scour depth will be carried out by performing physical experiments on a reduced scale (the results of Phase I and II will be used to properly design the physical experiments). The results of the physical experiments also provided useful information to derive predictive semi-empirical formulations to improve the prediction of the scour depth in future foundations.

This paper is a methodological paper in which the main assumptions made at each stage of the methodology are presented and discussed. In addition, some examples from previous experience are also included in the methodology section of each stage.

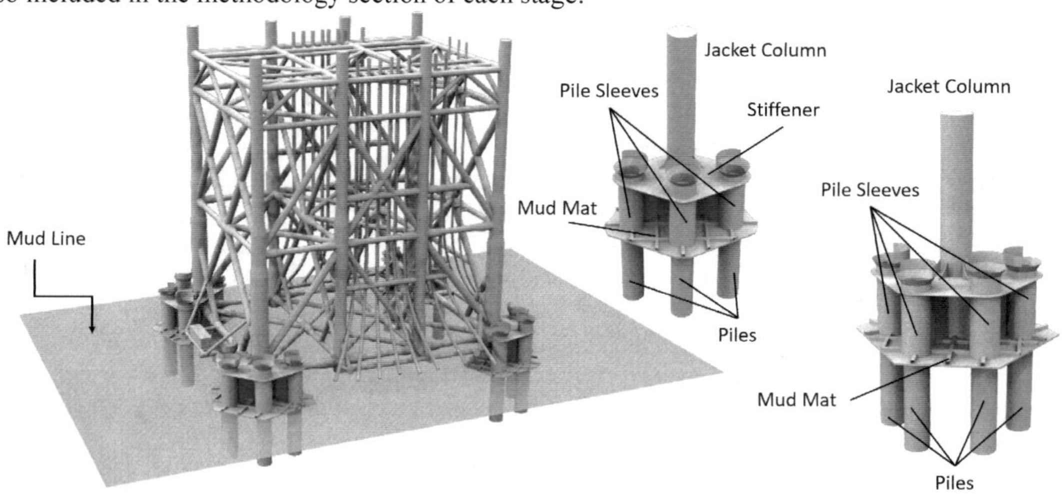


Figure 1. OSS geometry.

## 2. Methodology

The methodology proposed has been divided in three main phases:

a) Step I (early project stages): Advance numerical model simulations: Identification of the flow-structure interaction and the amplification coefficient around the foundation.
b) Step II (early project stages): Estimation of the scour based on the application of semi-empirical formulations.
c) Step III (final validation): Physical experiments at reduced scale (final validation).

In each section, different examples are presented to facilitate the understanding of the proposed methodology. In addition, the methodology presented has been applied to non-cohesive soils, where the maximum erodibility of the seabed material is expected [6].

### *2.1. Step I: Numerical Modelling (flow-structure interaction)*

Numerical model simulations are mainly carried out to estimate the flow amplification coefficient (ratio of the velocity around the foundation divided by the undisturbed flow velocity) around the CFL interaction is evaluated at this stage of the project.

Various numerical methods can be used to calculate the velocities around a jacketed structure exposed to currents and waves. The numerical method chosen must be able to deal with the viscosity of the water and the free surface. In the present case, a RANS method coupled with the VOF transport equation is used to simulate the water effects over the jacket structure.

The Reynolds Averaged Navier Stokes (RANS) method is one of the most widely used models for simulating fluid physics. This method includes the effects of viscosity and the effects of turbulence. The key point of the method is that the turbulence effects are averaged over time. This means that not all but some of the turbulence characteristics are captured. The time averaging of the equations generates additional terms that require additional equations to close the problem. These additional equations are the models used to simulate the turbulence properties. Depending on the model, the turbulence transport equations and their boundary conditions are different. In this case, the $k\omega{-}sst$ model is used. This model uses two transport equations to model the turbulence, one for the kinetic energy and one for the dissipation rate. This model assumes that the turbulence is isotropic.

The Volume of Fluid (VOF) method is used to transport the phase information along the domain. It allows the propagation and interaction of the water waves with the structure to be represented. The VOF method uses an additional equation for the propagation of the phase content along the domain.

There are other numerical alternatives capable of solving the fluid-structure interaction problem with much more detail (i.e. .LES, DNS), but RANS models are identified as a good choice in terms of the accuracy/cost ratio, which is one of the major constraints in most engineering problems.

The mesh is one of the key points in setting up the numerical model. The results of the RANS equations are usually quite mesh dependent. Therefore, in order to increase the accuracy of the simulation, a mesh convergence study is carried out. In this way it is possible to reduce the influence of the mesh properties on the final results. This does not mean that the results are correct, as the model used and the boundary conditions are still in the middle, but it allows the uncertainty of the simulation to be reduced.

Since the RANS equations are time-dependent, it is necessary to integrate the equations not only in space but also in time. To ensure that the time integration is correct, it is necessary to perform an additional convergence study for the time step. In this case, the simulations were performed using an adaptive time step algorithm. It relates the value of the time step to the CFL, which is also related to the mesh size. Therefore, the convergence study is not performed directly on the time step but on the CFL number.

The flow-structure interaction results provide very useful outputs in terms of identifying the maximum flow velocities and the extent of lee wakes (Figure 2) around the foundation. Figure 2 (left side) shows the flow amplification factor around a jacketed OSS under currents (maximum flow amplification factor combining the maximum values at each mesh node for different current directions). The coupling effect between flow amplification waves and jacket elements can lead to the development of global scour.

Based on previous experience with various foundations, the flow amplification factors are around 2 for currents and 3 for waves (in a few specific cases, maximum values have been reached up to 3.5-4 times the undisturbed wave-bottom velocity).

Journal of Physics: Conference Series **2745** (2024) 012020 doi:10.1088/1742-6596/2745/1/012020

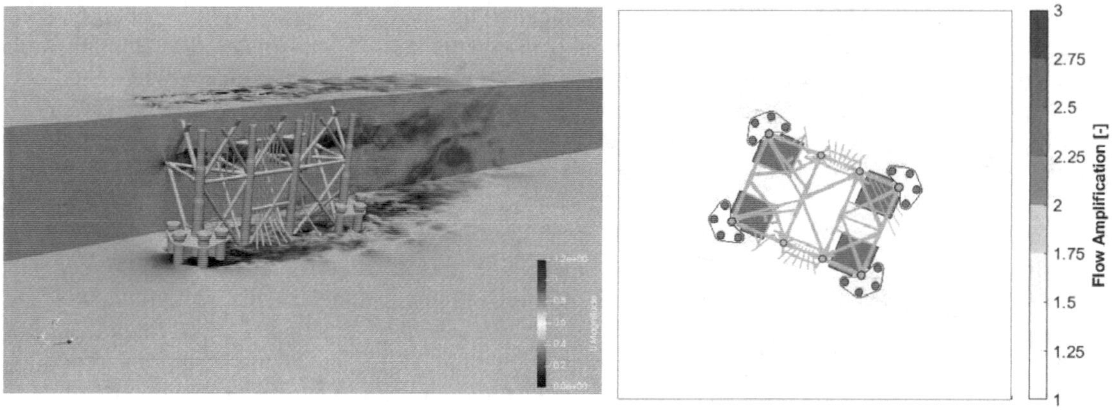

Figure 2. Left: Numerical Model Simulation (CFD, openFOAM). Right: Flow Amplification Coefficient.

### 2.2. *Step II: Estimation of the scour based on the application of semi-empirical formulations.*
In this section, both local and global scour around the foundation are estimated based on the application of semi-empirical formulations and the results of the numerical modelling. The total scour expected at the foundation is the sum of both quantities plus the expected seabed subsidence associated with the morphodynamic changes.

#### 2.2.1. *Local Scour*
Based on previous experience with jacket OSS, local scour is mainly located around the elements in contact with the seabed (mainly pile clusters and mud mats). The contribution to local scour from other elements, such as horizontal and vertical frames, J-tubes and upper foundation elements, can be considered negligible compared to the local scour observed around the pile clusters and mud mats.
Given the key assumption described in the previous paragraph, a simplification of the pile cluster geometry is proposed. The main idea is to replace the real geometry with an equivalent simplified geometry capable of producing an equivalent blocking effect. The pile clusters (elements in contact with the seabed) are simplified as an equivalent truncated vertical cylinder, where:

- Diameter: contour of the pile sleeves and jacket column (based on previous experience the equivalent diameter may vary from 8-16metres).
- Height (stick-up): From the ground to the top of the pile cluster (variable height depending on the pile cluster).

In general, the equivalent diameter of the truncated cylinder (simplified pile cluster geometry) is 3 to 5 times larger than the column diameter. Given these differences, local scour is estimated by considering only the contribution of the equivalent truncated cylinder (simplified geometry around the pile cluster). As the height of the equivalent vertical cylinder does not extend along the entire water column, the application of a height correction factor (Kh) is proposed to account for the limited height of the pile cluster with respect to the entire water column. Different height correction factors have been proposed for vertical cylinders (i.e. [7] and [8]).
The simplified geometry can be seen in Figure 3 (dashed red circles represent the simplified geometry).

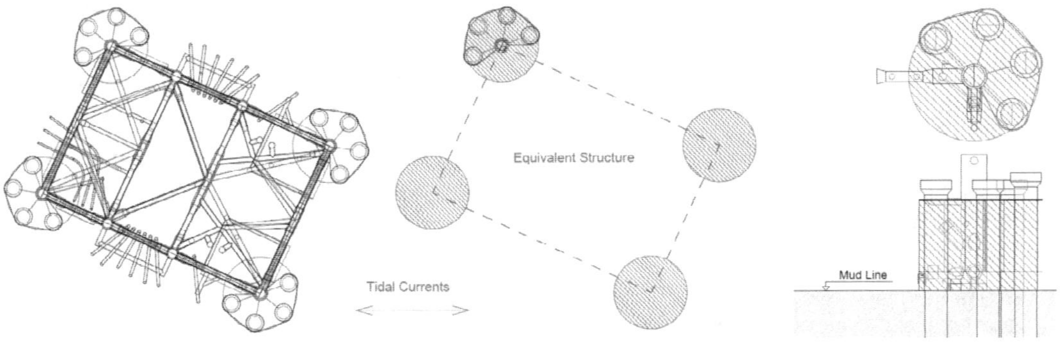


Figure 3. Simplified geometry.

Given the simplification suggested above, the equilibrium scour depth can be estimated individually using Equation 1:

$$S_{eq-f} = S_{eq} \cdot K_h \quad \text{Equation 1}$$

Where:

- $S_{eq-f}$ – Final Equilibrium scour depth.
- $Seq$ – Equilibrium scour depth for vertical cylinders.
- $K_h$- correction factor for a truncated cylinder, stick-up height.

It should be noted that the proposed simplification considers each pile cluster as an individual element, without considering the influence of the horizontal and vertical frames, as well as the clogging effect induced by all the jacket elements. In summary, local scour is estimated for individual elements by attempting to reproduce the clogging effect of a single pile cluster.
Based on the proposed simplification of the geometry, the local scour is calculated under steady-state conditions and the long-term analysis is simulated taking into account the metocean conditions associated with the lifetime of the foundation (different metocean databases are used to take into account the characteristics and duration of the events).

### *2.2.1.1 Steady Environmental Conditions*

Under steady uniform conditions (wave, currents or combined wave and currents), the evolution of the scour is often represented by Equation 2 ( [1] and [2]), where, S(t) is the evolution of the scour with time, $T_{char}$ (s) is the characteristic time, $S_{eq}$ (m) is the equilibrium scour depth and t (s) the time.

$$S(t) = S_{eq}[1 - e^{\left(-\frac{t}{T_{char}}\right)}] \quad \text{Equation 2}$$

In order to estimate the evolution of the scour depth along the time, the application of Equation 1 requires the knowledge of the equilibrium scour depth and the characteristic time:

a) Steady currents: According to [9], tidal currents are chosen as the driving parameter for scour evolution. The selection of maximum currents associated with the RP of 10y-50yr-100yr may give rather conservative results due to the limited duration of these events and combined with the presence of waves:
   - Live Bed Regime ($\theta > \theta_{cr}$): The parameters of Equation 2 can be estimated using the relationship proposed in [5] or [2]. An additional formulation derived from the physical experiments conducted at IHCantabria has been included in the methodology and will be presented soon.
   - Clear Water condition ($\theta < \theta_{cr}$): The parameters of Equation 2 can be estimated using the relationship proposed at [1] and [5].
b) Combine Waves and Currents: Formulations proposed at [10] and [2] (design sea states based on [11]).

In general, the results obtained at this stage are quite good compared to the results obtained during the physical experiments.

### *2.2.1.2 Long term Analysis*

In addition to the steady state, a model based on those proposed in [7]and [12] was also included in the methodology to estimate the final equilibrium scour depth around the OSS, taking into account the geometric simplification described above. The developed model considers the following inputs:

- Platform Geometry - simplified (as described in the previous section).
- Metocean Characteristics (Figure 4) - IHCantabria metocean databases considering waves, currents and water depth variations. Finally, the database time interval (dt) is 1 hour, so the scour estimated was updated hourly.
- Soil Characteristics (non-cohesive materials).
- Flow amplification coefficient (based on numerical model simulations described in the next section). To estimate the mobility parameter at the foundation.

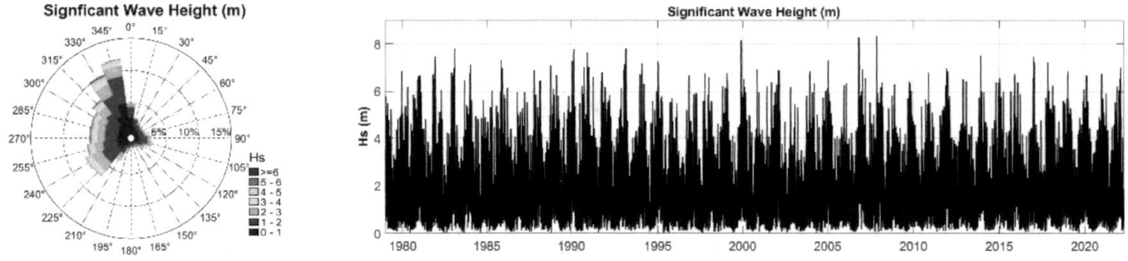


Figure 4. Example of the metocean characteristics: 1) Wave rose, 2) Time series.

The estimation of the scour depth at each time step follows the equation presented in [10]. The developed model allows the selection of different formulations (steady state) to estimate the scour and backfilling time and the equilibrium scour depth (live bed regime or clear water condition, depending on the sediment transport regime of each time step). Estimation of the backfilling time is not clear for this type of structure. For monopiles, according to [7] and [13], the backfilling characteristic time is often greater than the scour time (variable as a function of the KC number). The evolution model allows the backfilling characteristic time to be varied in order to assess the uncertainties associated with this parameter (variation of the backfilling characteristic time can reduce/increase the final equilibrium scour depth between 0.5-2m). Figure 5 shows an example of the results obtained after applying the evolution model. The blue lines show the evolution of scour over time, while the grey line shows the hourly significant wave height. The main difference between the blue lines is the characteristic time associated with the backfilling events (the backfilling time on the dashed line was considered to be much higher than on the solid line, an accurate estimation of the characteristic time is difficult due to the geometric simplification).

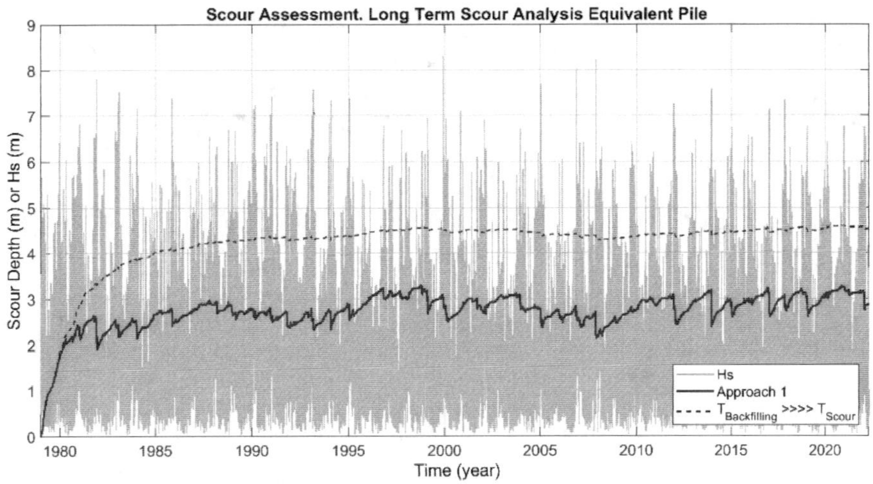


Figure 5. Left: Long term scour analysis. Right: Scour evolution under steady condition.

In general, the equilibrium scour depth reached after applying the long-term analysis gives a lower equilibrium scour depth than the values obtained for tidal steady currents. In many situations, backfilling events combined with the fluctuation of the tidal currents make it impossible to achieve the equilibrium achieved with steady currents.

### *2.2.2. Global Scour*

The estimation of global scour around the jacket foundation is quite difficult to predict. In many situations it may take several years to develop after the OSS has been installed. There is limited information on global scour.

To estimate the evolution of global scour, the formulation proposed in [14] can be used. Additionally, based on previous physical experiments conducted in-house at IHCantabria facilities, the global scour magnitude is estimated to be between 1-2m.

## *2.3. Step III: Physical Experiments*

Physical experiments allow to properly reproduce the interaction between flow, structure and soil, therefore the main objective of the physical experiments is the final estimation/validation of the equilibrium scour depth along the pile clusters, as well as the potential assessment of the global scour around the foundation.

The experimental methodology was developed at IHCantabria facilities (CCOB, equivalent to the basin tests described in [15]). Considering the main characteristics of the basin (CCOB) and the size of the platforms, the scale of the test can vary from 1/35 to 1/45. In addition, since the basin is wider rather than longer, two or even three OSS may be tested simultaneously (Figure 6 shows an example of the proposed setup). Froude scaling laws of similarity [16] and mobility similarity are applied to reproduce the environmental conditions and soil mobility (non-cohesive materials) at laboratory scale.

Figure 6. General view of the physical tests.

In order to assess the scour evolution over time, physical tests are performed in different steps, which are carried out in sequence. The test sequence is repeated until equilibrium is reached (the duration of the interval may vary depending on the cumulative test time). Between each step, laser scanner measurements are performed to measure the evolution of the scour (depth and extent) over time.

As an example of the results obtained from the basin tests, the next figures summarise the main outcomes obtained during a series of physical experiments around a typical OSS under steady uniform currents (tidal magnitude). Laser scanner measurements are taken around the entire area of the jacket foundation, so that both the local scour and the global erosion pattern can be determined after each time step. Figure 7 shows the laser scanner measurements after three different steps of the test sequences.

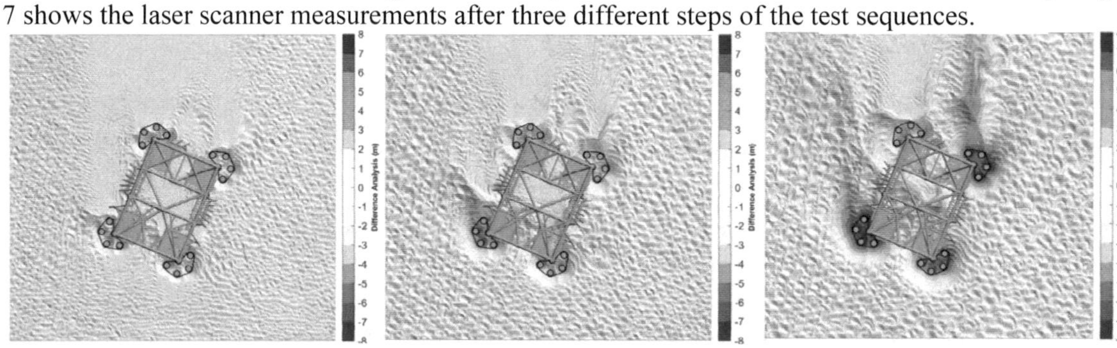


Figure 7. Laser Scanner measurements from the physicial experiments

With regard to local erosion close to the jacket piles, Figure 8 shows the evolution of scour along one of the front piles of the jacket foundation (Figure 8 left side, erosion magnitude after each time step, Figure 8 right side, longitudinal profiles from the centre of the pile). Figure 8 clearly shows how, for a single pile, the final equilibrium scour depth was reached by the end of the test.

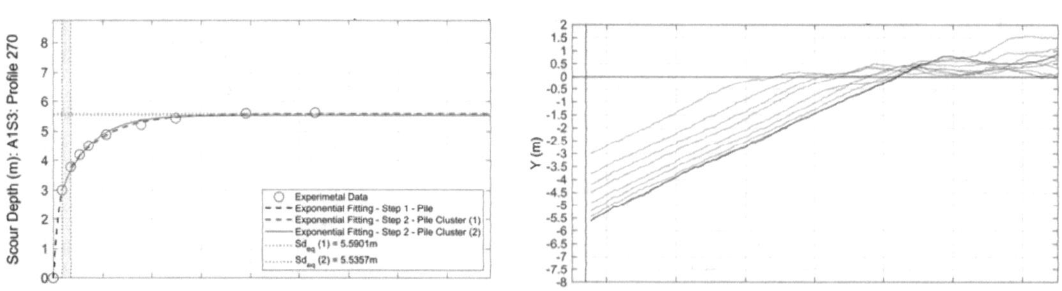


Figure 8. Left: evolution of the scour around a single pile. Right: Longitudinal Profiles from a single pile after each step of the sequence.

Finally, if the duration of the test is sufficiently long, global scour evolution patterns can also be observed during the physical experiments. For example, Figure 9 shows the global seabed lowering between one of the rows of OSS. This effect was observed when the front erosion wake reached the position of the rear pile cluster (Figure 9). The left side of Figure 9 shows, the longitudinal profiles between the two pile clusters (front and rear) after each time step. From this figure, it can be seen how from step 7 (blue line) to the last profile (black solid line) a general lowering of the seabed is generated along the jacket side. According to different results from physical experiments, the global seabed lowering varies between 1-2 metres.

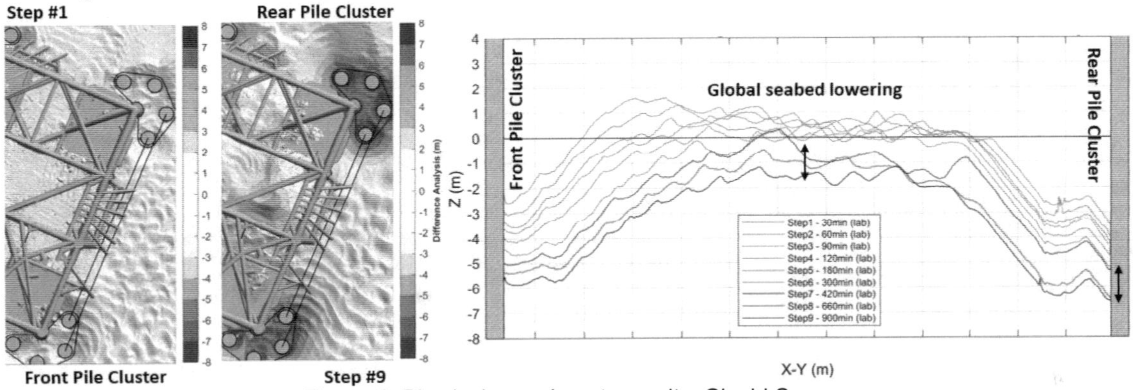


Figure 9. Physical experiments results: Gloabl Scour

Based on the methodology and the results presented above, the physical experiments allow us to estimate the global and local scour around complex structures such as the OSS. The execution of the tests in several steps, allows us to identify the global erosion patterns around the different elements of the OSS.

## 3. Conclusions

The paper summarises the methodology presented to estimate the scour around complex structures such as jacket OSS connected to the seabed by multiple piles. The methodology presented combines numerical model simulations, semi-empirical approaches and physical experiments with the aim of estimating the potential scour around complex structures from the early design stages. The main results are presented below:

- Local scour is mainly located in the elements in contact with the seabed (pile clusters and mud mats). The simplification of the pile clusters as vertical truncated cylinders allows us to estimate the scour depth using some formulations designed for monopiles (under steady conditions and long-term analysis). The application of the semi-empirical approaches allows to get a general idea of the local equilibrium scour depth expected at the pile clusters.
- 3D CFD (openFoam) simulations are performed to reproduce the full geometry of the foundation. Numerical model simulations are performed to estimate the flow amplification around the foundation. According to previous experience with jacketed OSS, the flow amplification factor is around: currents~2, waves~3 (in a few specific cases, the maximum values reached values up to 3.5-4 times the undisturbed bottom orbital velocity). Identification of the flow amplification factor around the foundation provides the potential eroded areas around the OSS. The coupling effect between erosion wakes and jacket elements can lead to the development of global scour.
- Physical experiments allow the identification of the global erosion patterns, the local scour equilibrium depth at the piles and the global scour.
- The comparison between the results obtained with the numerical and semi-empirical approaches and the results of the physical experiments gives quite reasonable results, taking into account the main assumptions made. Finally, the presented methodology has been

satisfactorily applied to various OSS already in operation, under construction or at an early design stage.

## List of Abbreviations and Symbols

- $K_h$- correction factor for a truncated cylinder, stick-up height.
- KC – Keulegan Carpenter.
- OSS - Offshore substation.
- OWF - Offshore Wind Farm.
- RANS - Reynolds Averaged Navier Stokes.
- $S(t)$– Scour depth in function of the time.
- $S_{eq-f}$ – Final Equilibrium scour depth.
- $Seq$ – Equilibrium scour depth for vertical cylinders.
- t – time.
- $T_{char}$ – Characteristic Time.
- VOF - Volume of Fluid.
- $\theta$ – Shields Number.
- $\theta_{cr}$ – Critical Shields Number.

## Acknowledgements

- The authors would like to thank DRAGADOS OFFSHORE for their collaboration in this research.
- R. Guanche also acknowledges Grant RYC-2017-23260 funded by MCIN/AEI/10.13039/501100011033 by "ESF Investing in your future".
- This study forms part of the ThinkInAzul programme and was supported by Ministerio de Ciencia e Innovación with funding from European Union NextGeneration EU (PRTR-C17.I1) and by Comunidad de Cantabria.

## References

[1] R. Whitehouse, Scour at Marine Structures: A Manual for Practical Applications, Thomas Telford Limited, 1998.

[2] B. M. Sumer y J. Fredsøe, The mechanics of scour in the marine environment, Advanced series on ocean engineering: River Edge, N.J, 2002.

[3] Deltares, Handbook of Scour and Cable Protection Methods, Deltf, 2023.

[4] H. N. C. Breusers, G. Nicollet y J. Shen, «Local Scour Around Cylindrical Piers,» *Journal of Hydraulic Research,* vol. 15:3, pp. 211-252, 1977.

[5] D. M. Sheppard, B. Melville y H. Demir, «Evaluation of Existing Equations forLocal Scour at Bridge Piers,» *Journal of Hydraulic Engineering,* vol. 140, nº 1, pp. 14-23, 2014.

[6] R. J. S. Whitehouse, «Scour at coastal structures (Invited lecture),» Gouda, The Netherlands, 2006.

[7] T. Raaijmakers y D. Rudolph, «Time-dependent scour development under combined current and waves conditions -laboratory experiments with only monitoring technique,» 4th International Conference on Scour and Erosion, 2008.

[8] R. J. Whitehouse y E. I. Stroescu, «Scour depth development at piles of different height under the action of cyclic (tidal) flow,» *Coastal Engineering,* vol. 179, p. 104225, 2023.

[9] T. Ugelvig Petersen, B. Mutlu Sumer, J. Fredsøe, , T. C. Raaijmakers, y J.-J. Schouten,, «Edge scour at scour protections around piles in the marine environment — Laboratory and field investigation,» *Coastal Engineering,* vol. 106, pp. 42-72, 2015.

[10] D. Rudolph, T. Raaijmakers y C. Stam, «Time-dependent scour development under combined current and wave conditions. Hindcast on field measurements,» 2008.

[11] DNVGL-OS-C101, «RECOMMENDED PRACTICE. Design of offshore steel structures, general - LRFD method,» 2019.

[12] T. Raaijmakers, T. Joon, M. Segeren y P. Meijers, «Scour: to protect or not to protect? That's the question. Feasibility of omitting scour protection,» de *Proc. EWEA Offshore, Frankfurt, Germany,* Proc. EWEA Offshore, 2013, Frankfurt, Germany, 2013.

[13] B. M. Sumer, T. U. Petersen, L. Locatelli, J. Fredsøe, R. E. Musumeci y E. Foti, «Backfilling of a Scour Hole around a Pile in Waves and Current,» *Journal of Waterway, Port, Coastal, and Ocean Engineering,* vol. 139, nº 1, 2012.

[14] A. Bolle, J. De Winter, W. Goossens, P. Haerens y G. Dewaele, «Scour monitoring around offshore jackets and gravity based foundations,» Paris, 2012.

[15] J. Sarmiento, R. Guanche, A. Iturrioz, T. Ojanguren, A. Ávila y C. Yanes, «Dynamic Rock Scour Protection in Morphodynamic Environments for Offshore Wind Jackets,» *Energies,* vol. 3379, nº https://doi.org/10.3390/en14123379, p. 14 , 2021.

[16] S. A. Hughes, Physical Models and laboratory techniques in coastal engineering, Singapore: World Scientific Publishing Co. Pte. Ltd, 1993.

[17] J. M. Harris, R. J. S. Whitehouse y T. Benson, «The time evolution of scour around offshore structures,» *Proceedings of the Institution of Civil Engineers,* vol. Maritime Engineering, nº 163, 2010.

Journal of Physics: Conference Series **2745** (2024) 012021 doi:10.1088/1742-6596/2745/1/012021

# Linking the microstructure with strain-life curves for improved utilization of the lightweight potential of thick-walled nodular cast iron

**C Pittel[1], J Niewiadomski[1], C Bleicher[1]**

[1]Fraunhofer Institute for Structural Durability and System Reliability LBF, Bartningstraße 47, 64289 Darmstadt, Germany

christian.pittel@lbf.fraunhofer.de

**Abstract**. The assessment of the fatigue life of critical, highly stressed component areas is an important aspect in the design of thick-walled components for wind turbines made of nodular cast iron. With the demand for resource efficiency through lightweight design combined with a certain capability to withstand extreme loads and special events, this issue is becoming increasingly important. However, there are still gaps in the description of local strain-based material behaviour. For the design and local strain-based fatigue assessment of large cast components, possibilities to determine the fatigue life as accurately as possible must be provided by standards and guidelines. The paper aims to provide a wide range of material and fatigue data from one common source for general design interests. During research projects at Fraunhofer LBF, material properties for the $P_{RAM}$-based fatigue life assessment and the elastic-plastic material behaviour have been derived from 9 nodular cast iron grades with 44 microstructural variations. To describe the influence of the technological size effect on cyclic stress-strain and $P_{RAM}$-N curves the quasi-static parameters (tensile strength and elongation at break) and the microstructure (nodularity and graphite particle density) are considered. In addition, suggestions for scatter bands in the strain-based design approach are given to the designers of thick-walled GJS components.

## Symbol directory

| | | |
|---|---|---|
| $R_\varepsilon$ | Strain ratio | - |
| $R_\sigma$ | Stress ratio | - |
| $N_{lim}$ | Limit number of cycles | - |
| f | Frequency | Hz = $s^{-1}$ |
| $\sigma_{a,n}$ | Nominal stress amplitude | MPa |
| k | Factor for mean stress sensitivity | - |
| $\sigma_m$ | Mean stress | MPa |
| $\varepsilon_{a,tl}$ | Total strain aplitude | % |
| E | Young's Modulus | MPa |
| $M_\sigma$ | Mean stress sensitivity | - |
| $P_{RAM}$ | Damage parameter | MPa |
| N | Number of cycles to crack initiation | - |
| $N_{k,1}$ | First knee point of damage parameter-life ($P_{RAM}$-N) curve | - |
| $N_{k,2}$ | Second knee point of $P_{RAM}$-N curve | - |

| | | |
|---|---|---|
| $P_{RAM,Z,WS}$ | Knee point (KP) of the $P_{RAM}$-N curve at $N_{k,1}$ | MPa |
| $P_{RAM,D,WS}$ | KP of the $P_{RAM}$-N curve at $N_{k,2}$ | MPa |
| $d_1$ | Slope of the $P_{RAM}$-N curve in the LCF range | - |
| $d_2$ | Slope of the $P_{RAM}$-N curve in the MCF range | - |
| $d_3$ | Slope of the $P_{RAM}$-N curve in the HCF range | - |
| $P_S$ | Probability of survival | % |
| P | Probability of occurrence | % |
| $T_\sigma$ | Scatter band of the cyclic stress-strain curve (P = 90 %) | - |
| $T_{PRAM}$ | Scatter band $P_{RAM}$-N curve ($P_S$ = 90 %) | - |
| $f_{2.5\%}$ | Factor for conversion to 2.5 % probability of failure ($P_S$ = 97.5 %) | - |
| $K'$ | Cyclic hardening coefficient | MPa |
| $n'$ | Cyclic hardening exponent | - |
| $\varepsilon_{a,pl}$ | Plastic strain amplitude | % |
| $\varepsilon_{a,el}$ | Elastic strain amplitude | % |
| $R_{p0,2\%}$ | Yield strength | MPa |
| $R_m$ | Tensile strength | MPa |
| $a_{K'}$ | Coefficient for determining the cyclic stress strain curve | MPa |
| $b_{K'}$;$a_{n'}$; $b_{n'}$; $n'_{limit}$ | Parameters for determining the cyclic stress strain curve | - |
| $n_{st}$ | Factor for the statistical size effect | - |
| $A_\sigma$ | Highly stressed specimen surface | $mm^2$ |
| $A_{ref}$ | Highly stressed reference specimen surface | $mm^2$ |
| $k_{st}$ | Weibull coefficient | - |
| $a_{P,Z}$; $a_{P,D}$ | Coefficients for determining the KP of the $P_{RAM}$-N curve | MPa |
| $b_{P,Z}$; $b_{P,D}$ | Exponents for determining the KP of the $P_{RAM}$-N curve | - |
| $a_{1/d1}$; $b_{1/d1}$ | Parameter for determining the slope of the $P_{RAM}$-N curve in the LCF range | - |
| A | Elongation at break | % |
| $\rho_{graphite}$ | Graphite particle density | $mm^{-2}$ |
| $n_{graphite}$ | Nodularity of the graphite particles | % |
| $A_{Z,0}$; $A_{Z,1}$; $A_{Z,2}$; $A_{Z,3}$ | Parameters for determining the first KP of the $P_{RAM}$-N curve | MPa |
| $A_{D,0}$; $A_{D,1}$; $A_{D,2}$; $A_{D,3}$ | Parameters for determining the second KP of the $P_{RAM}$-N curve | MPa |
| $N_{experimental}$ | Experimentally determined number of cycles | - |
| $N_{calculated}$ | Calculated number of cycles | - |

## 1. Introduction

The assessment of the fatigue life of critical, highly stressed component areas is an important aspect in the design of thick-walled components for wind turbines made of nodular cast iron. With the demand for resource efficiency through lightweight design combined with a certain capability to withstand extreme loads and special events, this issue is becoming increasingly important. The material class "nodular cast iron" (GJS) represents a wide range of different grades from ferritic microstructures, offering low tensile strength but therefore high elongation at break, to ausferritic microstructures (ADI) with inversed strength-ductility-ratio. Due to the casting process, the material further offers the possibility of producing components with a high freedom in design adopted to the specific load case.

The fatigue life assessment gains importance when notches or deviations in the local microstructure are present. However, there are still gaps in the description of local strain-based material behavior. New cast iron grades such as ADI have been little studied. However, they do offer the potential for lightweight design or increasing the maximum allowable load on the component. For the design and local strain-based fatigue assessment of large cast components, possibilities to determine the service life as accurately as possible must be provided by standards and guidelines [1, 2, 3]. While [3] and [2] only offer a stress-based fatigue assessment concept, [1] gives a guideline for a strain-based fatigue assessment concept using the damage parameter $P_{RAM}$, at least for smaller components and other groups of materials.

Recent investigations on thick-walled GJS have shown the influences on the fatigue strength resulting from changing microstructure for stress-based design purposes [4, 5, 6, 7, 8, 9]. In these works, mainly wind energy related ferritic GJS have been investigated based on stress-controlled fatigue tests while defects and size effects according to [10] were considered. In further works from the literature [11, 12, 13, 14, 15] and works performed at LBF [7, 8, 16, 17], the focus is also on strain-based approaches to better understand the cyclic deformation behavior as well as the strain-based fatigue behavior. The influence of locally different microstructures is not significant on the cyclic stress-strain behavior for large castings made of EN-GJS-400-18LT as well as on the material cyclic hardening [11, 12]. A cyclical hardening effect can be observed for all grades of nodular cast iron [7, 13, 16]. However, there is a lack of available strain-life curves that allow an evaluation for the fatigue life, which takes the local microstructure into account.

Based on the previous work at LBF and in the literature, the paper aims to provide an extension of the FKM guideline "Non-linear" [1]. The intention is to present the user of this guideline all the required parameters for the design of components made of thick-walled nodular cast iron. For the strain-based fatigue life assessment, material-specific $P_{RAM}$-N curves are essential. To derive these material-specific $P_{RAM}$-N curves and correlate them with the tensile strength, the damage parameter $P_{RAM}$-N curves of 9 different grades and different wall thicknesses have been summarized from several research projects at LBF. This leads to a fatigue assessment concept that follows the FKM "Non-linear" very closely and allows to derive $P_{RAM}$-N curves based only on the tensile strength.

The next step in the design of components using the FKM "Non-linear" is to determine the strain under plastic deformation at the highest stress area of the component. Therefore, the parameters of the deformation curve according to Ramberg-Osgood K' and n' are derived as a function of the tensile strength $R_m$, these correlations could be further confirmed with parameters from literature. This enables the designer of structural cast components to compute the local plastic deformations using an FE calculation with an elastic-plastic material model or to use the notch root approximation according to Neuber, as recommended by [1].

To obtain the design $P_{RAM}$-N curves from the material-specific $P_{RAM}$-N curves, the use of statistically verified safety factors is mandatory. A detailed safety assessment has shown that all available fatigue tests can be safely modelled using the in this paper specified factor for conversion to 2.5 % probability of failure $f_{2.5\%}$.

In addition, a second approach to the fatigue life assessment concept is presented based on more information. To describe the influence of the technological size effect more accurately, the characteristics of the microstructure in terms of graphite morphology and elongation at break A are also considered in the latter concept. A multiple linear regression is used to link the quasi-static and microstructural parameters with the parameters of the $P_{RAM}$-N curves and to statistically secure them.

## 2. Determination of $P_{RAM}$-N curves and cyclic stress-strain curves according to FKM19

The aim of this paper is to provide an extension of the FKM guideline "Non-linear" for components made out of thick-walled nodular cast iron based on previous work at LBF and in the literature. All the necessary parameters for the design of nodular cast iron components are presented to the user of this guideline. Material-specific damage parameter $P_{RAM}$-N curves are the main element of the strain-based fatigue life assessment according to [1]. The fatigue data of 9 different grades and wall thicknesses from several research projects at LBF were investigated to derive experimental $P_{RAM}$-N curves and cyclic stress-strain curves.

### *2.1. Definition of the damage parameter $P_{RAM}$*

The damage parameter $P_{RAM}$ is a modification of the damage parameter $P_{SWT}$ according to Smith, Watson and Topper [18] with consideration of the influence of mean stress according to Bergmann [19]. The damage parameter $P_{RAM}$ is used in the FKM guideline [1] „Rechnerischer Festigkeitsnachweis für Maschinenbauteile unter expliziter Erfassung nichtlinearen Werkstoffverformungsverhaltens" (engl., "Computational strength assessment for structural components with explicit consideration of non-linear material deformation behavior"). Equation (1) shows the formula for the damage parameter as a function of the mean stress $\sigma_m$ and the nominal stress amplitude $\sigma_{a,n}$.

$$P_{RAM} = \begin{cases} \left(\left(\sigma_{a,n} + k \cdot \sigma_m\right) \cdot \varepsilon_{a,tl} \cdot E\right)^{\frac{1}{2}}, \; for \left(\sigma_{a,n} + k \cdot \sigma_m\right) \geq 0 \\ 0, \; for \left(\sigma_{a,n} + k \cdot \sigma_m\right) < 0 \end{cases} \quad (1)$$

If there is a mean stress $\sigma_m$ present on the component that has to be assessed, the factor k must be determined. The factor k is then obtained from the mean stress sensitivity $M_\sigma$, which according to [2] results in a mean stress sensitivity $M_\sigma = 0.4$ independent of the tensile strength of the material which is assumed to be valid for all nodular cast iron grades. The factor k is therefore only dependent on the mean stress sensitivity and is thus calculated according to equation (2).

$$k = \begin{cases} M_\sigma \cdot (M_\sigma + 2) = 0.960, \; for \; \sigma_m \geq 0 \\ \frac{M_\sigma}{3} \cdot \left(\frac{M_\sigma}{3} + 2\right) = 0.284, \; for \; \sigma_m < 0 \end{cases} \quad (2)$$

*2.2. Determination of $P_{RAM}$-N curve from strain-controlled and stress-controlled fatigue tests*

The present series of experiments include ferritic, pearlitic, solid solution strengthened [20] and ausferritic [21] nodular cast iron grades. The materials investigated were thick-walled castings with wall thicknesses of over 75 mm and differing microstructures.

The strain-controlled fatigue tests were performed under constant amplitude loading with a strain ratio of $R_\varepsilon = -1$ on servo-hydraulic test rigs with an applied extensometer, which measures the strain as a controlled variable. The strain-controlled fatigue tests were performed until crack initiation (a force reduction of 10 % related to the stabilized hysteresis) or until the limit number of cycles $N_{lim} = 1 \cdot 10^7$. Specimens were tested in the frequency range of f = 0.1 and 25.0 Hz.

The stress-controlled fatigue tests were performed under constant amplitude loading with a stress ratio of $R_\sigma = -1$ on electric-resonance test rigs until crack initiation (a reduction in the frequency of 4 Hz) or until the limit number of cycles $N_{lim} = 1 \cdot 10^7$. These specimens were tested with a higher frequency and were mainly in the range of f = 100 to 200 Hz.

A more detailed overview of the specimen geometries used in the research projects [7, 22, 23, 24, 25, 26, 27], the geometry of the casted Y-blocks and casted components, sampling, the chemical composition, images of the microstructure and the test procedure can be found in the open source [28].

The experimental determination of the damage parameter $P_{RAM}$ is performed for strain-controlled tests with a strain ratio of $R_\varepsilon = 1$ by evaluating the nominal stress amplitude $\sigma_{a,n}$ and the total strain amplitudes $\varepsilon_{a,tl}$ from the hysteresis at half the number of cycles to crack initiation and is calculated according to equation (3).

For the force-controlled fatigue tests, with a stress ratio of $R_\sigma = -1$, the damage parameter $P_{RAM}$ is calculated using the nominal stress amplitude $\sigma_{a,n}$. The strain amplitude $\varepsilon_{a,tl}$ is also calculated from the stress amplitude using Hook's law and the average Young's modulus E under the assumption of strictly linear-elastic loading.

$$P_{RAM} = \left(\sigma_{a,n} \cdot \varepsilon_{a,tl} \cdot E\right)^{\frac{1}{2}} \quad (3)$$

In the FKM guideline "Non-linear" [1], a bilinear approach with a knee point at $N = 10^3$ cycles is used to describe the $P_{RAM}$-N curves. The main benefit of this approach is that the entire curve can be described using 3 parameters, a stress value for the knee point $P_{RAM,Z,WS}$ cycles as well as the slope of the LCF range $d_1$ and the MCF range $d_2$. For the analysis of the experimental fatigue data, an extended approach, adding an additional knee point to separate the LCF range, is used. This implies that two more parameters are required, the stress value of the second knee point $P_{RAM,D,WS}$ and the number of cycles of the second knee point $N_{k,2}$.

To determine the damage parameter $P_{RAM}$-N curve according to equation (4), a combined evaluation of the strain-controlled and the force-controlled tests is needed. This evaluation is carried out using maximum likelihood, including iterative definition of the second knee point $N_{k,2}$ for the maximum probability of occurrence of the fatigue tests with consideration of runouts. The evaluation results in the two knee points $P_{RAM,Z,WS}$ and $P_{RAM,D,WS}$ as well as the slope of the low cycle fatigue (LCF) range $d_1$ and the slope of the medium cycle fatigue (MCF) range $d_2$ , **Figure 1**. The first knee point $N_{k,1}$ is defined as $N_{k,1} = 10^3$ according to [1].

The approach for evaluating the damage parameter $P_{RAM}$-N curve is based on the approach for evaluating stress-life curves from force-controlled test series according to [29]. The slope of the high cycle fatigue (HCF) range $d_3$ is set to $1/d_3 = 44.89$ according to [30] for nodular cast iron. This is equal to a reduction of the damage parameter of 5 % per decade. In addition, the scatter band $T_{PRAM}$ is determined from the standard deviation of the fatigue tests in relation to the 50 % $P_{RAM}$-N curve. **Figure 1** shows an exemplary $P_{RAM}$-N curve.

$$N = \begin{cases} N_{k,1} \cdot \left(\frac{P_{RAM}}{P_{RAM,Z}}\right)^{\frac{1}{d1}}, \; for \; P_{RAM} \geq P_{RAM,Z} \\ N_{k,1} \cdot \left(\frac{P_{RAM}}{P_{RAM,Z}}\right)^{\frac{1}{d2}}, \; for \; P_{RAM,Z} > P_{RAM} \geq P_{RAM,D} \\ N_{k,2} \cdot \left(\frac{P_{RAM}}{P_{RAM,D}}\right)^{\frac{1}{d3}}, \; for \; P_{RAM} < P_{RAM,D} \end{cases} \tag{4}$$

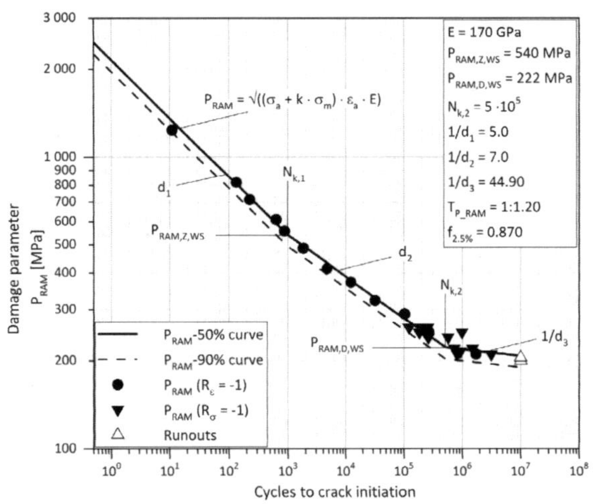


**Figure 1.** Combined evaluation of strain-controlled and force-controlled test series to determine the trilinear $P_{RAM}$-N curve. The $P_{RAM}$-N curve is shown for a probability of survival $P_S = 50$ % and $P_S = 90$ %, considering the scatter band $T_{PRAM}$.

### *2.3. Assessment of the fatigue life using the damage parameters*

To obtain the design $P_{RAM}$-N curves required for the numerical fatigue assessment, the material-specific $P_{RAM}$-N curves must be shifted by considering the statistical size effect [10] and converted from a 50% to a 97.5 % probability of survival using a safety factor. For the fatigue life assessment according to [1] the factor for conversion to 2.5 % probability of failure $f_{2.5\%}$ is used. This factor can be calculated from the scatter band of the damage parameter curve $T_{PRAM}$ according to equation (5). Furthermore, the elementary approach with a continuous slope after the second knee point $d_3 = d_2$ is applied to calculate the number of cycles in the HCF range according to [1], **Figure 2**.

$$f_{2.5\%} = 10^{\frac{1,960}{2 \cdot 1,282} \cdot log\,(T_{PRAM})} \tag{5}$$

If the highly stressed surface of the critical component area is not the same as the highly stressed reference surface $A_{ref}$ of the FKM guideline [1], the statistical size effect has to be considered. To compensate for the statistical size effect the factor $n_{st}$ is used to adjust the experimental damage parameter $P_{RAM}$-N curves according to equation (6). For this approach, the Weibull coefficient $k_{st}$ is set to $k_{st} = 10$ according to [2].

$$n_{st} = \left(\frac{A_{ref}}{A_\sigma}\right)^{\frac{1}{k_{st}}} \qquad A_{ref} = 500\,mm^2; k_{st} = 10 \tag{6}$$

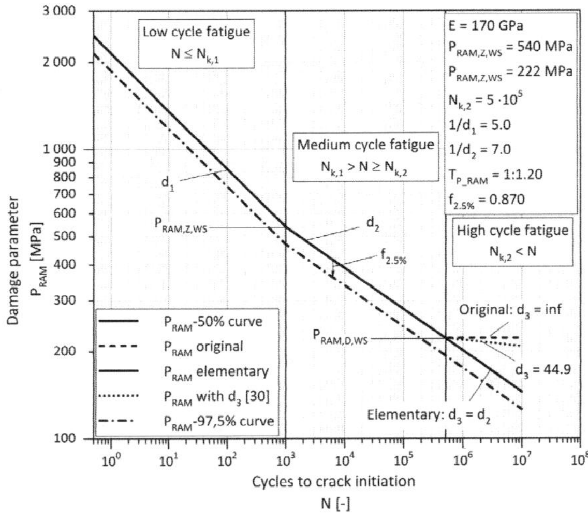


**Figure 2.** $P_{RAM}$-N curve diagram using the elementary approach in the HCF range. The $P_{RAM}$-N curve is shown for a survival probability of $P_S$ = 50 % and $P_S$ = 97.5 %, considering the factor for conversion to 2.5% probability of failure $f_{2.5\%}$.

### *2.4. Determination of the parameters of the cyclic stress-strain curve*

Above, the relationship between damage parameter and fatigue life was described using design $P_{RAM}$-N curves. Next, the relationship between an external load leading to plastic deformations in the highly stressed area on a component and the damage parameter is required. The value of the damage parameter has to be determined according to equation (1) considering the plastic strain amplitude $\varepsilon_{a,pl}$. The plastic strain amplitude $\varepsilon_{a,pl}$ has to be calculated either from FE simulation using a linear elastic material model and the notch root approximation by Neuber or by applying an elastic-plastic material model in the FE simulation. In both cases the relationship between an external load and the damage parameter is described by the cyclic stress-strain curve.

As recommended in [1], the Ramberg-Osgood approach [31] is applied for the definition of the cyclic stress-strain curve, equation (7). As described in [28], the cyclic stress-strain curve is defined with scatter bands to include variations in the hardening behavior and stress-strain curve of the investigated materials.

The Young's modulus E is calculated for each series of experiments as an average of all strain-controlled fatigue tests in the considered test series. The Young's modulus of each fatigue test is evaluated from the initial loading curve by linear regression in the range from 10 % to 40 % of the yield strength. The Young's modulus for the present test series is within a range of E = 155 - 172 GPa.

$$\varepsilon = \frac{\sigma}{E} + \left(\frac{\sigma}{K'}\right)^{\frac{1}{n'}} \tag{7}$$

The evaluation of the Ramberg-Osgood parameters K' and n' is performed using linear regression in the double-logarithmic diagram of the stress amplitude $\sigma_{a,n}$ versus the plastic strain amplitude $\varepsilon_{a,pl}$ according to [1] and only the strain-controlled tests are considered. The stress amplitude $\sigma_{a,n}$ and the plastic strain amplitude $\varepsilon_{a,pl}$ are derived from the hysteresis at half the cycles of crack initiation. The plastic strain amplitude $\varepsilon_{a,pl}$ results from the difference of the total strain amplitude $\varepsilon_{a,tl}$ and the elastic strain amplitude $\varepsilon_{a,el}$. The elastic strain amplitude $\varepsilon_{a,el}$ itself is calculated from the stress amplitude $\sigma_{a,n}$ using Hook's law and the average Young's modulus. Furthermore, the scatter band of the cyclic stress-strain curve $T_\sigma$ is also determined using the standard deviations of the linear regression. **Figure 3** shows an exemplary cyclic stress-strain curve.

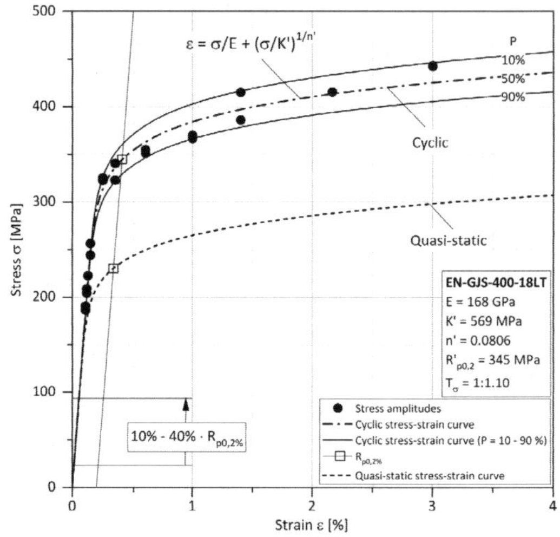


**Figure 3.** Determination of the cyclic stress-strain curve.

## 3. Extension of the FKM Non-Linear approach based on experimental data for thick-walled GJS

### *3.1. Present experimental PRAM-N curves and experimental cyclic stress-strain curves*

**Figure 4** shows exemplary the diagram of all test series of the EN-GJS-400-18LT. It shows that the damage parameter $P_{RAM}$ at $1 \cdot 10^7$ cycles and a probability of survival $P_S = 50$ % are within the range of 133 MPa to 238 MPa. This equals to a difference of 63 % within this nodular cast iron grade, which is widely used in many industrial sectors. This is presumably due to the different wall thicknesses and solidification times, which results in varying microstructures and therefore material properties. The damage parameter $P_{RAM}$-N curve from [24] has a significantly lower slope in the MCF range with $1/d_2 = 16.7$ while all other $P_{RAM}$-N curves have a slope in the range of $1/d2 = 5.4 – 8.4$. This is most likely because the cast component from which the specimens for the test series of [24] were taken was cast using the permanent mold casting process, whereas the other test series were cast using the sand molded casting process which results in a different pronounced local microstructure.

**Figure 5** shows the lognormal probability plot of the scatter bands of the damage parameter $P_{RAM}$-N curves and the cyclic stress-strain curves of the 44 analyzed test series. It appears that the two parameters mostly follow the lognormal distribution. From this graph, it is possible to determine that for a 90 % probability of occurrence, P = 90 %, the factor for conversion to 2.5 % probability of failure, $f_{2.5\%}$, can be set to $f_{2.5\%} = 0.803$. A factor for conversion to 2.5 % probability of failure, $f_{2.5\%} = 0.803$, corresponds to a scatter band $T_{PRAM}$ for a probability of survival $P_S = 90\%$ of $T_{PRAM} = 1:1.33$, following equation (5).

Applying the same approach on cyclic stress-strain curves for a 90 % probability of occurrence in the lognormal probability plot, the scattering band for the cyclic stress-strain curve can be determined to $T_\sigma = 1:1.12$. These characteristics provide a reference for the use on experimentally determined damage parameter $P_{RAM}$-N curves and cyclic stress-strain curves if experimental data is used for a numerical fatigue assessment.

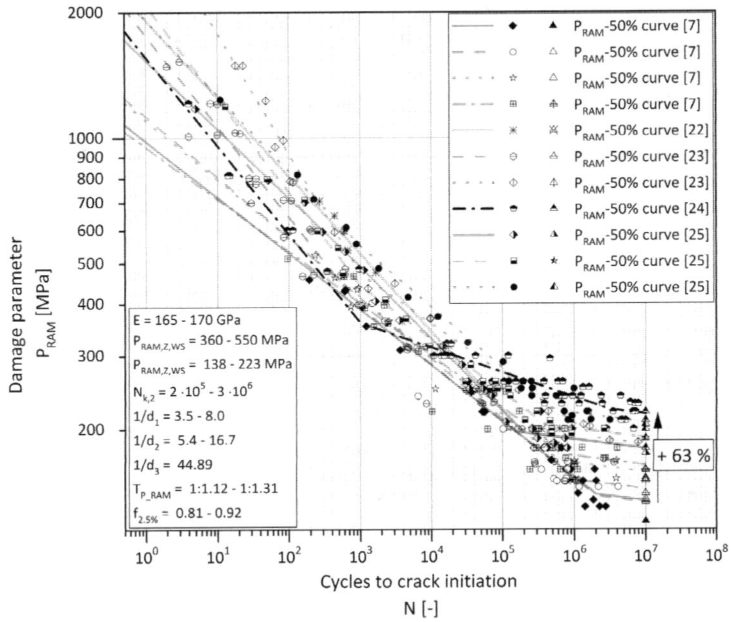


**Figure 4.** Diagram of the $P_{RAM}$-N curves of the EN-GJS-400-18LT test series.

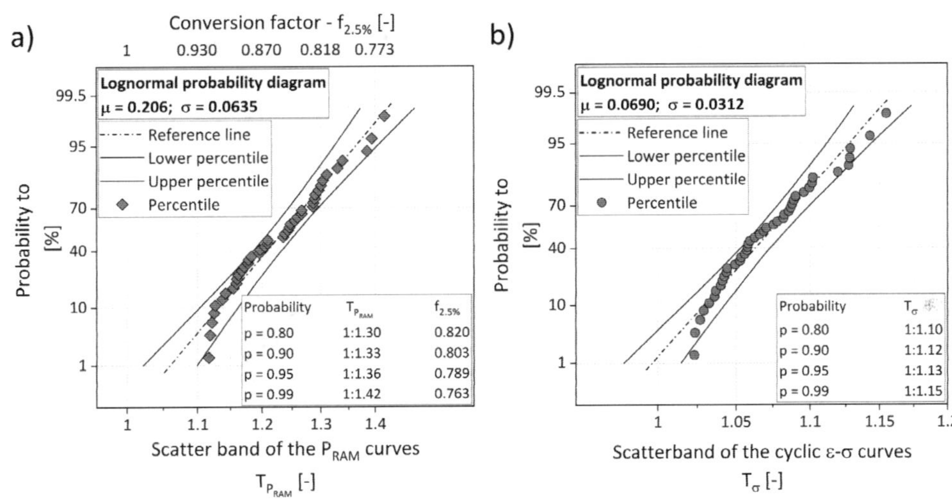


**Figure 5.** Lognormal probability plot of the scatter bands of the damage parameter $P_{RAM}$-N curves and the cyclic stress-strain curves.

### *3.2. Derivation of the fatigue assessment concept depending on the tensile strength*

The FKM guideline "Non-linear" [1] provides a fatigue assessment concept for the materials steel, cast steel and wrought aluminium alloys. The required parameters for the $P_{RAM}$-N curves and cyclic stress-strain curves of those materials are determined in the guidelines using only the tensile strength $R_m$. The exponential relationships used in the guideline are adopted and the required parameters are specified with updated values for nodular cast iron.

To enable the application of the above discussed approaches, it is necessary to show that all parameters required for the fatigue assessment of nodular cast iron components according to [1] can be determined with a statistical certainty using the tensile strength $R_m$. Furthermore, a safety concept is presented that follows closely the approach in [1] so that all available fatigue tests can be summarized.

 2745 (2024) 012021 doi:10.1088/1742-6596/2745/1/012021 

*3.2.1. Defining the parameters K' und n' for the determination of stress-strain curves.* To determine the parameters of the cyclic stress-strain curve in **Figure 6** and **Figure 7**, the cyclic hardening coefficient K' and the cyclic hardening exponent n', literature data [13, 14, 15, 32, 33, 34] is included in the evaluation alongside to the 44 present series of tests. This correlation has already been shown very similarly in [28] for nodular cast iron.

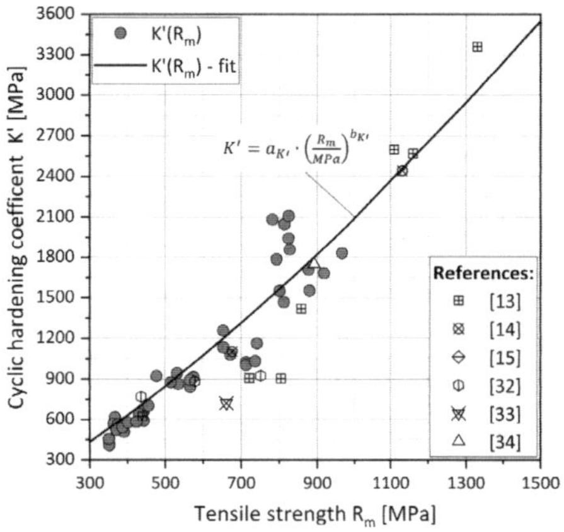


**Figure 6.** Cyclic hardening coefficients versus tensile strength.

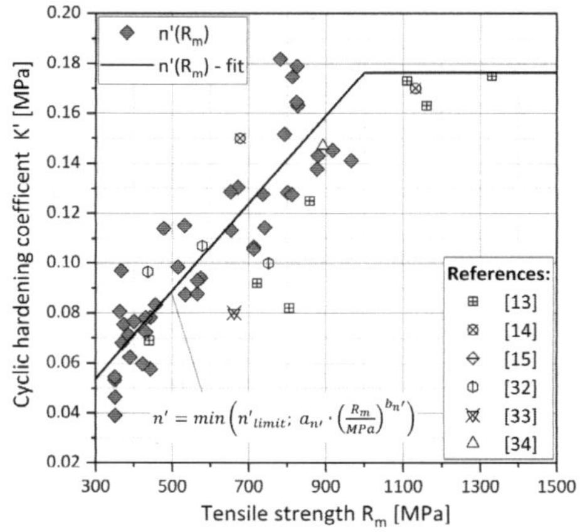


**Figure 7.** Cyclic hardening exponents versus tensile strength.

The correlation between the parameters of the cyclic stress-strain curve and the tensile strength $R_m$ can be derived from equation (8) and equation (9). Additionally, a threshold was set for the cyclic hardening exponents n' so that it is constant for a tensile strength of $R_m > 1000$ MPa. The resulting parameters of this correlation can be found in **Table 1**.

$$K' = a_{K'} \cdot \left(\frac{R_m}{MPa}\right)^{b_{K'}} \tag{8}$$

$$n' = min\left(n'_{limit};\ a_{n'} \cdot \left(\frac{R_m}{MPa}\right)^{b_{n'}}\right) \tag{9}$$

**Table 1.** Parameters for determining the cyclic hardening coefficients and cyclic hardening exponents from the tensile strength.

| $a_{K'}$ | $b_{K'}$ | $a_{n'}$ | $b_{n'}$ | $n'_{limit}$ |
|---|---|---|---|---|
| 0.256 | 1.30 | $1.93 \cdot 10^{-4}$ | 0.987 | 0.176 |

*3.2.2. Defining the parameters for the determination of $P_{RAM}$-N curves.* To determine the $P_{RAM}$-N curves based on the tensile strength $R_m$, the two knee points are the most important parameters for generating the design $P_{RAM}$-N curves. The correlation between the knee points of the damage parameter curve $P_{RAM,Z,WS}$ and $P_{RAM,Z,WS}$ and the tensile strength $R_m$ can be derived from equation (10) and equation (11). The resulting parameters of this correlation can be found in **Table 2**.

$$P_{RAM,Z,WS} = f_{2.5\%} \cdot a_{P,Z} \cdot \left(\frac{R_m}{MPa}\right)^{b_{P,Z}} \tag{10}$$

$$P_{RAM,D,WS} = f_{2.5\%} \cdot a_{P,D} \cdot \left(\frac{R_m}{MPa}\right)^{b_{P,D}} \tag{11}$$

Furthermore, in contrast to the other material groups in [1], no constant value is assumed for the slope of the LCF range $d_1$, but instead this parameter is described as a function of the tensile strength $R_m$ for this approach, equation (12). The slope of the MCF range $d_2$ results from the two knee points of the damage parameter curve $P_{RAM,Z,WS}$ and $P_{RAM,Z,WS}$ and is therefore also dependent on the tensile strength $R_m$, equation (13).

$$\frac{1}{d_1} = a_{1/d1} \cdot \left(\frac{R_m}{MPa}\right)^{b_{1/d1}} \tag{12}$$

$$\frac{1}{d_2} = \frac{log\left(\frac{N_{k,2}}{N_{k,1}}\right)}{log\left(\frac{P_{RAM,Z,WS}}{P_{RAM,D,WS}}\right)} \tag{13}$$

To shift all the experimental $P_{RAM}$-N curves with different specimen geometries used for the fatigue tests to a same level, the factor for statistical size effect $n_{st}$ has to be considered according to equation (6). In this case, the aim is to compensate for the statistical size effect [10] so that all experimental $P_{RAM}$-N curves are related to the same highly stressed reference surface $A_{ref}$ = 500 mm². Therefore, the reciprocal of the factor for statistical size effect $n_{st}$ is used and multiplied with the knee points $P_{RAM,Z,WS}$ and $P_{RAM,Z,WS}$.

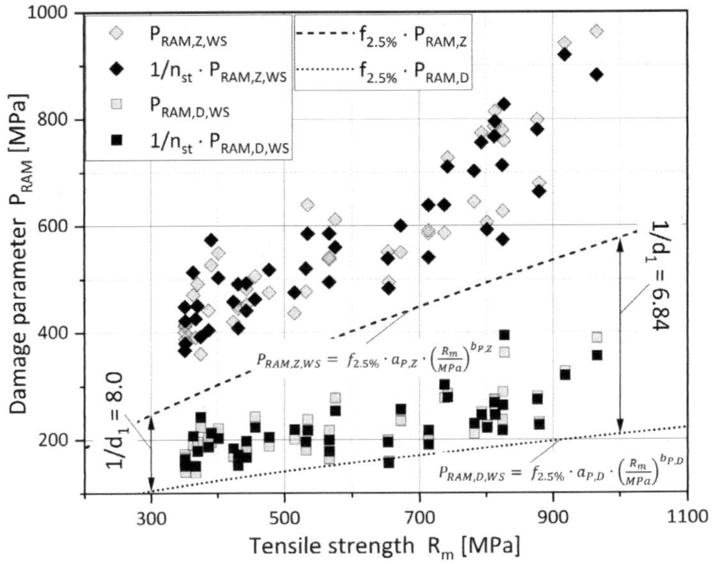


**Figure 8.** Knee points of the damage parameter $P_{RAM}$ versus the tensile strength $R_m$.

The relationship between the product of the experimentally determined knee points $P_{RAM,Z,WS}$ and $P_{RAM,D,WS}$, the inverse factor for the statistical size effect $1/n_{st}$ and the tensile strength $R_m$ is shown in **Figure 8**. Moreover, the dependency of the resulting slope of the MCF range $d_2$ according to equation (13) is depicted. Furthermore, the factor for conversion to 2.5% probability of failure $f_{2.5\%}$ is applied to the regression line. The factor for conversion to 2.5% probability of failure $f_{2.5\%}$ is determined below and must be selected so that a safe design based on the tensile strength is achieved.

*3.2.3. Defining the factor for the conversion of material-specific $P_{RAM}$-N curves to design $P_{RAM}$-N curves with a 2.5 % probability of failure.* To determine the factor for conversion to 2.5 % probability of failure $f_{2.5\%}$ for this fatigue assessment concept, the experimentally determined cycles $N_{experimental}$ of all fatigue tests are compared with the calculated cycles to crack initiation $N_{calculated}$ in a scatterplot, **Figure 9**. Moreover, the logarithmic ratio of the calculated to the experimental fatigue strength $\log(N_{experimental}/N_{calculated})$ is displayed in a normal probability plot with 95 % confidence bounds, **Figure 10**.

The logarithmic ratio of experimental to calculated fatigue strength $\log(N_{experimental}/N_{calculated})$ is equal to 0 if the experimental fatigue strength is equal to the calculated fatigue strength. The factor for conversion to 2.5 % probability of failure $f_{2.5\%}$ is hereby chosen so that in the normal probability plot for $\log(N_{experimental}/N_{calculated}) = 0$ a 2.5 % probability of occurrence, P = 2.5 %, with the use of the lower percentile of the confidence bounds is achieved. This approach leads to the factor for conversion to 2.5 % probability of failure $f_{2.5\%}$ = 0.71 and secures a robust and conservative design using the parameters derived from the experimental fatigue data.

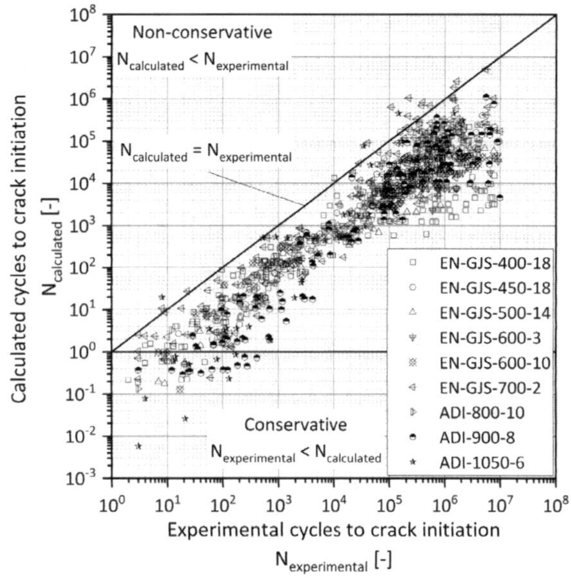


**Figure 9.** Scatterplot of the calculated cycles to crack initiation, $N_{experimental}$, versus the experimental cycles to crack initiation, $N_{calculated}$, of all fatigue tests.

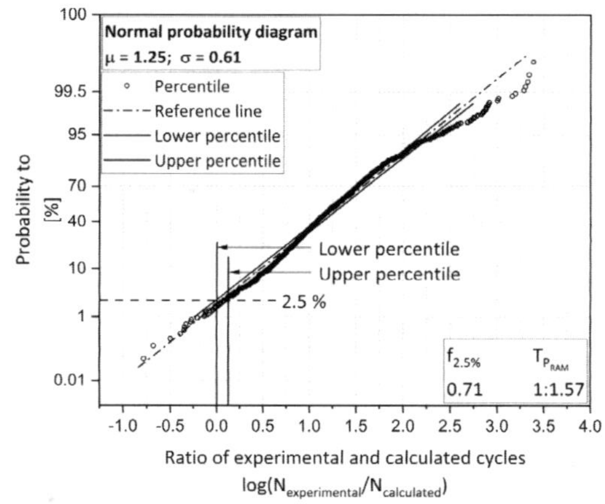


**Figure 10.** Logarithmic ratio of the calculated to the experimental cycles to crack initiation in a normal probability plot.

**Table 2.** Parameters for determining the knee points of the damage parameter curve and the slope of the LCF range from the tensile strength.

| $a_{P,Z}$ | $b_{P,Z}$ | $a_{P,D}$ | $b_{P,D}$ | $a_{1/d1}$ | $b_{1/d1}$ | $1/d_2$ | $f_{2,5\%}$ |
|---|---|---|---|---|---|---|---|
| 6.40 | 0.703 | 5.52 | 0.578 | 0.0239 | 0.901 | 6.8 – 8.0 | 0.71 |

## 4. Improvement of the fatigue strength assessment concept with consideration of the tensile strength, the elongation at break and the microstructure

Since a low factor for conversion to 2.5 % probability of failure $f_{2.5\%}$ must be chosen if the damage parameter $P_{RAM}$-N curve is only determined from the tensile strength, another fatigue life assessment concept is presented in the following based on two more variables. In this case the knee points of the damage parameter curve at the two knee points $P_{RAM,Z,WS}$ and $P_{RAM,Z,WS}$ are determined using not only the tensile strength $R_m$ but also the elongation at break A and the graphite particle density $\rho_{graphite}$ to account for the microstructure. The latter parameter is to be determined according to [35].

The relationship between the experimental parameters is determined with a multiple linear regression (MLR). The statistical significance of the parameters on the model is evaluated using the t-value and the statistical significance of the entire model is evaluated using the F-value. Furthermore, the adjusted coefficient of determination $R^2_{adj}$, which considers how many independent variables are added, is also taken as a measure of the performance of the MLR. The experimental knee points of the damage parameter $P_{RAM,Z,WS}$ and $P_{RAM,Z,WS}$ are shown in **Figure 11** and **Figure 12** in comparison to the calculated knee points of the damage parameter using the MLR.

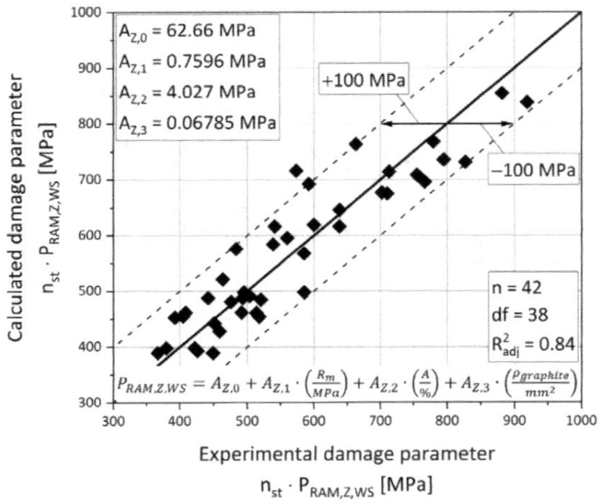


**Figure 11.** Calculated knee point at $N_{k,1}$ of the damage parameter curve $n_{st} \cdot P_{RAM,Z,WS}$ versus the experimentally determined knee point of the damage parameter curve.

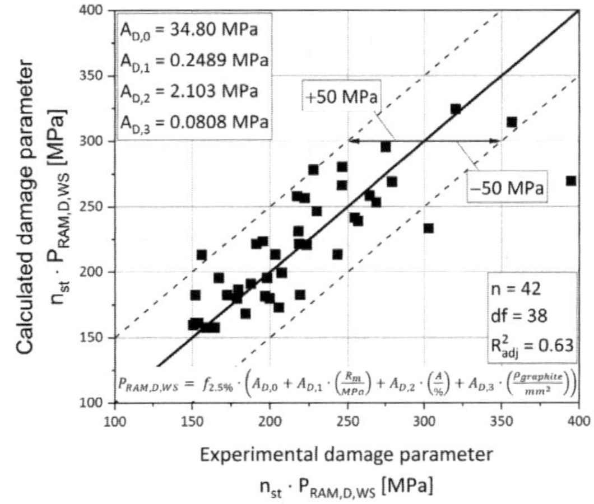


**Figure 12.** Calculated knee point at $N_{k,2}$ of the damage parameter curve $n_{st} \cdot P_{RAM,D,WS}$ versus the experimentally determined knee point of the damage parameter curve.

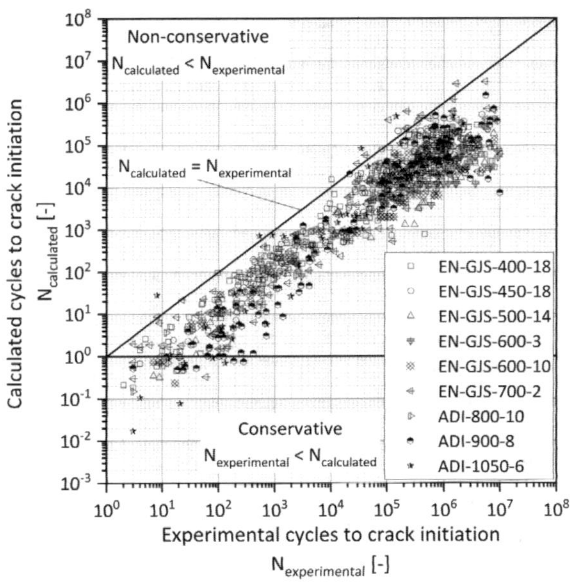


**Figure 13.** Scatterplot of the calculated $N_{experimental}$ versus the experimental cycles to crack initiation $N_{calculated}$ of all fatigue tests.

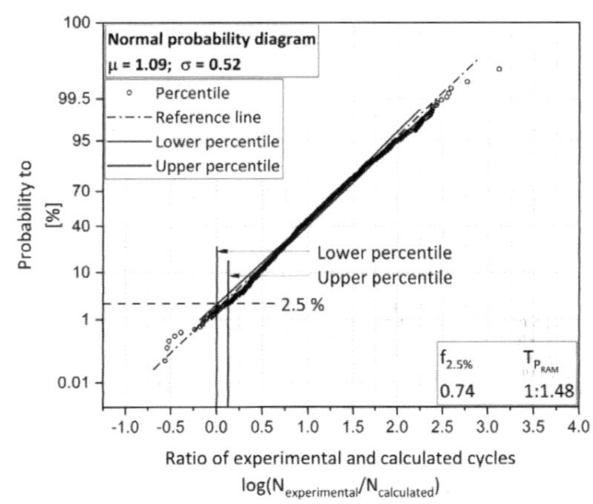


**Figure 14.** Logarithmic ratio of the calculated to the experimental cycles to crack initiation in a normal probability plot.

**Table 3.** Parameters for determining the knee points of the damage parameter curve and the slope of the LCF range from the quasi-static parameters and the microstructure

| $A_{Z,0}$ | $A_{Z,1}$ | $A_{Z,2}$ | $A_{Z,3}$ | $A_{D,0}$ | $A_{D,1}$ | $A_{D,2}$ | $A_{D,3}$ | $1/d_2$ | $f_{2,5\%}$ |
|---|---|---|---|---|---|---|---|---|---|
| 62.66 | 0.7596 | 4.027 | 0.06785 | 34.80 | 0.2489 | 2.103 | 0.0808 | 6.7 – 10.4 | 0.74 |

The resulting relationship between the knee points of the damage parameter curve and the tensile strength $R_m$, elongation at break A, graphite particle density $\rho_{graphite}$ can be found in equation (14) and equation (15). The resulting parameters of this correlation can be found in **Table 3**.

$$P_{RAM,Z,WS} = f_{2.5\%} \cdot \left(A_{Z,0} + A_{Z,1} \cdot \left(\frac{R_m}{MPa}\right) + A_{Z,2} \cdot \left(\frac{A}{\%}\right) + A_{Z,3} \cdot \left(\frac{\rho_{graphite}}{mm^2}\right)\right) \quad (14)$$

$$P_{RAM,D,WS} = f_{2.5\%} \cdot \left(A_{D,0} + A_{D,1} \cdot \left(\frac{R_m}{MPa}\right) + A_{D,2} \cdot \left(\frac{A}{\%}\right) + A_{D,3} \cdot \left(\frac{\rho_{graphite}}{mm^2}\right)\right) \quad (15)$$

The factor for conversion to 2.5% probability of failure $f_{2.5\%}$ is likewise chosen so that in the normal probability plot, **Figure 14** for $\log(N_{experimental}/N_{calculated}) = 0$ a 2.5 % probability of occurrence P = 2.5 % with the use of the lower percentile of the confidence bounds is achieved. This approach leads to the factor for conversion to 2.5% probability of failure $f_{2.5\%} = 0.74$.

## 5. Discussion and Conclusion

The work presented gives the designer of thick-walled GJS components an overview of the strain-life material behaviour and provides methods and results for the assessment of components under cyclic loading. In the present paper, the material properties for the damage parameter $P_{RAM}$-N curves and the elastic-plastic material behaviour of 9 different GJS grades, EN-GJS-400-18LT, EN-GJS-450-18, EN-GJS-500-14, EN-GJS-600-3, EN-GJS-700-2 and ADI were summarised. The materials investigated differ in their chemical composition, the specimen geometry, the relevant wall thickness as well as the casting process. From these process parameters, 44 different microstructures with different quasi-static properties and graphite morphologies were obtained. The tensile strengths of the materials investigated were in the range of 351 MPa to 966 MPa with elongations at break in the range of 1 % to 27 %. The damage parameter $P_{RAM}$-N curves in the LCF range at $N = 1 \cdot 10^7$ cycles were between 133 MPa to 362 MPa. The microstructures were ferritic, pearlitic, solid solution strengthened and ausferritic and showed a nodularity in the range of 31 % to 94 % and graphite particle density in the range of 30 $mm^2$ to 663 $mm^2$.

Furthermore, suggestions for scatter bands in the strain-based design approach are given to the designers of thick-walled GJS components, if experimental fatigue data is used for the design process. For a 90 % probability of occurrence, the factor for conversion to 2.5 % probability of failure $f_{2.5\%}$ can be set to $f_{2.5\%} = 0.803$ and the scatter band of the cyclic stress-strain curve to $T_\sigma = 1.12$. However, it should be noted that the factor for conversion to 2.5 % probability of failure $f_{2.5\%}$ of the experimental damage parameter curves was at least $f_{2.5\%} = 0.74$ and the maximum scatter band of the cyclic stress-strain curve was $T_\sigma = 1.16$.

Based on the approach in [1], which does not provide a fatigue assessment concept for nodular cast iron, a corresponding solution was determined from the present fatigue data. This fatigue assessment concept determines all required parameters of the cyclic stress-strain curve according to Ramberg Osgood and the $P_{RAM}$-N curve from the tensile strength $R_m$. The parameters of the cyclic stress-strain curves according to Ramberg-Osgood K' and n' enables the designer of a cast component to calculate the local plastic strain $\varepsilon_{a,pl}$ using an FE simulation with an elastic-plastic material model or to use the notch root approximation according to Neuber, as explicitly recommended in [1]. This fatigue assessment concept, which only requires the tensile strength $R_m$ as an input variable, can be used very conveniently for the design process. Furthermore, the reduced tensile strengths from standardization [20, 21] for thick-walled nodular cast irons can be used as input variable for the fatigue assessment concept since most of the material data used meet these standards.

The design concept including tensile strength $R_m$, elongation at break A and graphite particle density $\rho_{graphite}$ allows a more precise assessment compared to the simpler approach. This is accomplished with a statistically equally secured assessment. Considering the microstructure in the use of strain-based concepts has not yet been applied in industry, as no practical approaches for nodular cast iron have been found in research in this area either.

A direct comparison of the two design concepts presented in the scatterplot of the experimental versus the calculated fatigue strength, **Figure 15**, shows that the calculated cycles of the fatigue assessment concept depending on the tensile strength and the microstructure are closer to the line for $N_{experimental} = N_{calculated}$. From this it can be summarized that a more precise determination of the damage parameter curve can be achieved by considering multiple parameters.

A less conservative design concept is obtained due to the better assessment of the damage parameter curve compared to the approach that only considers the tensile strength. By reducing conservativity in the design process, the relevant wall thickness and thus the total component weight could be reduced.

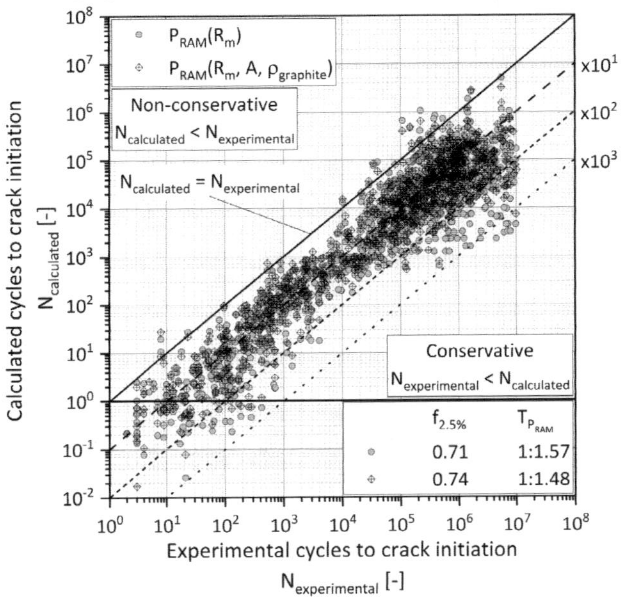


**Figure 15.** Comparison of the scatterplots of the fatigue assessment concept depending on the tensile strength with the fatigue assessment concept depending on the tensile strength and the microstructure.

The advantage of the proposed approach offers potential in terms of lightweight construction and resource efficiency. The reduction of conservativity in the design process combined with a statistically validated safety concept allows the component weight to be reduced, while simultaneously considering extreme loads and special events with elastic-plastic loadings based on a calculated damage parameter curve. Considering the immense challenges in the production of components with unit weights of several tonnes and the associated costs for transport and assembly, cost and energy savings can be achieved along the entire life cycle of wind turbines through better design methods.

## 6. Funding and Acknowledgement

The results presented in this paper were derived during the research project „GaßnerWind" (Grant number 0325707) [23], "Lunkerfest" (Grant number 0325239) [22], "Gusswelle" (Grant number 0324329) [24], "Gusszahnrad" (Grant number 03ET1555) [27], "SiGuphit-B³" (Grant number 03EE2007) [26], "DNAguss³" (Grant number 033EE3018) [25]. For the funding of this project, sincere thanks are given to the German Federal Ministry of Economic Affairs and Climate Action (BMWK). All project partners are thanked for their participation and support to complete this project successfully.

**Table-A1.** Summary of the results of the quasi-static tensile tests, metallography and highly stressed specimen surface.

| Material[a] | Reference | $R_m$ [MPa] | $R_{p0.2\%}$ [MPa] | A [%] | $\rho_{graphite}$ [$mm^{-2}$] | $n_{graphite}$ [%] | $A_\sigma$ [$mm^2$] |
|---|---|---|---|---|---|---|---|
| 400-18LT | [7] | 352 | 233 | 15.9 | 65 | 30.5 | 1178 |
| | [7] | 351 | 235 | 13.6 | 86 | 42.1 | 1178 |
| | [7] | 352 | 233 | 15.9 | 65 | 30.5 | 207.3 |
| | [7] | 351 | 235 | 13.6 | 86 | 42.1 | 207.3 |
| | [22] | 390 | 253 | 22 | - | - | 1178 |
| | [23] | 367 | 261 | 10.6 | 136 | 85.3 | 1178 |
| | [23] | 363 | 225 | 26.6 | 226 | 74.4 | 1178 |
| | [23] | 374 | 239 | 15.1 | 663 | 85.6 | 1178 |
| | [25] | 370 | 245 | 21.6 | 178 | 60.6 | 207.3 |
| | [25] | 386 | 248 | 21.6 | 182 | 63.5 | 207.3 |
| | [25] | 401 | 256 | 25.5 | 311 | 81.3 | 207.3 |
| 450-18 | [7] | 431 | 318 | 16.8 | 63 | 52.1 | 1178 |
| | [7] | 443 | 325 | 20.4 | 91 | 58.4 | 1178 |
| | [7] | 431 | 318 | 16.8 | 63 | 52.1 | 207.3 |
| | [7] | 443 | 325 | 20.4 | 91 | 58.4 | 207.3 |
| | [26] | 424 | 333 | 8.9 | 117 | 78.4 | 1178 |
| | [26] | 456 | 344 | 22.7 | 304 | 84.1 | 207.3 |
| 500-14 | [26] | 477 | 412 | 5.6 | 93 | 57.5 | 1178 |
| | [26] | 535 | 434 | 20.1 | 262 | 71.9 | 207.3 |
| 600-3 | [24] | 672 | 408 | 6.6 | 287 | 88.0 | 1178 |
| 600-10 | [26] | 515 | 449 | 4.7 | 118 | 36.6 | 1178 |
| | [26] | 575 | 478 | 18.9 | 296 | 69.8 | 207.3 |
| 700-2 | [7] | 566 | 492 | 0.9 | 30 | 59.6 | 1178 |
| | [7] | 713 | 537 | 1.9 | 64 | 70.0 | 1178 |
| | [7] | 566 | 492 | 0.9 | 30 | 59.6 | 207.3 |
| | [7] | 713 | 537 | 1.9 | 64 | 70.0 | 207.3 |
| | [23] | 532 | 411 | 2.7 | 109 | 75.9 | 1178 |
| | [23] | 737 | 418 | 4.8 | 65 | 82.5 | 1178 |
| | [27] | 742 | 544 | 3.2 | 530 | 85.5 | 395.8 |
| | [27] | 653 | 533 | 1.6 | 281 | 75.1 | 395.8 |
| | [27] | 654 | 535 | 1.5 | 153 | 74.2 | 395.8 |
| | [27] | 877 | 766 | 1.6 | 483 | 86.7 | 395.8 |
| | [27] | 801 | 724 | 1.1 | 247 | 77.8 | 395.8 |
| 800-10 | [23] | 782 | 614 | 2.5 | 148 | 90.9 | 1178 |
| 900-8 | [23] | 827 | 596 | 7.3 | 166 | 92.2 | 1178 |

| | | | | | | | |
|---|---|---|---|---|---|---|---|
| | [27] | 918 | 632 | 11.8 | 453 | 86.8 | 395.8 |
| | [27] | 880 | 643 | 5.4 | 162 | 75.5 | 395.8 |
| | [27] | 812 | 680 | 1.2 | 172 | 75.6 | 395.8 |
| | [27] | 793 | 590 | 6.5 | 253 | 74.0 | 395.8 |
| | [27] | 813 | 592 | 8.4 | 316 | 82.3 | 395.8 |
| | [27] | - | - | - | 214 | 65.5 | 395.8 |
| | [27] | 966 | 649 | 11.2 | 195 | 93.6 | 207.3 |
| 1050-6 | [27] | 824 | 621 | 4.5 | 104 | 91.3 | 207.3 |
| | [27] | 825 | 618 | 4.9 | 101 | 86.3 | 207.3 |

[a] Abbreviation of the material without the terminator "EN-GJS-" according to [20] and [21]

**Table-A2.** Summary of the results of the evaluation of the Young's modulus and the Ramberg-Osgood parameters.

| Material[a] | Reference | E [GPa] | K' [MPa] | n' [-] | $T_\sigma$ [-] |
|---|---|---|---|---|---|
| 400-18LT | [7] | 168 | 421.3 | 0.0465 | 1.05 |
| | [7] | 169 | 453.9 | 0.0533 | 1.03 |
| | [7] | 169 | 408.0 | 0.0391 | 1.04 |
| | [7] | 165 | 455.1 | 0.0545 | 1.06 |
| | [22] | 166 | 507.1 | 0.0623 | 1.04 |
| | [23] | 168 | 618.6 | 0.0970 | 1.16 |
| | [23] | 168 | 568.7 | 0.0806 | 1.10 |
| | [23] | 170 | 563.0 | 0.0754 | 1.09 |
| | [25] | 168 | 523.5 | 0.0680 | 1.06 |
| | [25] | 168 | 541.4 | 0.0713 | 1.06 |
| | [25] | 169 | 577.9 | 0.0767 | 1.08 |
| 450-18 | [7] | 166 | 618.9 | 0.0724 | 1.07 |
| | [7] | 169 | 663.5 | 0.0782 | 1.04 |
| | [7] | 163 | 641.8 | 0.0781 | 1.05 |
| | [7] | 163 | 590.9 | 0.0576 | 1.04 |
| | [26] | 167 | 584.1 | 0.0597 | 1.05 |
| | [26] | 169 | 704.6 | 0.0833 | 1.08 |
| 500-14 | [26] | 167 | 923.3 | 0.1140 | 1.13 |
| | [26] | 170 | 865.5 | 0.0874 | 1.06 |
| 600-3 | [24] | 171 | 1079.8 | 0.1305 | 1.10 |
| 600-10 | [26] | 169 | 876.3 | 0.0984 | 1.04 |
| | [26] | 170 | 916.4 | 0.0940 | 1.03 |
| 700-2 | [7] | 166 | 843.6 | 0.0877 | 1.02 |
| | [7] | 163 | 1022.8 | 0.1066 | 1.07 |

| | | | | | |
|---|---|---|---|---|---|
| | [7] | 166 | 893.8 | 0.0929 | 1.09 |
| | [7] | 163 | 1004.8 | 0.1056 | 1.04 |
| | [23] | 166 | 943.3 | 0.1152 | 1.06 |
| | [23] | 162 | 1030.9 | 0.1277 | 1.04 |
| | [27] | 172 | 1163.2 | 0.1144 | 1.06 |
| | [27] | 170 | 1257.5 | 0.1286 | 1.08 |
| | [27] | 171 | 1134.0 | 0.1133 | 1.03 |
| | [27] | 169 | 1706.3 | 0.1378 | 1.09 |
| | [27] | 169 | 1550.5 | 0.1284 | 1.14 |
| 800-10 | [23] | 163 | 2081.1 | 0.1818 | 1.10 |
| 900-8 | [23] | 155 | 1858.0 | 0.1631 | 1.07 |
| | [27] | 164 | 1681.6 | 0.1452 | 1.09 |
| | [27] | 166 | 1552.1 | 0.1430 | 1.12 |
| | [27] | 167 | 1466.7 | 0.1276 | 1.13 |
| | [27] | 161 | 1785.6 | 0.1517 | 1.08 |
| | [27] | 166 | 2046.7 | 0.1747 | 1.10 |
| | [27] | 163 | 1583.2 | 0.1374 | 1.02 |
| 1050-6 | [27] | 160 | 1833.1 | 0.1411 | (1.33)[b] |
| | [27] | 160 | 1942.0 | 0.1644 | 1.13 |
| | [27] | 160 | 2108.1 | 0.1790 | 1.09 |

[a] Abbreviation of the material without the terminator "EN-GJS-" according to [20] and [21]
[b] Excluded from the evaluation of the scattering bands, as some fatigue tests showed a very high hardening effect.

**Table-A3.** Summary of the results of the evaluation of the damage parameter $P_{RAM}$-N curve with scatter bands.

| Material[a] | Reference | $P_{RAM,Z,WS}$ [MPa] | $P_{RAM,D,WS}$ [MPa] | $1/d_1$ [-] | $1/d_2$ [-] | $N_{k,2}$ [-] | $T_{PRAM}$ [-] |
|---|---|---|---|---|---|---|---|
| 400-18LT | [7] | 388 | 139 | 7.4 | 7.4 | $2 \cdot 10^6$ | 1.15 |
| | [7] | 412 | 151 | 6.9 | 6.9 | $1 \cdot 10^6$ | 1.24 |
| | [7] | 415 | 168 | 7.1 | 7.1 | $6 \cdot 10^5$ | 1.16 |
| | [7] | 401 | 174 | 8.0 | 8.0 | $8 \cdot 10^5$ | 1.30 |
| | [22] | 527 | 196 | 5.3 | 5.3 | $2 \cdot 10^5$ | 1.16 |
| | [23] | 391 | 138 | 4.6 | 7.3 | $2 \cdot 10^6$ | 1.31 |
| | [23] | 472 | 190 | 3.5 | 8.4 | $2 \cdot 10^6$ | 1.18 |
| | [23] | 360 | 223 | 4.7 | 16.7 | $3 \cdot 10^6$ | 1.20 |
| | [25] | 492 | 196 | 6.1 | 5.8 | $2 \cdot 10^5$ | 1.14 |
| | [25] | 442 | 205 | 4.5 | 7.8 | $4 \cdot 10^5$ | 1.12 |
| | [25] | 550 | 222 | 5.5 | 6.9 | $5 \cdot 10^5$ | 1.12 |

| | | | | | | | |
|---|---|---|---|---|---|---|---|
| 450-18 | [7] | 451 | 158 | 4.9 | 7.2 | $2 \cdot 10^6$ | 1.17 |
| | [7] | 453 | 182 | 3.3 | 8.3 | $2 \cdot 10^6$ | 1.18 |
| | [7] | 446 | 166 | 4.5 | 7.7 | $2 \cdot 10^6$ | 1.27 |
| | [7] | 482 | 182 | 4.1 | 7.8 | $2 \cdot 10^6$ | 1.21 |
| | [26] | 421 | 169 | 5.6 | 7.3 | $8 \cdot 10^5$ | 1.25 |
| | [26] | 506 | 244 | 4.2 | 8.5 | $5 \cdot 10^5$ | 1.13 |
| 500-14 | [26] | 475 | 189 | 6.6 | 8.2 | $2 \cdot 10^6$ | 1.33 |
| | [26] | 639 | 238 | 6.1 | 8.4 | $4 \cdot 10^6$ | 1.20 |
| 600-3 | [24] | 551 | 236 | 5.2 | 10.4 | $7 \cdot 10^6$ | 1.29 |
| 600-10 | [26] | 437 | 201 | 6.4 | 9.8 | $2 \cdot 10^6$ | 1.14 |
| | [26] | 612 | 278 | 4.7 | 7.9 | $5 \cdot 10^5$ | 1.12 |
| 700-2 | [7] | 537 | 164 | 6.7 | 6.7 | $3 \cdot 10^6$ | 1.29 |
| | [7] | 586 | 200 | 6.0 | 6.0 | $6 \cdot 10^5$ | 1.39 |
| | [7] | 541 | 218 | 7.1 | 7.1 | $6 \cdot 10^5$ | 1.24 |
| | [7] | 591 | 209 | 6.0 | 6.0 | $5 \cdot 10^5$ | 1.30 |
| | [23] | 478 | 181 | 9.6 | 6.9 | $8 \cdot 10^5$ | 1.29 |
| | [23] | 586 | 278 | 5.6 | 7.6 | $3 \cdot 10^5$ | 1.13 |
| | [27] | 727 | 286 | 12.0 | 8.6 | $3 \cdot 10^6$ | 1.29 |
| | [27] | 552 | 200 | 9.4 | 8.2 | $4 \cdot 10^6$ | 1.16 |
| | [27] | 495 | 160 | 10.0 | 7.5 | $5 \cdot 10^6$ | 1.25 |
| | [27] | 798 | 282 | 13.7 | 8.2 | $5 \cdot 10^6$ | 1.30 |
| | [27] | 606 | 227 | 9.0 | 6.1 | $4 \cdot 10^5$ | 1.27 |
| 800-10 | [23] | 645 | 211 | 11.4 | 6.1 | $9 \cdot 10^5$ | 1.17 |
| 900-8 | [23] | 759 | 362 | 6.2 | 6.2 | $1 \cdot 10^5$ | 1.24 |
| | [27] | 941 | 328 | 18.9 | 6.6 | $1 \cdot 10^6$ | 1.26 |
| | [27] | 679 | 233 | 8.3 | 6.3 | $8 \cdot 10^5$ | 1.42 |
| | [27] | 785 | 275 | 12.5 | 6.2 | $7 \cdot 10^5$ | 1.34 |
| | [27] | 773 | 252 | 13.7 | 6.0 | $8 \cdot 10^5$ | 1.20 |
| | [27] | 814 | 252 | 12.0 | 6.5 | $2 \cdot 10^6$ | 1.21 |
| | [27] | 770 | 168 | 12.5 | 4.5 | $1 \cdot 10^6$ | 1.38 |
| 1050-6 | [27] | 963 | 389 | 9.3 | 6.6 | $4 \cdot 10^5$ | 1.18 |
| | [27] | 779 | 237 | 9.0 | 6.7 | $3 \cdot 10^6$ | 1.25 |
| | [27] | 627 | 289 | 7.7 | 5.9 | $1 \cdot 10^5$ | 1.49 |

[a] Abbreviation of the material without the terminator "EN-GJS-" according to [20] and [21]

**References**

[1] FKM-Guideline: Richtlinie nichtlinear, 1th. Ed. 2019, ISBN 978-3-8163-0729-7
[2] FKM-Richtlinie – Rechnerischer Festigkeitsnachweis für Maschinenbauteile aus Stahl, Eisenguss- und Aluminiumwerkstoffen, 7. Auflage 2020, Forschungskuratorium Maschinenbau (FKM), Frankfurt am Main, 2020
[3] DNV GL: "Machinery for wind turbines, DNVGL-ST-0361," DNV GL AS, September 2016.
[4] M. Shirani: Probabilistic and defect tolerant fatigue assessment of wind turbine casting, Dissertation, Norges teknisk-naturvitenskapelige universitet NTNU Trondheim, 2011
[5] Kaizinger, P.; Guster, Ch.; Severing, M.; Ballmes, H.: Bewertung des Einflusses von Defekten auf die Schwingfestigkeit von Gusseisen mit Kugelgraphit. 40. Tagung des DVM Arbeitskreises Betriebsfestigkeit, Herzogenaurach, 9. Und 10. Oktober, 2013
[6] Kaufmann, H.: Zur schwingfesten Bemessung dickwandiger Bauteile aus GGG-40 unter Berücksichtigung gießtechnisch bedingter Gefügeungänzen. Fraunhofer Institute LBF, Darmstadt, LBF report No. FB-214, 1998
[7] Ch. Bleicher: Ein Beitrag zur Beurteilung der Schwingfestigkeit von Gusseisen mit Kugelgraphit mit besonderer Berücksichtigung der Auswirkungen von Lunkern auf die Bauteillebensdauer. Dissertation, Technische Universität Darmstadt, Fraunhofer Institute LBF, Darmstadt, LBF report No. FB-246, ISBN 978-3-8396-1048-0, 2016, 299 pages.
[8] J. R. Bisping: Entwicklung einer Bemessungsmethode für Mahlanlagen in der Zementindustrie am Beispiel von Vertikalrollenmühlen. Dissertation, Technische Universität Darmstadt, Fraunhofer Institute LBF, Darmstadt, LBF report No. FB-242, 2014, 178 pages
[9] Niewiadomski J., Bleicher C., Kaufmann H.: Influence of the chill casting process on the cyclic material behavior of nodular cast iron, 2021, pp. 2404 – 2411
[10] Kloos, K.-H., Einfluss des Oberflächenzustandes und der Probengröße auf die Schwingfestigkeitseigenschaften, VDI-Berichte, 268, Darmstadt, VDI, 63-76, 1976
[11] P. Čanžar, Z. Tonković, N. Drvar, A. Bakić, J. Kodvanj, J. Sorić, Experimental investigation and modelling of fatigue behaviour of nodular cast iron for wind turbine applications, in EURODYN 2011, Proceedings of the 8th International Conference on Structural Dynamics, 2011, pp. 3252 – 3257.
[12] P. Čanžar, Z. Tonković, J. Kodvanj: Microstructure influence on fatigue behaviour of nodular cast iron, Materials Science and Engineering: A, Volume 556, 2012, Pages 88-99, ISSN 0921-5093, https://doi.org/10.1016/j.msea.2012.06.062.
[13] B. Atzori, G. Meneghetti, M. Ricotta – Unified material parameters based on full compatibility for low-cycle fatigue characterisation of as-cast and austempered ductile irons, International Journal of Fatigue, Volume 68, 2014, Pages 111-122, ISSN 0142-1123, https://doi.org/10.1016/j.ijfatigue.2014.05.012.
[14] Hübner, P., Pusch, G. and Krodel, L.: Fatigue behaviour of cast iron with globular graphite. Advanced Engineering Materials, 6(7), pp. 541-544, 2004
[15] Scacco, F., Campagnolo, A., Franceschi, M.; Meneghetti, G.: Strain-Controlled Fatigue Behavior of a Nodular Cast Iron in Real Off-Highway Axles: Effects of Casting Skin and Strain Ratio. Metals, 12(3), p. 426, 2022.
[16] Ch. Bleicher: Influence of Different Load Histories on the Cyclic Material Behavior of Nodular Cast Iron for Thick-walled Application. International Journal of Offshore and Polar Engineering. Transactions of The ISOPE, Volume 28, Number 3, September 2018, ISSN: 1053-5381, S. 294 –302.
[17] Ch. Bleicher, R. Wagener, H. Kaufmann, T. Melz: Fatigue behavior of large cast components under variable amplitude loading with overloads. WCX SAE World Congress Experience, 9.-11. April 2019.
[18] K. N. Smith, P. Watson, and T. H. Topper. A stress-strain function for the fatigue of metals. Journal of Materials, 5(4):767-778, Dec 1970.
[19] J.W. Bergmann. Zur Betriebsfestigkeitsbemessung gekerbter Bauteile auf der Grundlage der

örtlichen Beanspruchungen. 1983. Dissertation, Heft 37.
[20] Normenausschuss Gießereiwesen (GINA) im DIN Deutsches Institut für Normung e.V.: DIN EN 1563: Gießereiwesen – Gusseisen mit Kugelgraphit. Deutsche Fassung EN 1563:2019, ICS 77.080.10, Beuth Verlag, Berlin, April 2019
[21] Normenausschuss Gießereiwesen (GINA) im DIN Deutsches Institut für Normung e.V.: DIN EN 1564: Gießereiwesen – Ausferritisches Gusseisen mit Kugelgraphit. Deutsche Fassung EN 1564:2012, ICS 77.080.10, Beuth Verlag, Berlin, Januar 2012
[22] A. Heinrietz, C. Bleicher, J. Kurz, J. Rabe, Ermüdungsfestigkeitsnachweis auf Basis zerstörungsfreier Prüfungen an dickwandigen Eisengussbauteilen in der Windenergie, Fraunhofer Institute LBF, Darmstadt, Germany, Technical Report, Grant No. 0325239, Jun. 2014.
[23] C. Bleicher, J. Hesseler, R. Wagener, J. Baumgartner, H. Kaufmann, Vereinfachte Bemessung von Eisengussbauteilen für Windenergieanlagen unter Berücksichtigung von Betriebsbelastungen und Sonderereignissen, Fraunhofer Institute LBF, Darmstadt, Germany, Technical Report, Grant No. 0325707, Feb. 2019.
[24] J. Kirsch, H. Kyling, J. Niewiadomski, Werkstoff- und Bauteiloptimierung für leistungsfähigere Gusskomponenten im Antriebsstrang von Windenergieanlagen durch den Einsatz von Kokillenguss, Fraunhofer Institute LBF, Darmstadt, Germany, Technical Report, Grant No. 0324329A, Aug. 2022.
[25] F. Reissner, C. Prüfer, N. Adamchuck, M. Thomä, C. Wulf, G. Hartmann, M. Ziegler, Durchgängige numerische Auslegung von Gussbauteilen entlang der Prozesskette, Technical Report, Grant No. 033EE3018, Jan. 2024.
[26] C. Bleicher, J. Niewiadosmki, Bestimmung der Leistungsfähigkeit von hochsiliziumhaltigem Gusseisen mit Kugelgraphit für die sichere Bauteilbemessung nach Betriebsfestigkeits- und Bruchmechanikaspekten, Technical Report, Grant No. 03EE2007, Apr. 2024.
[27] C. Bleicher, K. Lipp, C. Pittel, M. Gietzen, A. Merten, Optimierte und energieschonende Herstellung von Großzahnrädern aus Gusseisen mit Kugelgraphit, Fraunhofer Institute LBF, Darmstadt, Germany, Technical Report, Grant No. 03ET1555, Jan. 2022.
[28] C. Bleicher, C. Pittel, A. Kansy, J. Niewiadomski, H. Kaufmann, On the strain-life behavior of thick-walled nodular cast iron, , Materials Testing, 2024.
[29] Störzel, K.; Baumgartner, J.: Statistical Evaluation of Fatigue Tests Using Maximum Likelihood, Materials Testing, De Gruyter, 63 (2021) No. 8
[30] Sonsino, C. M.: Course of SN-curves especially in the high-cycle fatigue regime with regard to component design and safety. International Jour-nal of Fatigue, Nr. 29, 2007, S. 2246 – 2258.
[31] Ramberg, W; Osgood, W. R.: Description of stress–strain curves by three parame-ters. NACA Techn. Rep. 902, NACA, 1943
[32] H. U. Leutwein, Einfluss von Autofrettage auf die Schwingfestigkeit innendruckbelasteter Bauteile aus Kugelgraphitguss, Ph. D. dissertation, Technische Universität Ilmenau, Ilmenau, Germany, 2008.
[33] B. Šamec, I. Potrč, M. Šraml, Low cycle fatigue of nodular cast iron used for railway brake discs, Engineering Failure Analysis, vol. 18, no. 6, pp. 1424 – 1434, 2011, DOI: https://doi.org/10.1016/j.engfailanal.2011.04.002.
[34] T. Engelke, Beitrag zur Ermüdungsfestigkeit von ADI-Gussbauteilen, Ph.D. dissertation, Technische Universität Clausthal, Clausthal-Zellerfeld, Germany, 2019.
[35] DIN EN ISO 945-1:2019-10: Mikrostruktur von Gusseisen - Teil 1: Graphitklassifi-zierung durch visuelle Auswertung (ISO 945-1:2019); Deutsche Fassung EN ISO 945-1:2019, Beuth Verlag, Berlin, 2019

# Designing Large Segmented Flexible Conical Plain Bearings as Wind Turbine Main Bearings

**T Jakobs, G Jacobs, A Rolink, J Euler, C Hollas, and J Röder**

Chair for Wind Power Drives, RWTH Aachen University, 52074 Aachen, Germany

Corresponding author: timm.jakobs@cwd.rwth-aachen.de

**Abstract**. Main bearings are failure-prone components in wind turbines, significantly increasing the LCOE of wind energy. Therefore, segmented plain bearings are being discussed as replacement for the currently used rolling bearings. Segmented plain bearings allow an up-tower replacement of faulty segments without the need to dismantle the WT drivetrain, as is required for rolling bearing replacements. One such bearing is the segmented, flexible, conical "FlexPad" plain bearing as alternative to existing double row tapered roller bearings. Designing the FlexPad bearing manually is very complex and thus time-consuming. Hence, during the research project a new holistic design method for large and highly loaded, conical plain bearings was developed. To prove the suitability of the method, two bearing prototypes were built and tested. The prototype, designed with the new method, outperformed the first, manually designed bearing, in terms of the design objectives. Most importantly, the friction losses were reduced by 32.1%. The project results show that the new method is able to efficiently design safely operating FlexPad bearing variants for wind turbines.

## 1. Introduction

Wind turbines (WT) are a key technology towards a carbon neutral energy production worldwide. The EU aims to be climate neutral till 2050 and Germany set itself a target of a fully carbon neutral electricity production till 2035 [1–3]. To increase the competitiveness of wind energy, the levelized cost of electricity (LCOE) needs to be lowered further. Currently about 30% of the LCOE is caused by service and maintenance activities [4]. The main rolling bearings are critical components in terms of maintenance costs due to their failure probability of 15-30% within their 20-year design lifetime [5]. Main bearing replacements are expensive procedures as an external crane is needed which costs about 200.000 € per day for offshore WTs [6]. Hence, a discussed alternative for rolling bearing as main bearings in the wind industry are segmented plain bearings. Segmented plain bearings allow an up-tower sub-component wise replacement of faulty segments without the need to dismantle the WT drivetrain, as it is needed for rolling bearing replacements.

The use of segmented plain bearings for WT is still in the prototype stage and subject of ongoing research. The large and highly loaded structures of WTs deform under load. Plain bearings are sensitive towards these deformations as their clearance is typically only about 1‰ of the bearing's diameter. Exceeding the plain bearings capacity to endure the deformation will lead to edge loading and mixed friction, resulting in excessive wear and thus premature bearing failure. The FlexPad bearing, shown in **Figure 1**, was designed with these challenges in mind and fulfils the requirement of an easy to maintain WT main bearing. Its flexible support structure for the segments ensures they stay parallel to the shaft even under deformation or displacement of such. The first prototype of the FlexPad bearing was built in the *WEA-GLiTS* project and the feasibility of the bearing concept was shown [7, 8].

However, designing the FlexPad bearing manually, as done in the *WEA-GLiTS* project, is very complex and time-consuming. The conical and inherent flexible design of the FlexPad plain bearing precludes the usage of established methods or guidelines, such as DIN [9]. Hence, a new holistic design method for large and highly loaded conical plain bearings needed to be developed. This objective was subject of research in the successor of the *WEA-GLiTS* project called *FlexPad* project. To achieve the research goal, the modelling strategy of the FlexPad bearing was revised first and an automated simulation workflow was implemented. Based on the new modelling approach a new design method was developed [10, 11]. The method starts with multiple simulations of different designs which are translated into a surrogate model. To derive the best bearing design an optimization algorithm utilizes the surrogate model. Finally, two FlexPad prototypes were built during the project. The first design (V1) is a scaled down version of the FlexPad bearing prototype developed manually in the previous project. The second bearing (V2) was designed and optimized with the design method. Both prototypes underwent excessive testing with various load cases typical for WT. The experimental results were used to validate the simulation models as well as the new design method.

This paper presents key results from the *FlexPad* project. Firstly, the developed design method and its application to the FlexPad bearing is explained. Afterwards, the experimental setup used to validate the simulation models as well as the design method itself is presented. Finally, a short summary of the experimental results is given.

## 2. Method

The FlexPad bearing is defined by ten key design parameters (see **Figure 1**). Five parameters describe the bearing global parameters: outer and inner diameters ($D_o$ and $D_i$), span (S), inclination ($\alpha$) and number of segments ($N_{segm}$). The segments are supported by a flexible arm, giving each segment the ability to adapt to the shaft's deflection. Each arm is defined by five parameters: arm thickness ($s_{arm}$), dimension and position of the groove ($t_{groove}$, $x_{groove}$, $b_{groove}$), and the normalized arm length ($l_{arm,norm}$). Each of the ten parameters affects the performance of the bearing individually. Furthermore, there are cross-influences between the parameters affecting performance. Thus, finding an optimal design manually is very difficult. Therefore, an optimization method is needed to design this bearing efficiently.

The modelling and simulation of the FlexPad bearing for one parameter set during the *WEA-GLiTS* project required about eight hours, including a lot of error-prone manual user interaction. Improvement to the simulation model and automatization of the entire process, from parameter definition, modelling and finally running the simulation reduced this time to about one hour, without the need of user interaction.

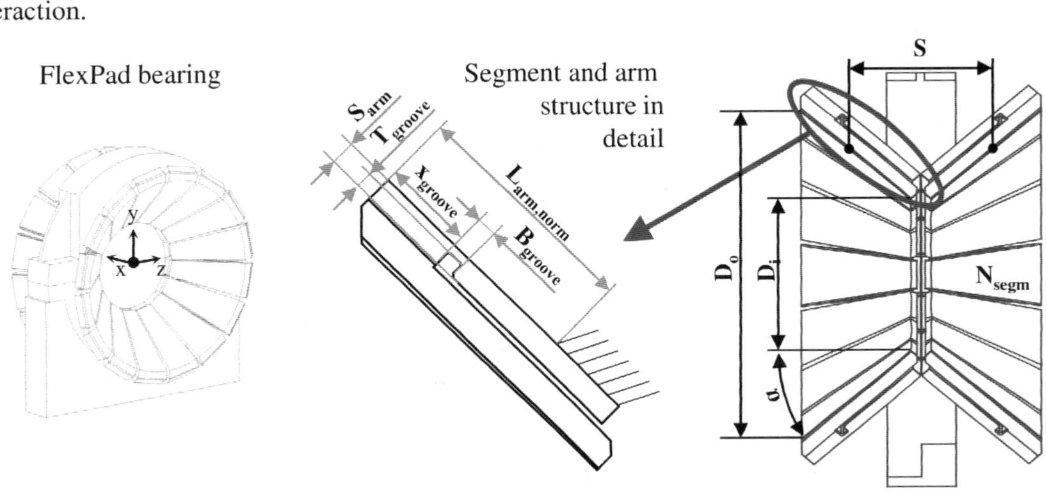


**Figure 1.** The FlexPad bearing and its parameters.

With this improved modelling approach a new design method was developed, first described by Rolink et al. [11]. Further finetuning of this method leads to the approach depicted in Figure 2 and consists of five steps:

1. Definition of design parameter space
   Each parameter can vary in a defined range. These ranges need to be set by preliminary studies (e.g. spot sampling) or can be defined by experienced engineers.
2. Sampling
   Within the defined parameter space, discrete and unique parameter sets are defined. Sampling can be carried out utilizing a sampling algorithm, e.g. Latin Hypercube Sampling.
3. Simulation
   Each of the sampled parameter set is simulated. For the FlexPad, this is done by using an in-house developed toolchain. The new modelling approach automatically links the necessary steps to perform the elasto-hydrodynamic-multi-body simulations.
4. Surrogate model
   The simulation results are post-processed and used to derive surrogate models for every objective stated in table of the next step. For the FlexPad bearing, a linear regression function with first and second order terms was used. This approach is further described by Rolink et al. [10].
5. Optimization
   In the last step, surrogate models are used to find an optimized bearing design which bests fulfills the design objectives. Regarding the FlexPad prototype, the following design objectives were optimized:

| **Objective** | **Target** | **Explanation** |
|---|---|---|
| Friction power | Minimize | Frictional losses of bearing need to be minimized for an efficient bearing |
| Pressure area ratio | Maximize | Area on the segment with a hydrodynamic pressure over 0.1 MPa. Maximized for good sliding area utilization |
| Maximum hydrodynamic pressure | Maximize to limit | Maximize to limit of sliding material with safety margin for a high-power density of bearing |

A bearing designed and optimized with the design method (V2) is depicted at the bottom of Figure 2. All objectives, except for hydrodynamic pressure, could be optimized in the direction of the target by using the new design method (cf. Table 1). The hydrodynamic pressure was reduced by 4%, despite the target being maximizing to a limit. This objective is of secondary importance and ensures that the sliding material strength is not exceeded. As the limit has not been exceeded, this objective can also be stated achieved.

The V1 design was developed without utilizing the new design method and is a scaled down version of the FlexPad bearing prototype developed in the *WEA-GLiTS* project [8, 12]. The outer diameter of V1 was reduced from 926 mm to 503 mm to decrease the prototype manufacturing costs. Both V1 and V2 were built and extensively tested at the Chair for Wind Power Drives (CWD). Within the *FlexPad* project the V1 bearing was used to validate the simulation models implemented in the aforementioned toolchain and as a baseline for optimizing the FlexPad's design. The second prototype (V2) was designed using the design method presented in this chapter. It was used to validate the design method itself and demonstrate that the computed optimization results can also be measured on the test bench.

**Table 1.** Comparison of the reference FlexPad design with the optimized design.

| **Objective** | **V1 (reference)** | **V2** |
|---|---|---|
| Friction power [W] | 550.0 | 373.3 (-32.1%) |
| Pressure area ratio RS [%] | 62.1 | 70.4 (+13.4%) |
| Pressure area ratio GS [%] | 61.7 | 62.9 (+1.9%) |
| $p_{max,hyd}$ [MPa] | 24.6 | 23.6 (-4%) |

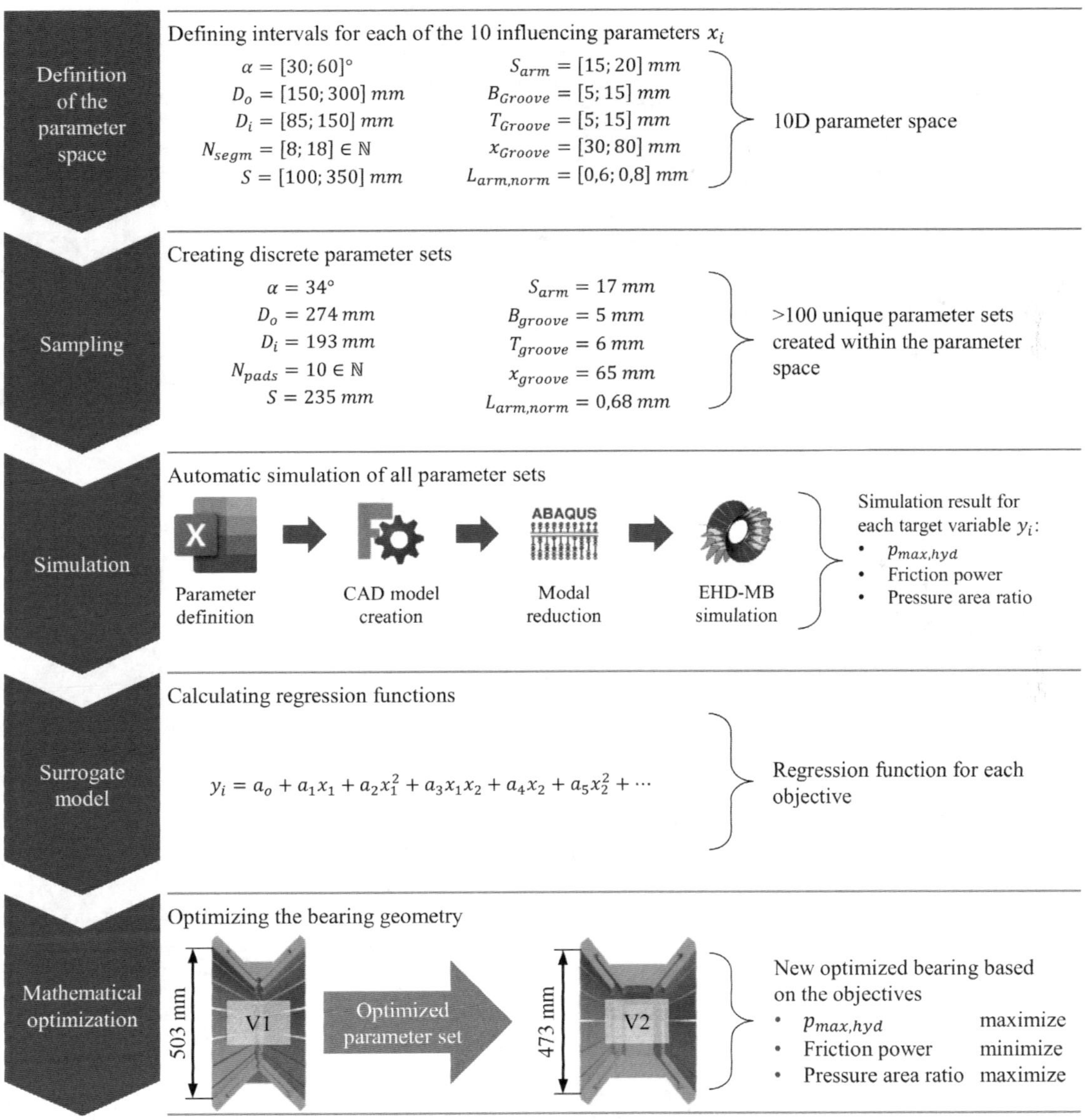


**Figure 2.** Steps of the design method

Journal of Physics: Conference Series **2745** (2024) 012022 doi:10.1088/1742-6596/2745/1/012022

## 3. Test bench setup

Both prototypes, V1 and V2, were tested on the same test bench at the CWD. The experiments were used to validate the improved modelling approach and the new design method for the FlexPad bearing. Also, the general qualification to design a safely operating plain bearing for WT with the design approach should be proven. Figure 3 (left) shows the setup of the V1 prototype on the test bench. The following section describes the measurement system setup used on the test bench and afterwards the applied load cases during testing.

**Figure 3.** FlexPad bearing (V1) on the test bench (left) | FlexPad in operation with the applied measurement system in close-up (centre) | Comparison of the two prototypes tested (right)

### *3.1. Temperature and lubrication gap height measurement*

To measure the temperature distribution, two PT100 temperature sensors were integrated into each pad, as shown in Figure 4. The sensors' measuring tip is located 0.5 mm below the pad sliding surface, allowing for precise measurement of the oil temperature in the lubrication gap. This enables precise determination of the steady-state temperature of the bearing, which is crucial in assessing the temperature safety of the plain bearing. Figure 4 right shows a detailed section of the temperature measurement system. The temperature sensors are identifiable by the white cables.

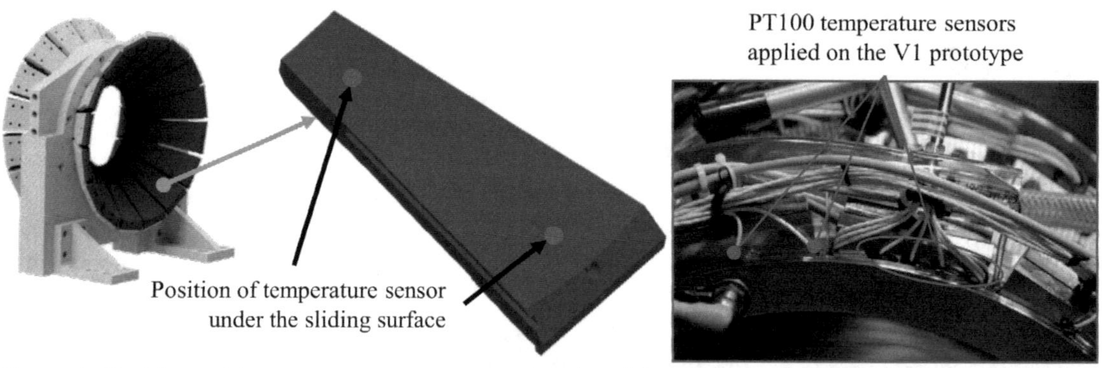


**Figure 4.** Position of the temperature sensors under the sliding surface (left) and their application holes in the cutaway view (right)

Each cone of the FlexPad bearing was equipped with two eddy current sensors. These sensors allow to measure the distance between the shaft and the sliding surface, and therefore the lubrication gap height. The measurement is unaffected by the oil in the lubrication gap. The sensors were positioned flush with the cone running surfaces, as shown in Figure 5.

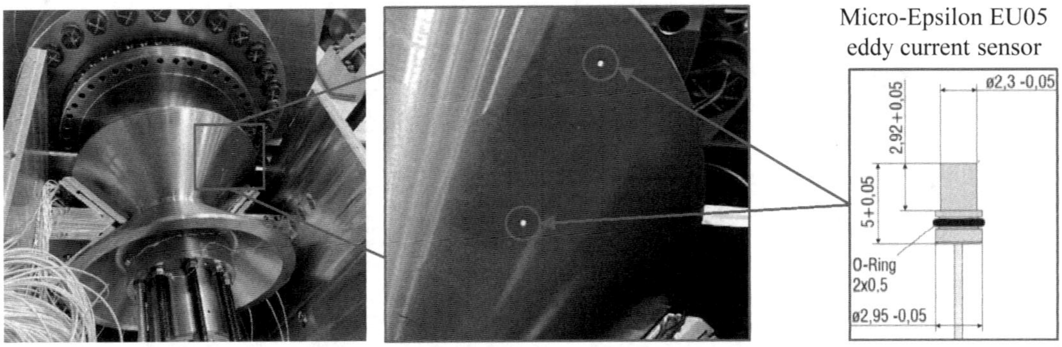


**Figure 5.** Lubrication gap measurement system applied to the main shaft cone on the test bench

### *3.2. Test load cases*

Both prototypes were tested on the test bench shown in Figure 3 left, which allowed load application in the following directions at variable shaft speeds:

- Thrust and vertical forces
- Bending moments around the horizontal and vertical axes

All load cases shown in the following are derived from multi-body simulations of an 850 kW turbine. Each conical FlexPad test specimen underwent a similar test program. The prototypes were tested under constant load cases, dynamic load cases, idling, and start-stop operation to demonstrate the bearings' overall feasibility as a WT main bearing. The bearing design was performed with the rated operation load case shown in Table 2. These loads were applied for 12 hours to determine the steady-state temperature of the test bearing.

**Table 2.** Exemplary production and idling load case with constant values

| Load Case | $F_x$ [N] | $F_z$ [N] | $M_y$ [Nm] | $M_z$ [Nm] | n [$min^{-1}$] | Load time [s] |
|---|---|---|---|---|---|---|
| Production | 29218 | -26004 | 22111 | 21322 | 28 | 1800 |
| Idling | 2684 | -23848 | -13981 | 659 | 0.6 | 3600 |

Figure 6 displays one of the dynamic load cases, with clear force and moment oscillations in the four-second detail section on the right.

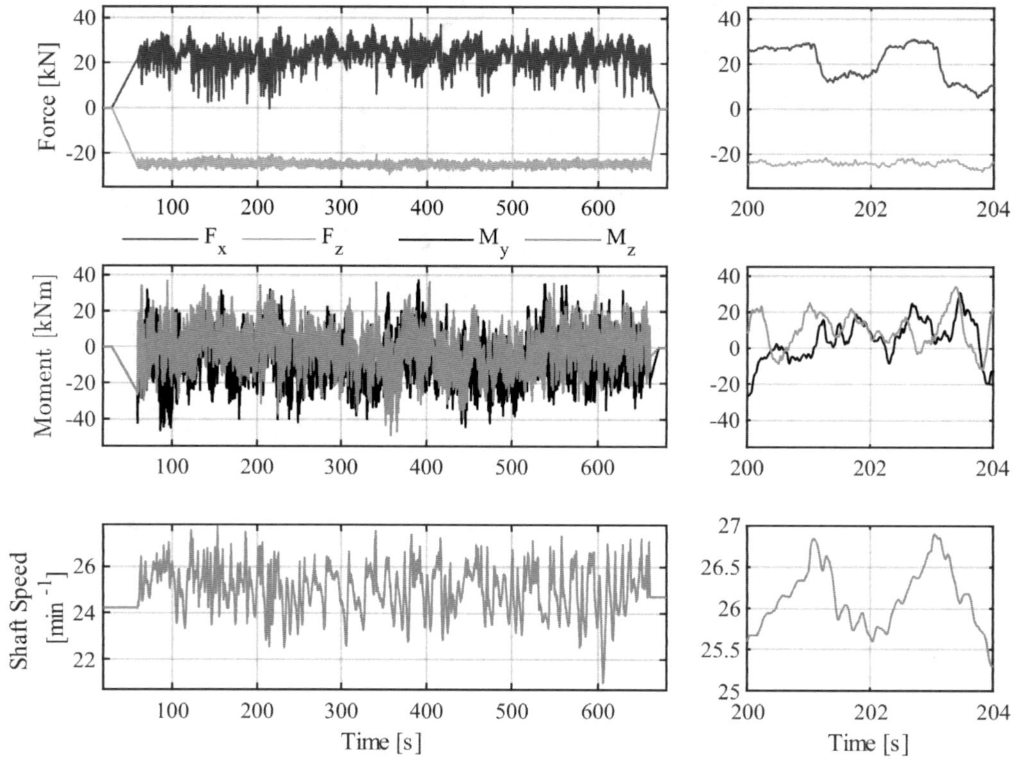


**Figure 6.** Exemplary dynamic load case and shaft speed with detailed 4 s section (right)

The start-stop load case is a critical load case for plain main bearings due to solid-state friction between the shaft and bearing segments during WT start-up, resulting in inevitable wear. Each bearing prototype underwent approximately 10,000 start-stop cycles of 30 s each, as shown in Figure 7. During the first five seconds, of the steady forces are applied without rotation. In the following ten seconds, the shaft speed and thrust force ($F_x$) increase. This operating point is held for five seconds. The loads then are ramped down to zero and finally, the shaft rotation is ramped down to stand still. After another five seconds the cycle restarts. Each bearing variant successfully completed 10,000 start-stop cycles without any damage or significant wear beyond the initial run-in wear.

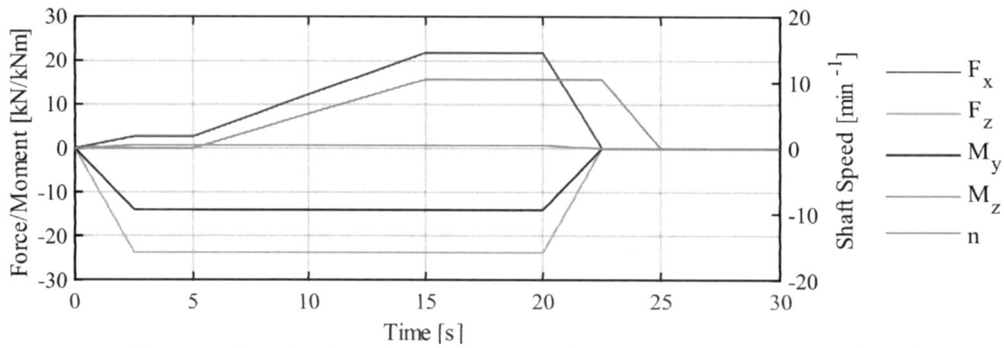


**Figure 7.** Load and shaft speed curves of exemplary start-stop load cycle

Idling is another critical load case for plain main bearings. Forces and torques act on the bearing even at low speeds (see Table 2). As the shaft speed is only 0.6 $min^{-1}$, the bearing does not operate fully in the hydrodynamic regime and runs continuously in mixed friction. Each prototype variant was operated

Journal of Physics: Conference Series **2745** (2024) 012022 doi:10.1088/1742-6596/2745/1/012022

for approximately 100 h under this load conditions. These 100 h only represent a fraction of the expected operation in idling of a WT in the field, but the test bench time was limited and the start-stop operation is the more critical operation condition.

## 4. Results and Discussion

The project's design method relies on simulation data, making a validated simulation model crucial for its reliability. The validation of the simulation model is shown below exemplarily for prototype V1.

### *4.1. Validation of the simulation model*

The validation was based on temperature measurements and thermo-elasto-hydrodynamic (TEHD) simulations conducted by the project partner IST with the simulation software FIRST. The friction losses in the TEHD model are converted into a temperature distribution in the lubrication gap, as shown in Figure 8 bottom left. To validate the simulation model, the measured surface temperatures of the pads at the test bench were compared to the simulated ones (Figure 8). The measured values, shown in blue, were gathered after ten hours of operation with static 100% production loads (cf. Table 2). The simulated values of the oil temperature at the sensor position, shown in red, resulted from simulating the same load case. Overall, the simulation model and the measurement show very good agreement, with a maximum deviation of less than 5%.

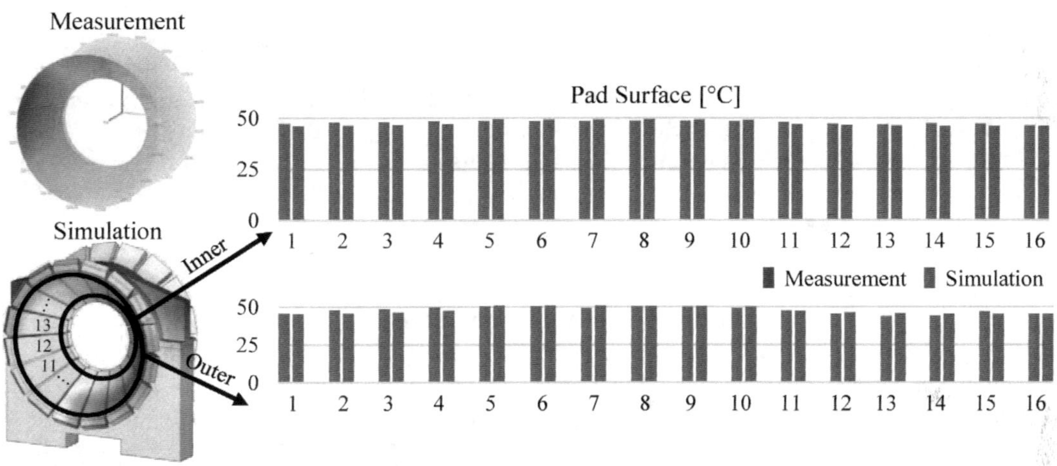


**Figure 8.** Validation of the TEHD simulation model via test bench measurements of the pad temperature distribution on the gearbox side of prototype V1

The validation of the lubrication condition was based on the measurement of the lubrication gap height. Figure 9 compares the results of the simulation and measurement for the static production load case in Table 2. The measured lubrication gap height corresponds qualitatively to the simulated one, although the absolute height of the lubrication gap does not match. This discrepancy is due to the complicated application and calibration of the sensors in the installed state. The simulation model is considered validated due to the good quantitative agreement of the temperatures and the qualitative agreement of the lubrication gap height.

Journal of Physics: Conference Series **2745** (2024) 012022 doi:10.1088/1742-6596/2745/1/012022

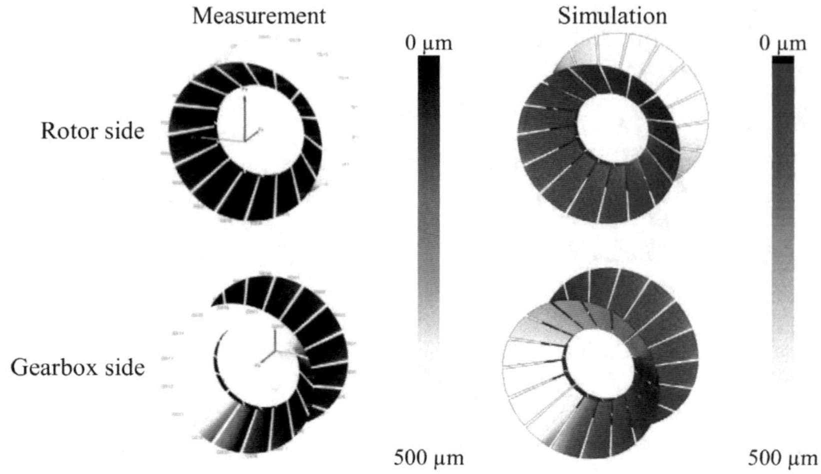


**Figure 9.** Comparison of lubrication gap height between measurement and simulation

*4.2. Validation of the design method*

Both tested prototypes successfully completed the test bench program consisting of static and dynamic loads of rated operation for a WT. In addition, each bearing withstood 10,000 start-stop cycles and 100 h of idling. These two load cases are critical for plain bearings due to the continuous mixed friction operation. During the tests no bearing failure occurred. A visual inspection of all sliding segments after testing showed no extensive damage and an intact sliding surface. Therefore, the general qualification of the new design method to design a safely operating plain bearing for WT is proven.

To validate the effectiveness of the design method the measured frictional torque of the two bearing variants is compared. The friction torque measurement during an eleven-hour production load case test is shown in Figure 10, left. The decrease in frictional torque at the beginning is a common occurrence in plain bearings and is referred to as the running-in process. To obtain a comparative value, the measured frictional torque on the test bench was averaged over six minutes after 10.5 h (Figure 10, red line). The table on the right in Figure 10 compares this value with the simulation result for both tested bearing variants. The deviation between the measurement and simulation is approximately 43 Nm for both variants. It is expected that the deviation in measurements is a result of the friction in the rolling bearing of the test bench located between the drive and load unit (cf. Figure 3 left). Due the design of the bench, the torque measurement includes the losses in the rolling bearing. As the same load case was used for both V1 and V2, the frictional torque of the rolling bearing in the load unit should remain constant. Since the deviation in simulated and measured frictional torque also remains constant, it can be concluded that this deviation is a result of friction in the rolling bearing. Disregarding this offset it can be seen, that the friction torque for V2 was reduced by 32.1% with the design method compared to V1 (based on the simulated values). This validates the effectiveness of the design method to optimize for the defined objectives.

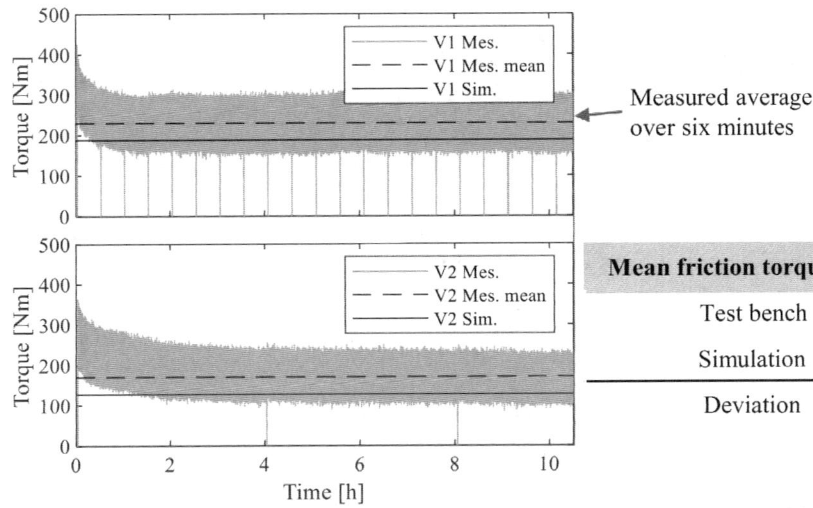


| Mean friction torque [Nm] | V1 | V2 |
|---|---|---|
| Test bench | 229.7 | 170.6 |
| Simulation | 187.6 | 127.3 |
| Deviation | 42.1 | 43.3 |

**Figure 10.** Validation of the design methodology by comparing the frictional torque between test bench tests and simulations of V1 and V2 for the same load case

## 5. Conclusion

WTs could use segmented plain bearings as main bearings in the future. Currently, this concept is still subject to ongoing research. The large and highly loaded structures of a WT deform under load. Plain bearings are sensitive towards these deformation as their clearance is only about 1‰ of the bearing's diameter. Exceeding the plain bearings capacity to endure the deformation will lead to excessive wear and thus premature bearing failure. The FlexPad bearing was designed with these challenges in mind and to be a replacement for tapered roller bearings.

However, designing the FlexPad bearing manually is very complex and thus time-consuming. To handle the complexity a new design method was developed and validated during the *FlexPad* project. The method uses a detailed elasto-hydrodynamic-multibody simulation model. Different designs of the FlexPad have been simulated and translated into a surrogate model. The surrogate model allows for the utilization of a mathematical optimization algorithm to design a new FlexPad bearing.

Two prototypes were built to validate the design method. The V1 design is a scaled down version of the FlexPad bearing prototype developed manually in a previous project. The second bearing (V2) was designed and optimized with the design method presented herein. The optimization objectives were to minimize friction power, maximize pressure area ratio, and maximize hydrodynamic pressure (to a limit). Both bearing designs were built and extensively tested at the CWD under realistic load conditions for WT. Both successfully completed the test bench program consisting of static and dynamic loads of rated operation for a WT. In addition, each bearing withstood 10,000 start-stop cycles and 100 h of idling. These two load cases are critical for plain bearings due to the continuous mixed friction operation. Simulations showed that the bearing designed using the new method had 32.1% less friction power during rated operation. The reduction of friction between the two designs was validated on the test bench. The project results show that the new method is able to design a FlexPad bearing variant with the optimization targets achieved.

In summary, the new method enables to design safely operating FlexPad bearings while meeting specific optimization targets. The method's current limitation to the FlexPad bearing can be expanded in future research by adapting it to other segmented plain bearing designs for WT.

**Affiliations**

The investigations presented in this paper are based on the results of the joint project 'FlexPad - Development of a design methodology for an innovative bearing concept', which was funded by the German Federal Ministry for Economic Affairs and Energy (grant ID 03EE2002B). The authors would like to express their gratitude to the ministry for their support of the 'FlexPad' project. Additionally, the authors would like to express their gratitude to their project partners, Miba Industrial Bearings Germany GmbH (Göttingen, Germany) and IST Ingenieurgesellschaft für Strukturanalyse und Tribologie mbH (Aachen, Germany), for their outstanding cooperation and support.

Supported by:

Federal Ministry
for Economic Affairs
and Energy

on the basis of a decision
by the German Bundestag

**References**

[1] Bundestag 2022 *Gesetz zu Sofortmaßnahmen für einen beschleunigten Ausbau der erneuerbaren Energien und weiteren Maßnahmen im Stromsektor*

[2] Bundestag 2022 *Zweites Gesetz zur Änderung des Windenergie-auf-See-Gesetzes und anderer Vorschriften*

[3] European Commission *The European Green Deal* vol 11.12.2019

[4] Stehly T and Duffy P 2022 *2021 Cost of Wind Energy Review* (National Renewable Energy Laboratory)

[5] Hart E, Turnbull A, Feuchtwang J, McMillan D, Golysheva E and Elliott R 2019 Wind turbine main-bearing loading and wind field characteristics *Wind Energy* **22** 1534–47, DOI: 10.1002/we.2386

[6] Dalgic Y, Lazakis I, Turan O and Judah S 2015 Investigation of optimum jack-up vessel chartering strategy for offshore wind farm O&M activities *Ocean Engineering* **95** 106–15, DOI: 10.1016/j.oceaneng.2014.12.011

[7] Schröder T N 2021 Konisches Gleitlager für die Rotorlagerung einer Windenergieanlage *Dissertation* Chair for Wind Power Drives, Rheinisch-Westfälischen Technischen Hochschule

[8] Schröder T, Rolink A, Mehmet Ö and Wietheger W 2019 *Thermisch gespritzte Gleitlagerbeschichtungen für Hauptlager von Windenergieanlagen: Abschlussbericht* (TIB - Technische Informationsbibliothek Universitätsbibliothek Hannover), DOI: 10.2314/KXP:1686138954

[9] DIN Deutsches Instltut fur Normung e. V. 1991 Hydrodynamische Axial-Gleitlager im stationären Betrieb: Berechnung von Axialsegmentlagern *(*31653-1*)*

[10] Rolink A, Jacobs G, Pérez A, Bosse D and Jakobs T 2022 Sensitivity analysis of geometrical design parameters on the performance of conical plain bearings for use as main bearings in wind turbines *J. Phys.: Conf. Ser.* **2265** 32010, DOI: 10.1088/1742-6596/2265/3/032010

[11] Rolink A, Jacobs G, Schröder T, Keller D, Jakobs T, Bosse D, Lang J and Knoll G 2021 Methodology for the systematic design of conical plain bearings for use as main bearings in wind turbines *Forsch Ingenieurwes* **85** 629–37, DOI: 10.1007/s10010-021-00452-3

[12] Schröder T, Jacobs G, Rolink A and Bosse D 2019 "FlexPad": Innovative conical sliding bearing for the main shaft of wind turbines *J. Phys.: Conf. Ser.* **1222** 12026, DOI: 10.1088/1742-6596/1222/1/012026

# Assessing aerodynamic influences on offshore foundation design for large wind farms

**K Z Wen[1], D Dehtyriov[1], B W Byrne[1]**

[1] Department of Engineering Science, University of Oxford, Parks Road, Oxford OX1 3PJ, UK.

E-mail: kathy.wen@spc.ox.ac.uk

**Abstract.** An evaluation of the impact of aerodynamic interactions on offshore wind turbine (OWT) monopile design in large wind farms is presented. The interactions between turbines within a wind farm, and between the atmosphere and the entire wind-farm, act to reduce the mean effective loads across the farm. This reduction impacts the operational performance of OWT foundations in two notable ways: a decrease in turbine structural loads which affects design to serviceability limit state and a shift in excitation frequencies of the passing blades which impacts fatigue performance. To assess the implications on monopile design, we conduct static and dynamic analyses using a 1-D finite element model (FEM). We show that the farm-atmosphere interaction effect leads to marked reduction in monopile lengths between $\sim 5-25\%$ across an entire wind farm. The dynamic analysis reveals a competing balance between shifting frequency bands and reduced wind loads on the fatigue response. The monopile design is found to be strongly dependent on wind farm size and ratio of wind to wave loading, with wave loading coupled to the farm-atmosphere interaction effect. Exploiting these interactions plays a pivotal role in reducing the levelised cost of wind energy and ensuring robust design of OWTs.

## 1. Introduction

Global wind turbine development is rapidly growing to meet renewable energy demands. As the wind energy industry has expanded, wind farm clusters have moved from onshore to offshore, and the size, number of turbines and mean water depth of these farms is expected to continue increasing over the coming decades. A driving factor contributing to this expansion is the falling cost of wind energy through increases in energy production efficiency and decreases in capital and operational expenditure. Optimisation of the design methods and technology for offshore wind farms is therefore a key ingredient in meeting both low carbon and energy security objectives.

Each turbine in a wind farm comprises a rotor and nacelle to extract power and a tower and foundation to transfer loads to founding soil. Modelling and optimising wind farms requires a detailed understanding of each sub-system and the ways in which they interact. We focus on the influence of aerodynamic interaction on the design of foundations for large offshore wind farms.

The majority of current offshore developments make use of a monopile foundation, which makes up one of the largest portions of capital expenditure. These foundations are individually designed for each turbine in the farm, with upwards of 40% variation in foundation mass across a single farm [1]. There are three variables which require optimisation for each turbine; length, $L_p$, wall-thickness, $t_p$ and diameter, $D_p$. These variables depend on location specific meteorological conditions, turbine loads, sea-depth and soil properties. The limit states which decide these

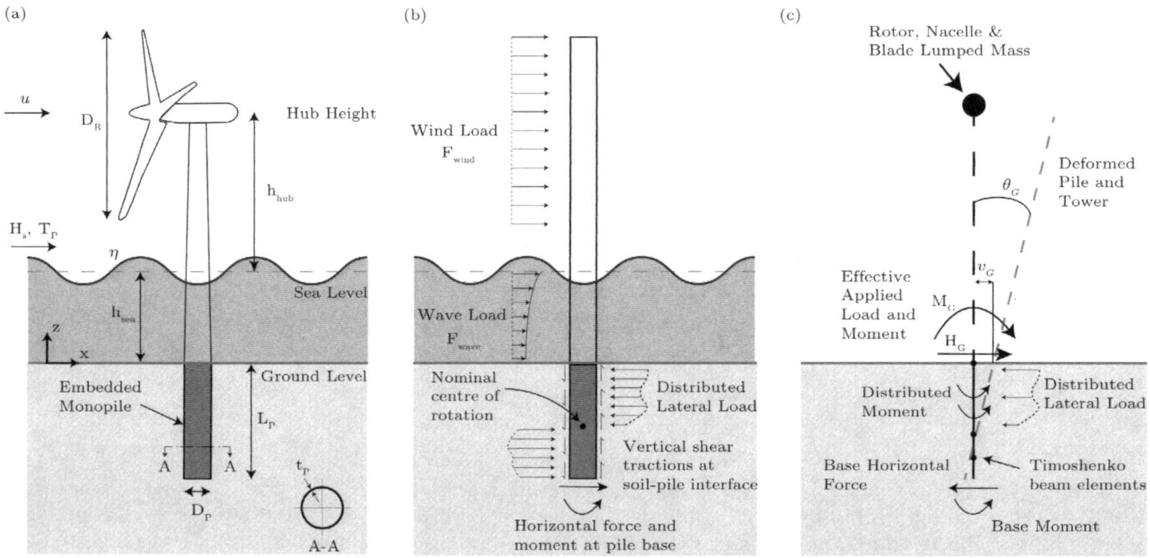


Figure 1: Schematic of the coupled load-monopile design problem. (a) Geometry of an OWT with monopile.(b) Simplified free-body diagram showing the distributed loads, and the balancing distributed soil-structure loads. (c) 1-D modelling of the system based on the PISA method.

designs are governed by installation, extreme weather events and loads under 'operational' conditions [2]. These loads include environmental loads i.e. wind and wave loading and cyclic operational loads caused by the rotation of the blades passing [3].

The characterisation of environmental and operational loads is incomplete due to aerodynamic interaction between individual wind turbines and between the wind farm and the atmospheric boundary layer. Each operating turbine locally reduces the wind speed and increases the turbulence in its wake. These wakes interact with other turbines, leading to power losses, upwards of 40% across a farm [4]. This effect, known as the wake effect, has been extensively explored, with models developed to quantify its impact. A recent observation has additionally identified a 'farm-atmosphere interaction' effect, where flow resistance caused by large wind farms reduces the wind speed upstream and within the farm [5]. Neglecting the farm-atmosphere interaction effect has historically led to a non-negligible bias in the wind resource assessment and has further implications for foundation design which forms the focus of this paper.

Engineering wake models approximating the relevant aerodynamics are extensively applied to estimate the net power production of a wind farm, determine wind farm control strategies [6], solve the wind farm layout optimisation problem [7], and estimate project costs and profit. Recently, the effects of turbine-wake interactions on foundation loads has additionally been identified [8]. However, an evaluation of the effects of farm-atmosphere interaction on foundation design is not presented in the literature thus far. In this paper, we investigate the implications of turbine-wake and farm-atmosphere interaction effect on monopile design by considering a simple wind turbine optimisation problem illustrated in figure 1. Both turbine-wake and farm-atmosphere interaction effects act to reduce the mean loads seen by turbines which subsequently influences their optimal design. We demonstrate that considering farm-atmosphere interaction effects significantly influences monopile design under both serviceabilty limit state (SLS) and fatigue limit state (FLS) design, especially as wind farms grow in size.

Table 1: IEA 15MW turbine parameters [11], herein used as the benchmark case.

| | | |
|---|---|---|
| Rotor Diameter | $D_R$ (m) | 240 |
| Hub Height | $h_{\text{hub}}$ (m) | 150 |
| Mean Sea Level | $h_{\text{sea}}$ (m) | 30 |
| Rated Wind Speed | $u_0$ (m/s) | 10.59 |
| Design Thrust Coefficient | $C_T$ | 0.804 |
| Design Power Coefficient | $C_P$ | 0.489 |
| Design Tip Speed Ratio | TSR | 9 |
| Monopile diameter | $D_P$ (m) | 10 |
| Monopile wall thickness | $t_P$ (mm) | 55 |

Table 2: IEA 15MW lumped mass structural properties [11]

| | | **Nacelle** | **Hub** |
|---|---|---|---|
| Lumped mass | $m$ (t) | 821 | 196 |
| Offset from lumped mass about x-axis | $x_{off}$(m) | 5.486 | 10.6 |
| Offset from lumped mass about z-axis | $z_{off}$(m) | 3.978 | 5.462 |
| Moment of inertia x-axis | $I_{xx}$ (tm$^2$) | 12607 | 303000 |
| Moment of inertia y-axis | $I_{yy}$ (tm$^2$) | 21434 | 152000 |
| Moment of inertia z-axis | $I_{zz}$(tm$^2$) | 18682 | 152000 |

## 2. Methodology

Consider the monopile design problem illustrated in figure 1a. The OWT rotor sees an incoming wind speed $u$ across the rotor diameter $D_R$ at height $h_{\text{hub}}$. The significant wave height $H_S$ and peak spectral period $T_P$ govern wave dynamics, with local sea free surface elevation $\eta$ and mean sea-level elevation $h_{\text{sea}}$. The rotor is supported by a tower and embedded in the soil is a monopile. A simplified load representation of this problem is presented in figure 1b. The wind load is assumed uniformly distributed across a region bounded by the rotor, centered about $h_{\text{hub}}$. The wave load is distributed over the tower base between the ground and sea level $\eta$, assumed to act in the same direction as the wind load. The monopile resists loads under soil-structure interaction. We further simplify the load state to the 1-D representation illustration in figure 1c. The nacelle, hub and blades are modelled by a lumped mass. The monopile is modelled as a 1-D discretised beam. At each node, the soil-structure interaction is modelled as an independent non-linear spring, calibrated against 3-D finite element simulations and validated through field tests, see [9, 10] for details. We evaluate aerodynamics influences on monopile design considering the IEA 15MW reference turbine detailed in [11], with relevant parameters in table 1. Although specific loading cases, atmospheric conditions, wind farm size, soil conditions, blade designs and layouts determine precise design, the broader conclusions are independent of the reference case.

### *2.1. Load determination*

To estimate the static wind load, we assume that the OWTs are operating in region II (above cut-in and below rated $u$) of the wind speed curve, representative of operating conditions, where the wind load varies quadratically with $u$

$$F_{\text{wind}} = \frac{1}{2}\rho u^2 A_T C_T. \tag{1}$$

Journal of Physics: Conference Series **2745** (2024) 012023 doi:10.1088/1742-6596/2745/1/012023

where $\rho$ is the air density, $A_T$ is the rotor swept area and $C_T$ the design rotor thrust coefficient. Dynamic analysis requires the calculation of turbulent wind kinematics which are added to the mean $u$, based on a Kaimal spectrum following the methodology presented in[12].

To estimate the static wave load, we approximate $H_S$ and $T_S$ by linearly interpolating wave conditions in [11] to the relevant $u$. The wave load estimation is based on Morrison's equation [13], where linear wave (Airy) theory is used to approximate the wave geometry. The mean horizontal velocity component of a wave propagating is

$$v_x = \frac{H_S g k}{2\omega} \frac{\cosh\left(k(z + h_{\text{sea}})\right)}{\cosh\left(k h_{\text{sea}}\right)} \cos\left(kx - \omega t\right), \tag{2}$$

where $k$ is the wavenumber, determined from $\omega^2 = gk \tanh k h_{\text{sea}}$, $\omega = 2\pi/T_P$ is the wave frequency. The wave force on each element of the tower is determined by Morrison's equation

$$dF = dF_D + dF_I = \frac{1}{2} C_D \rho D v|v| dz + C_M \rho \frac{\pi D^2}{4} \dot{v} dz, \tag{3}$$

where the drag and inertial coefficients are assumed $C_D = 1$ and $C_M = 2$ respectively. The maxima of drag and inertial force terms are out of phase, we evaluate the two separately and sum them linearly. The forces are found by integrating equation 3

$$F_{\text{wave}} = \int_{-h_{\text{sea}}}^{\eta} dF_D + \int_{-h_{\text{sea}}}^{\eta} dF_I, \tag{4}$$

$$M_{\text{wave}} = \int_{-h_{\text{sea}}}^{\eta} dF_D (h_{\text{sea}} + z) + \int_{-h_{\text{sea}}}^{\eta} dF_I (h_{\text{sea}} + z), \tag{5}$$

where the local maxima of inertial load occurs for $t = 0$, $\eta = 0$, and for drag load for $t = T_P/4$, $\eta = H_S/2$. Aerodynamic interaction affects wind and wave loads differently, see sections 3-4. As such, We additionally consider the limiting case where $F_{\text{wave}} = M_{\text{wave}} = 0$ such that $F_{\text{wind}}/F_{\text{wave}} \rightarrow \infty$ to quantify the impact of variations in this ratio.

For the dynamic analysis, wave kinematics calculation based on [14] are undertaken. The Jonswap spectrum is used to estimate $H_S$ and $T_P$ based on $u$ and estimated fetch [15].

Operational loads, 1P and 3P are caused by mass and aerodynamic imbalance of the rotor and by wind shear and tower shadow effects as blades pass the tower respectively. The 1P frequency ranges ($f_{1P}$) of the IEA 15MW turbine are determined from the tip speed ratio (TSR)

$$f_{1P} = \frac{\lambda u}{\pi D_R}, \tag{6}$$

where $\lambda$ is the TSR. The 3P freqeuncy ranges is $f_{3P} = 3 f_{1P}$. The magnitude of the tangential forces associated 1P and 3P $m_b u \lambda \approx 6$MN in region II, where the blade mass $m_b = 65$t.

### *2.2. Soil profile*

The primary soil profile considered herein is Dunkirk sand, a dense marine sand profile heavily characterised for several field testing campaigns [16]. The sand stiffness profile is modelled by the relationship proposed by [17]

$$G_0 = \frac{B p'_{\text{ref}}}{0.3 + 0.7 e_0^2} \sqrt{\frac{p'}{p'_{\text{ref}}}} = m\sqrt{z}, \tag{7}$$

where the $p'$ is mean effective stress variation with soil depth, $e_0 = 0.629$ (for a relative density of $D_{\text{Rel}} = 75\%$), $\gamma' = 10.09\text{kN/m}^3$, $B = 875$ and $p'_{\text{ref}} = 101.3$kPa. As a soil sensitivity study, we consider variations in $m$ (equivalent to variations in $B$), and comparisons to an overconsolidated glacial clay till profile, with properties described in [18, 19].

*2.3. Design methodology*
We are concerned with design of OWT monopiles under SLS and FLS. Typically, SLS limits $D_P/10$ to the ground displacement $v_G$, and $0.5°$ to ground rotation $\theta_G$, which are applied heriein. FLS requires the natural frequency $f_0$ of the OWT to be desined designed to sit between rotor vibration ($f_{1P}$) and blade passing ($f_{3P}$) excitation frequency ranges, known as the "soft-stiff" region to avoid resonance and structural degradation [20].

The loads and soil profile are inputs to an in-house 1-D finite element model (FEM), referred to as OxDyn, described in[21]. The nacelle, hub and blades are modelled by a lumped mass with properties in table 2. The monopile is represented as a series of interconnected 1-D Timoshenko beam elements, (see figure 1c). The resultant finite-element equations are solved for $v_G$ and $\theta_G$ of each beam element. This approach is described in [9, 10]. The dynamic fatigue assessment further requires the time integration calculation of the OWT response to a time varying load excitation (from wind, waves and 1P/3P loads). The dynamic response is solved by forward integration using the non-linear Newmark-beta method, obtaining the time-series load and displacement distribution at each beam element.

Turbine-wake and farm-atmosphere interaction effects decrease mean $u$ and loads seen by OWTs. The aerodynamic load scales quadratically with reduction in $u$ through equation 1, where $u = \beta u_0$, $\beta$ is the wind speed reduction factor. For turbine-wake interaction, the wave load is assumed unaffected, for farm-atmosphere interaction effects, updated wave loads are computed for new $u$ through equations 4-5.We assume a fixed $D_p$ and vary $L_p$ in response to changes in load. The 1-D FEM is solved iteratively for minimum $L_p$ to satisfy SLS conditions.

FLS is constrained by $f_0$, determined by eigenvalues analysis, solving the following simplified characteristic equation for values of undamped natural frequencies, $f_n$

$$(\mathbf{K} - (2\pi f_n)^2\mathbf{M})x \sin(2\pi f_n t) = 0, \tag{8}$$

where $\mathbf{K}$ and $\mathbf{M}$ are the stiffness and mass matricies respectively, $x$ is the displacement vector, $f_0$ is the first dominant modal frequency. The proximity of $f_0$ to $f_{1P}$ and $f_{3P}$ dictates the amount of fatigue damage. Based on the assumption of linear cumulative damage, the fatigue life of structures may be calculated based on the S-N fatigue approach (Palmgren-Miner rule) [22]. The accumulated damage $D$, represents the ratio of number of cycles accumulated for a given stress range to the total number of cycles to failure for the same stress range and is determined by equation 9 described in[23]

$$D = \sum_{i=1}^{k} \frac{n_i}{N_i} = \frac{1}{\bar{a}} \sum_{i=1}^{k} n_i (\Delta\sigma_i)^m, \tag{9}$$

where $\bar{a}$ is the intercept of the S-N curve with the $\log N$ axis, $m$ is the negative inverse slope of the S-N curve, $k$ is the number of stress blocks, $n_i$ is the number of stress cycles in each block $i$, computed from the time-series input from dynamic analysis and $N_i$ is the number of cycles to failure at constant stress range $\Delta\sigma_i$. Failure is assumed to occur in the butt welds at ground level where bending moment is the highest, with the S-N curves provided in DNV-RP-C203 [23].

## 3. Turbine wake interaction effect
Across a wind farm, the first column of turbines see the upstream meteorological $u$, however, turbines deeper in the farm experience a significant reduction in $u$ and load due to turbine-wake interactions. For large farms, the majority of turbines will experience similar levels of reduced load, except the first few columns.

We focus on expected normalised wake losses at the Horns Rev wind farm based on experimentally measured results presented in [4], although engineering wake models such as

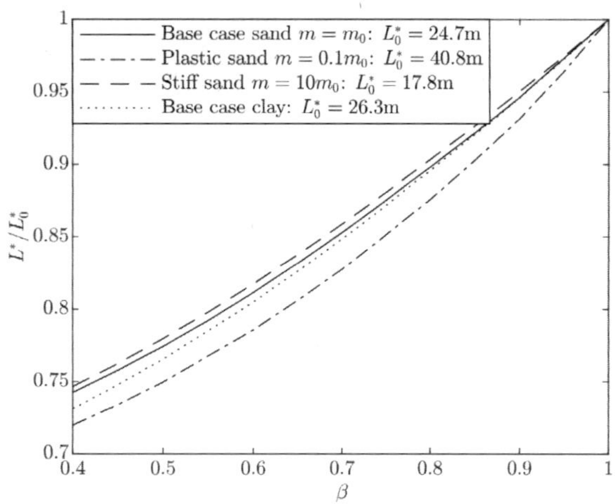


Figure 2: Soil sensitivity study relating the aerodynamic impact, characterised by $\beta$ on $L_p$ variation $L^*/L_0^*$ for various soil profiles. The base case sand is compared to a clay profile [19], and the proportionality constant $m$ for the base case sand (see equation 7) is varied in both directions and compared. Over a large range of $\beta$ and soil strengths, the mean $L_p$ reductions are relatively insensitive to the soil profiles.

the Jensen model can be used as a first-order prediction instead. In particular, we focus on the mean power reduction across the range of wind directions considered in [4], and changes in mean $u$ seen by turbines as a function of their depth within the farm.

Since far-wake velocity deficits are unaffected by specifics of OWT blade design [24], and primarily depend on $C_T$, we assume normalised power losses in [4] are reasonable approximations of expected turbine-wake losses and apply the same $\beta$ for the IEA 15MW turbine presented herein, assuming turbine-wake interactions have no impact on the wave load.

Figure 2 presents the results for the IEA 15MW. The difference in soil profiles impact nominal $L_p$, however changes to $L_p$ is insensitive to soil type. For a two order magnitude change in soil stiffness, and $\beta = 0.4$, the difference in relative $L_0^*$ remains below 3%. The influence of turbine-wake effects on monopile design is considered insensitive to realistic variations in soil conditions.

Across wind directions (figure 3a), mean reduction to $L_p \approx 5\%$. Reductions are not uniformly distributed, and depend on the 'depth' of each turbine in the farm and wind direction. For wind direction corresponding to the least wake interaction (figure 3b), $L_p$ decreases near linearly with distance into the farm. For the worst-case scenario (figure 3c) where the turbine array is aligned with wind direction, turbines in the first wind-facing column require an increased $L_p$, and mean $L_p$ reduction for other turbines is $\approx 7\%$. As wind farms grow, plateaus in $L_p$ are expected for all wind directions; for medium sized farms, considering wake interaction in monopile design requires additional coupling to $u$ and direction joint probability distributions [25].

The ratio of wind to wave loads emerges as a critical parameter that controls the relative impact of turbine-wake interactions on monopile design. It is anticipated that the significance of these impacts will increase as OWTs grow. Likewise, if operating in a sea-state with low wave loading, aerodynamic interaction effects increase in significance. As demonstrated in figure 3, in the limiting case where the wave loads are negligible compared to the wind load, the mean $L_p$ reduction across the farm falls from 5% to 8% (figure 3a), with larger variation across changing wind directions. Turbines 'deeper' in the farm see a significant reduction in $L_p$ (figures 3b-c), with reduction approaching 15% for both best and worst case wind alignment scenarios.

## 4. Farm-atmosphere interaction effect

As the size of wind farms increase, changes to mean $u$ and thus monopile design become governed by farm-atmosphere interaction effects. A numerical weather prediction model simulation used to compare atmospheric flow with and without a wind farm present is presented in [26], and illustrates velocity slow down across an entire wind farm as response to the presence of a wind farm. The $\beta$ is independent of turbine layout and instead depends on interaction between

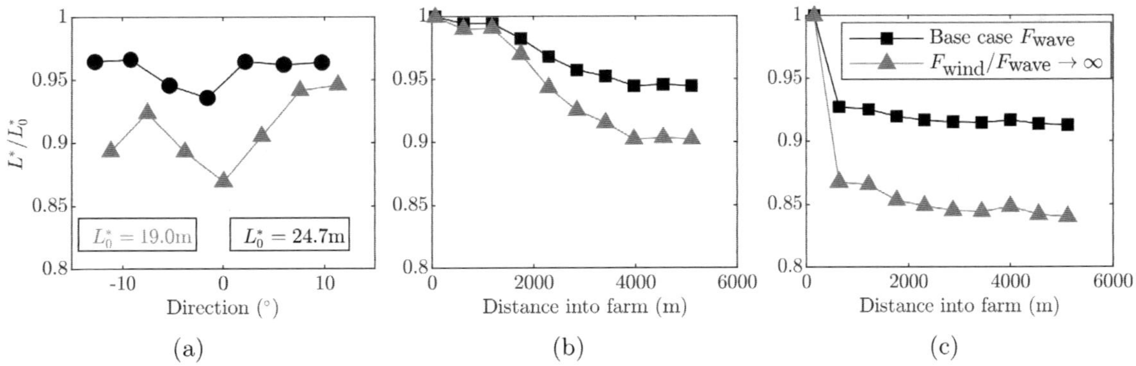


Figure 3: Reduction in optimal $L_p$ for two wind to wave loading ratio cases, due to turbine-wake interaction effects measured at the Nysted wind farm, with larger reductions for a given loading corresponding to increased wake-turbine interactions. (a) The mean farm-wide reduction variation with wind direction. (b), (c) $L_p$ variation with distance into the farm for the wind directions with minimum wake interaction and maximum wake interaction respectively.

the wind farm and atmospheric boundary layer. A two-scale theoretical model predicting the velocity slowdown for fully developed wind farms was presented by [27], after it was observed to impact power predictions for current wind farms [5]:

$$4\alpha(1-\alpha)\frac{\lambda}{C_{f0}}\beta_f^2 + \beta_f^\gamma = M, \tag{10}$$

where we assume $\gamma = 2$ based on [27], with the results relatively insensitive to this parameter, and the momentum availability factor $M$ is modelled as a linear approximation which has been shown to work well over a practical range of $\beta_f$ [26]

$$M = 1 + \zeta(1-\beta_f). \tag{11}$$

Equation 10 relates the turbine-scale flow physics, 'internal' conditions (e.g, the turbine layout and operating conditions) on the left-hand side, to the interaction between the wind farm and atmospheric boundary layer, 'external' conditions, on the right-hand side. Equations 10-11 are solved for farm flow velocity slowdown due to farm-atmosphere interaction $\beta_f$ assuming turbines operate at maximum efficiency ($C_P = 4\alpha^2(1-\alpha)\beta_f^3$). We note turbine operating points can be adjusted to mitigate some of the farm-atmosphere interaction effects [28] which could be considered in a detailed design.

The theoretical model predicts the global velocity deficit based on; the momentum response parameter $\zeta$, an empirically determined parameter which relates the extracted momentum from the atmosphere due to the presence of the farm with the mean velocity reduction, and the effective array density $\lambda/C_{f0}$, where $\lambda = nA_R/S_F$ ($n$ is the number of turbines, $A_R$ the swept area of the turbine and $S_F$ the wind farm area) and $C_{f0}$ is the sea-surface friction coefficient (typically $C_{f0} \approx 0.002$). For an infinitely large wind farm, $\zeta = 0$, $\zeta$ increases exponentially with decreasing wind farm size. Further details of the theoretical modelling are presented in [27].

For a specific wind farm site and layout, numerical weather simulations are required to determine precise extractability factor and turbine-wake interaction estimates; here, we seek a first order estimate to expected impacts, and to demonstrate their significance. To understand the joint effect of turbine-wake interaction and farm-atmosphere interaction effects, [29] simulated wind farms varying layouts over the $\zeta$-$\lambda/C_{f0}$ space. The relative impact of the

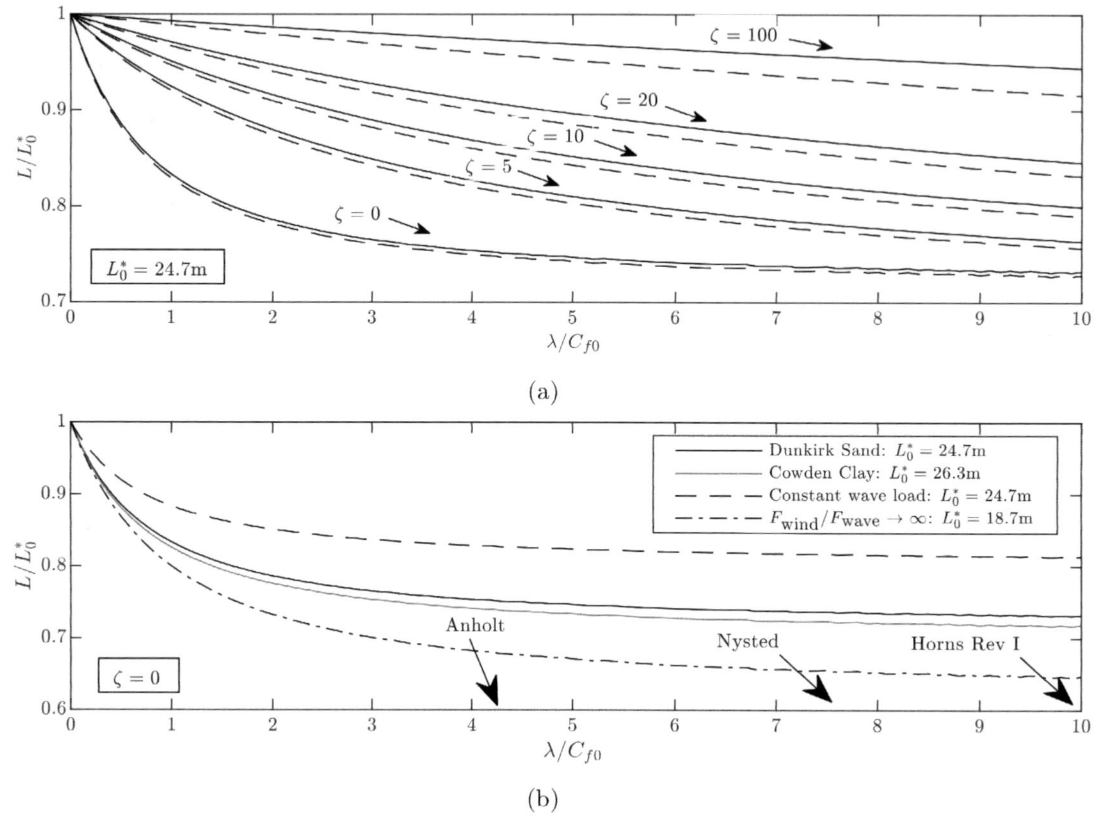


Figure 4: Reduction in optimal $L_0^*$ under aerodynamic coupling, parametrised against the momentum extractability factor $\zeta$, and effective array density $\lambda/C_{f0}$ (approximate $\lambda/C_{f0}$ for wind farms indicated). (a) The base case, assuming wave loads are altered by farm-atmosphere interaction, for a range of $\zeta$. The solid lines show farm-atmosphere interaction effect only, the dashed lines show combined turbine-wake and farm-atmosphere effect impact. (b) Solutions for an infinitely large wind farm ($\zeta = 0$). The solid curves show the base reference case, assuming wave loads are affected by farm-atmosphere interaction, with black red lines indicating Dunkirk sand and Cowden clay respectively. The dashed curve shows the solution assuming farm-atmosphere interaction effect does not affect wave loading, and the dash-dotted line assumes the wave loading is negligible compared to the wind loading.

turbine-wake effect is estimated by averaging the results presented in [29] for $\lambda/C_{f0} \leq 10$ over the range of $\zeta$ presented, which exhibits a near linear relationship. This linear fit is used to estimate turbine-wake impact, the theoretical model in [27] is employed to estimate the farm-wide impact.

The farm-scale aerodynamic interaction decreases the mean $u$ within and upstream of the entire wind farm to a larger scale compared to turbine-wake interactions. Wind-generated waves are assumed to adjust in $H_S$ and $T_S$ based on the new mean $u$, according to section 2. Consequently, the farm-atmosphere effect reduces both the wind and wave loads.

Figure 4a presents the results demonstrating significant farm-atmosphere interaction impact on monopile design. In the limiting case of a very large wind farm ($\zeta = 0$), and for realistic $\lambda/C_{f0}$, the mean $L_p$ is reduced by upwards of 25%. Here, farm-atmosphere interaction dominates the turbine-wake effect, and monopile designs are insensitive to turbine layout and wind direction.

As $\zeta$ increases, and wind farms become smaller, the relative importance of turbine-wake

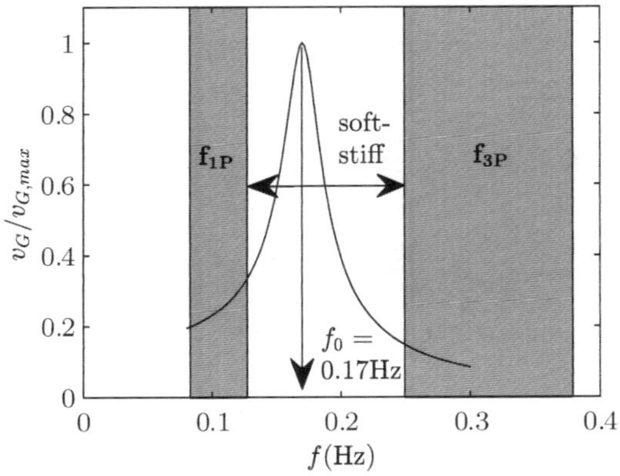

(a)

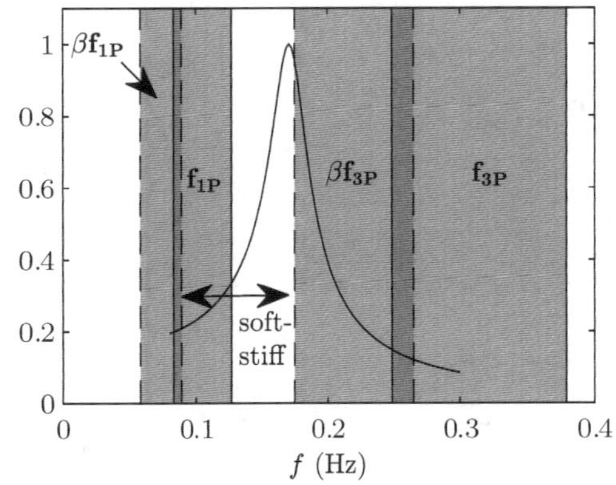

(b)

Figure 5: $v_G$ response of the IEA 15MW turbine ($L_0^* = 24.7m$) with excitation frequency (a) The reference case $\beta = 1$ (b) $\beta = 0.7$. The shaded regions represent the $f_{1P}$ and $f_{3P}$ ranges.

aerodynamics grow, at $\zeta = 100$, the impact of farm-atmosphere and turbine-wake effects approximately balance. For small wind farms, load considerations are predominantly influenced by turbine-wake interaction, however, as wind farms expand, the governing factor shifts to farm-atmosphere interaction.

The results show that the convexity of the relationship between $\lambda/C_{f0}$ and $L_p$ reduction changes with farm size. For small farms, $\lambda/C_{f0}$ has a near linear relationship with $L_p$ reduction, therefore doubling the number of turbines approximately doubles the relative $L_p$ reduction. For larger farms, increasing the number of turbines at low densities will have an appreciable impact on the permissible change in mean $L_p$. As more turbines are added to the farm, $L_p$ decreases until the addition of further turbines no longer impacts monopile design.

Soil conditions do not significantly influence $L_p$ reduction (figure 4b). Instead, the ratio of wind to wave loads is critical in evaluating the aerodynamic impact on monopile design. We explore two modifications to the base assumptions; the first where the $u$ reduction under farm-atmosphere interaction has no impact on $H_S$ and $T_S$ (and therefore wave loads); and second where wave loads are negligible compared to wind loads. These scenarios highlight the importance of considering the combined effect of wind and wave loads on monopile design.

The ratio of wind to wave loading decreases with increasing $\lambda/C_{f0}$ as only wind loads decrease. The plateau in $L_p$ drops from 25% to near 15%. Contrariwise, for a negligible wave load, this plateau increases to 35%. A detailed understanding of the response of wave loads to the presence of a wind farm is thus critical to the optimisation of offshore monopile design.

As both wind farms and wind turbines continue to increase, the farm scale aerodynamic physics are expected to have a more pronounced impact on reduction on mean $u$ and hence in the optimisation of monopile design. It has been shown that farm-atmosphere interaction effects have substantial implications regarding reduction in $L_p$ for large wind farms.

## 5. Natural Frequency and Fatigue

The implication of turbine-wake interaction and farm-atmosphere interaction on the fatigue performance of OWTs are now discussed. The $v_G$ response of the IEA 15MW turbine across a range of excitation frequencies is shown in figure 5a.

As anticipated, the excitation frequency at $f = f_0$ results in the largest $v_G$. The optimal TSR for power production depends on the turbine diameter and blade lift-to-drag ratios and is constant in region II, where the turbine operates the majority of the time. In the other

Journal of Physics: Conference Series **2745** (2024) 012023 doi:10.1088/1742-6596/2745/1/012023

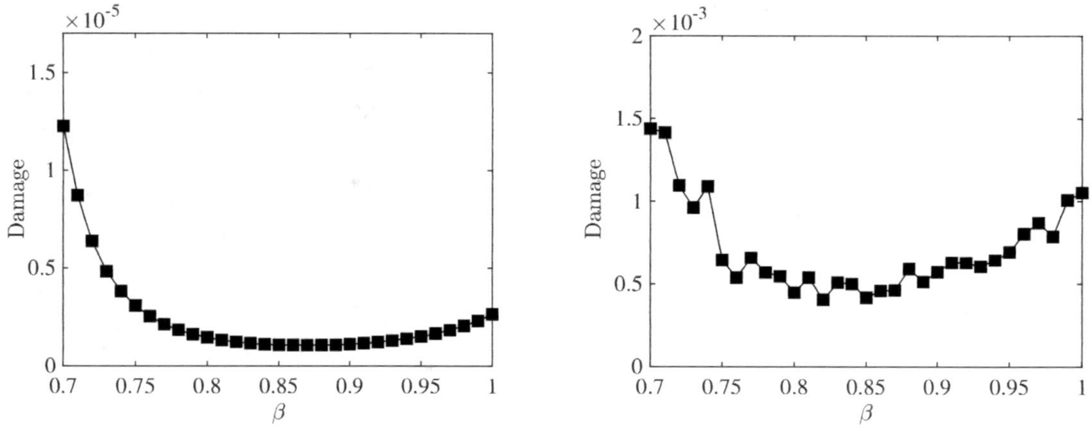


(a) (b)

Figure 6: Damage response as a function of $\beta$ calculated for a 10 minute time interval by (a) 1P and 3P loads and (b) combined environmental and operational loads.

operational regions, the TSR is varied to limit power production and rotation speed and avoid damage to the turbine [11]. For scope, we only consider changes in $\beta$ on fatigue impact, however we note that the turbine-wake interaction additionally increases the mean turbulence levels across the farm. For a reduction in mean $u$ across the farm and constant TSR, the $f_{1P}$ and $f_{3P}$ bands shift by a factor $\beta$ (see equations 6) to maintain optimal aerodynamic efficiency. This shift is illustrated in figure 5b, where as $\beta$ decreases, the "soft-stiff" region also decreases both in nominal value and in range. As the "soft-stiff" region shifts to lower frequencies, the amount of $v_G$ (a proxy for damage) increases as represented by the area under the curve.

The influence of the reduced $u$ on the damage response caused by isolated 1P/3P loads and combined environmental and operational loads are shown in figures 6(a, b) respectively.

Here it is demonstrated that reduction in $u$ largely influences the damage response caused by 1P/3P loads, see Figure 6a. Damage by 1P/3P increases rapidly as $\beta$ reduces due to the reduction of the 3P region interfering with $f_0$ as shown in figure 6b. However, the influence of $\beta$ on the combined damage response from all loads produces a varied response. Under all loading, the damage to the OWT decreases for small for $\beta \leq 0.85$, this is attributed to the decrease in damage from the wind load from reduced $u$. Further reduction in $\beta$ however shows that damage from the 1P/3P loads counteracts the beneficial effect of decreased wind load. As the 3P region shifts closer to $f_0$ greater damage to the OWT is observed. For $u$ reductions associated with large wind farms (farm-atmosphere interaction considerations), a re-design of the monopfile may be necessary to maintain the fatigue performance of the OWT. Eigenvalue analysis indicates that changes to $L_p$ do not greatly influence $f_0$ within typical $L/D$ ratios of OWTs, corroborated by [30]. Variation in $t_p$ is therefore necessary to counteract the aerodynamic effects ensuring $f_0$ remains within the "soft-stiff" region. The optimal value of $t_p$ will depend on the site specific atmospheric characteristics, soil profile and turbine attributes. OWT foundation designers need to consider all these factors to ensure robust fatigue design for large wind farms.

## 6. Conclusion

The implications of aerodynamic interactions on monopile design for large wind farms has been explored in this study. Two aerodynamic phenomena are known to impact the mean loads and excitation frequencies seen by wind turbines across a farm; turbine-wake interaction physics where turbulent wakes behind turbines interact with other turbines in the farm, and farm-atmosphere interaction, where the whole wind farm resists the atmospheric flow. The changes to mean loading is coupled to a 1-D FEM where the static and dynamic response of an OWT is considered to investigate the magnitude of expected impact of aerodynamics on monopile design.

The turbine design and environmental data is taken from the IEA 15MW OWT [11]. Two soil profiles are considered; a cohesionless sand and a cohesive clay, representative of typical offshore soil profiles and characterised for the PISA project [16]. To understand the sensitivity of our results to soil stiffness, we varied soil stiffness by an order of magnitude in both directions. To quantify the impact of turbulent wake interaction, we used experimental data from the Horns Rev wind farm [4]. For the farm-atmosphere scale impact, we coupled our monopile design methodology with the two-scale flow theory presented in [27]. Finally, we considered the combined impact of both aerodynamic effects using simulation data presented in [29]. The first-order analysis assumes a fully developed wind farm with uniform unsheared flow.

For any given loading condition, $L_p$ is minimised to satisfy SLS. Dynamic analyses are performed to evaluate the effect of turbine-wake and farm-atmosphere interaction on $f_0$ and subsequently the fatigue performance of OWTs. The results indicate that accounting for the turbine and farm-atmosphere aerodynamic effects significantly reduces the required mean $L_p$ to satisfy SLS by $\sim$ 5% upwards 25%. These reductions become more pronounced for large wind farms where farm-scale aerodynamics dominate the turbine-wake interaction effects. Simultaneously fatigue analysis demonstrates that a re-design of $t_p$ may be necessary to ensure the $f_0$ of the system remains between the $f_{1P}$ and $f_{3P}$ to maintain adequate fatigue performance. However this heavily depends on wind farm specific characteristics. For small changes to incoming mean wind speed such as those induced by turbine-wake interactions, the fatigue performance of the OWT increases directly as a results of wind load reduction. For large wind farms, where farm-atmosphere aerodynamics dominate, the fatigue performance decreases due to the $f_{3P}$ region interfering with $f_0$, necessitating a re-design of $t_p$. The farm-atmosphere interaction effect is additionally expected to reduce $H_S$ as the mean atmospheric $u$ is reduced by the farm-atmosphere interaction. The ratio of wind to wave loading is a critical parameter in estimating the influence of the changes in environmental load on monopile design.

Neglecting these aerodynamic interactions not only results in overly conservative monopile design but also can lead to unaccounted structural fatigue damage. These effects are expected to become more pertinent as individual wind turbine sizes increase and wind farms grow in size. Establishing a framework which integrates the aerodynamic interactions with foundation design is crucial for robust performance of OWT and ongoing reductions in the levelised cost of energy. Turbine manufacturer's will need to further collaborate with wind farm developers to ensure that these components of wind farm design are appropriately modelled in future.

## References

[1] Kallehave D, Byrne BW, LeBlanc TC, Mikkelsen KK. Optimization of monopiles for offshore wind turbines. Philosophical Transactions of the Royal Society A: Mathematical, Physical and Engineering Sciences. 2015;373(2035):20140100.

[2] Det Norske Veritas and Germanischer Lloyd. Support structures for wind turbines. DNV GL; 2016. 0126.

[3] Nikitas G, Vimalan N, Bhattacharya S. An innovative cyclic loading device to study long term performance of offshore wind turbines. Soil Dynamics and Earthquake Engineering. 2016 03;82:154-60.

[4] Barthelmie RJ, Pryor SC, Frandsen ST, Hansen KS, Schepers JG, Rados K, et al. Quantifying the impact of wind turbine wakes on power output at offshore wind farms. Journal of Atmospheric and Oceanic Technology. 2010;27(8):1302-17.

[5] Bleeg J, Purcell M, Ruisi R, Traiger E. Wind Farm Blockage and the Consequences of Neglecting Its Impact on Energy Production. Energies. 2018 Jun;11(6):1609.

[6] Shapiro CR, Starke GM, Meneveau C, Gayme DF. A Wake Modeling Paradigm for Wind Farm Design and Control. Energies. 2019;12(15).

[7] Park J, Law KH. Layout optimization for maximizing wind farm power production using sequential convex programming. Applied Energy. 2015;151:320-34.

[8] Larsen G, Larsen T, Pedersen M, Enevoldsen K, Aagaard Madsen H. Wake affected offshore tower and foundation loads. No. 131 in DTU Wind Energy E. Denmark: DTU Wind Energy; 2016.

[9] Byrne BW, Houlsby GT, Burd HJ, Gavin KG, Igoe DJP, Jardine RJ, et al. PISA design model for monopiles for offshore wind turbines: application to a stiff glacial clay till. Géotechnique. 2020;70(11):1030-47.

[10] Burd HJ, Taborda DMG, Zdravković L, Abadie CN, Byrne BW, Houlsby GT, et al. PISA design model for monopiles for offshore wind turbines: application to a marine sand. Géotechnique. 2020;70(11):1048-66.

[11] Evan G, Rinker J, Sethuraman L, Zahle F, Anderson B, Barter G, et al. Definition of the IEA 15-Megawatt Offshore Reference Wind. Golden, CO, USA: National Renewable Energy Laboratory; 2020. NREL/TP-5000-75698.

[12] International Electrotechnical Commission. Wind energy generation systems - Part 1: Design requirements. IEC; 2019. IEC 61400-1.

[13] Morison J, Johnson JW, Schaaf SA. The force exerted by surface waves on piles. Journal of Petroleum Technology. 1950;2(05):149-54.

[14] International Electrotechnical Commission. Wind energy generation systems - Part 3-1: Design requirements for fixed offshore wind turbines. IEC; 2019. IEC 61400-3.

[15] Hasselmann K, Barnett TP, Bouws E, Carlson H, Cartwright DE, Enke K, et al. Measurements of wind-wave growth and swell decay during the Joint North Sea Wave Project (JONSWAP). Ergaenzungsheft zur Deutschen Hydrographischen Zeitschrift, Reihe A. 1973.

[16] Zdravković L, Jardine RJ, Taborda DMG, Abadias D, Burd HJ, Byrne BW, et al. Ground characterisation for PISA pile testing and analysis. Géotechnique. 2020;70(11):945-60.

[17] Hardin BO, Black WL. Vibration Modulus of Normally Consolidated Clay. Journal of the Soil Mechanics and Foundations Division. 1968;94(2):353-69.

[18] Byrne BW, McAdam RA, Burd HJ, Beuckelaers WJAP, Gavin KG, Houlsby GT, et al. Monotonic laterally loaded pile testing in a stiff glacial clay till at Cowden. Géotechnique. 2020;70(11):970-85.

[19] Zdravković L, Taborda DMG, Potts DM, Abadias D, Burd HJ, Byrne BW, et al. Finite-element modelling of laterally loaded piles in a stiff glacial clay till at Cowden. Géotechnique. 2020;70(11):999-1013.

[20] Arany L, Bhattacharya S, Macdonald J, Hogan SJ. Design of monopiles for offshore wind turbines in 10 steps. Soil Dynamics and Earthquake Engineering. 2017 jan;92:126-52.

[21] Ryan GV, Adcock TAA, McAdam RA. Influence of soil plasticity models on offshore wind turbine response. Wind Energy. 2024;27(2):131-51.

[22] Miner MA. Cumulative Damage in Fatigue. Journal of Applied Mechanics. 2021 03;12(3):A159-64.

[23] Veritas DN, Lloyd G. Fatigue design of offshore steel structures. DNV GL; 2019. DNV-RP-C203.

Journal of Physics: Conference Series **2745** (2024) 012023 doi:10.1088/1742-6596/2745/1/012023


[24] Porté-Agel F, Bastankhah M, Shamsoddin S. Wind-Turbine and Wind-Farm Flows: A Review. Boundary-Layer Meteorology. 2020 jan;174(1):1-59.

[25] Feng J, Shen WZ. Modelling Wind for Wind Farm Layout Optimization Using Joint Distribution of Wind Speed and Wind Direction. Energies. 2015;8(4):3075-92.

[26] Patel K, Dunstan TD, Nishino T. Time-Dependent Upper Limits to the Performance of Large Wind Farms Due to Mesoscale Atmospheric Response. Energies. 2021;14(19).

[27] Nishino T, Dunstan TD. Two-scale momentum theory for time-dependent modelling of large wind farms. Journal of Fluid Mechanics. 2020;894.

[28] Legris L, Pahus ML, Nishino T, Perez-Campos E. Prediction and Mitigation of Wind Farm Blockage Losses Considering Mesoscale Atmospheric Response. Energies. 2023;16(1).

[29] Kirby A, Nishino T, Dunstan TD. Two-scale interaction of wake and blockage effects in large wind farms. Journal of Fluid Mechanics. 2022;953:1-28.

[30] Alkhoury P, Soubra AH, Rey V, Aït-Ahmed M. A full three-dimensional model for the estimation of the natural frequencies of an offshore wind turbine in sand. Wind Energy. 2021;24(7):699-719.

# AUTHOR INDEX

**Institute of Physics**
Dirac House, Temple Back
Bristol BS1 6BE UK

ISSN: 1742-6588
ISBN 978-1-7138-9733-0

CURRAN ASSOCIATES INC.
proceedings.com